ADVANCES IN BIOPHYSICS, Volume 14, 1981

Vol. 14, 1981

ADVANCES IN BIOPHYSICS

Masao Kotani, Editor
Haruhiko Noda, Associate Editor

Editorial Board
Setsuro Ebashi, Kazutomo Imahori, Yasuji Katsuki,
Jin-ichi Nagumo, Fumio Oosawa, Mitsuru Takanami,
Ei Teramoto, Akiyoshi Wada

Secretary
Koscak Maruyama
Faculty of Science, Chiba University
Yayoi, Chiba

JAPAN SCIENTIFIC SOCIETIES PRESS
Tokyo
UNIVERSITY PARK PRESS
Baltimore
1981

Supported in part by the Ministry of Education, Science and Culture under the Publication Grant-in-Aid.

Published jointly by
JAPAN SCIENTIFIC SOCIETIES PRESS
Tokyo

and

UNIVERSITY PARK PRESS
Baltimore
ISBN 0-8391-0111-2
Library of Congress Catalogue Card No. 76-118663

PREFACE

Biophysics is an interdisciplinary field of study, in which scientists and students from various disciplines work together, and in which research results obtained are often of much interest to wide circles in the life sciences and sometimes to those in physical sciences too. In such a field, a journal devoted to presentation of important achievements of authors in a personal review form so as to be accessible and understandable to wide circles of readers may contribute to meeting the needs of the scientific community. Such consideration led to the birth of this journal, *Advances in Biophysics*, in 1968 with the aid of an opportune offer by University of Tokyo Press and the scientific backing of the Biophysical Society of Japan with University Park Press cooperating in its distribution outside Japan. Since 1978, *Advances in Biophysics* is published by Japan Scientific Societies Press.

Thus, *Advances in Biophysics* includes mainly well-organized and well-documented contributions which present complete accounts of the authors' works on specified subjects that will be understandable to interested readers without the necessity of their referring to other papers by the authors and others. Pertinent background information and a short review of related studies by other scientists are included.

During the childhood of the journal all papers were contributed by Japanese authors. We wish to invite, however, contributions from scientists in other countries also, in the hope that our journal will enjoy worldwide recognition, not only from the point of view of distribution, but also from a large variety of authors of many nationalities.

Advances in Biophysics will publish papers which the Editorial Board has solicited from contributors. It is not the intention of the Editor and the Editorial Board, however, to exclude unsolicited contributors. The Editorial Board welcomes voluntary contributions suitable for publication in the journal; contributors wishing to submit material are requested to write in advance to the Editor. Prospective authors should carefully read the " Suggestions to Authors " at the end of this issue.

Masao Kotani, Editor
Haruhiko Noda, Associate Editor

CONTENTS

TWO-DIMENSIONAL NMR SPECTROSCOPY: AN APPLI-
CATION TO THE STUDY OF FLEXIBILITY OF PROTEIN
MOLECULES   K. Nagayama 139

THE CAP STRUCTURE IN EUKARYOTIC MESSENGER
RNA AS A MARK OF A STRAND CARRYING PROTEIN IN-
FORMATION   K. Miura 205

THE IONIC MECHANISM OF EXCITATION IN INTESTINAL SMOOTH MUSCLE CELLS

T. Suzuki and H. Inomata 239

Adv. Biophys., Vol. 14, pp. 1–35 (1981)

SUBUNIT ASSEMBLY OF *ESCHERICHIA COLI* RNA POLYMERASE

AKIRA ISHIHAMA

Department of Biochemistry Institute for Virus Research, Kyoto University, Kyoto, Japan

DNA-dependent RNA polymerase (ribonucleoside 5′-triphosphate: RNA nucleotidyltransferase [EC 2. 7. 7. 6]) is the key enzyme for the transcription of genetic information in *Escherichia coli*. The RNA polymerase holoenzyme is a complex enzyme of molecular weight 500,000 and is composed of at least four different polypeptide chains, α, β, β', and σ, the molecular weight of each subunit being 38,500, 155,000, 165,000, and 87,000, respectively (*1, 2*). On the basis of the known molecular weights and the relative contents of these components, the holoenzyme is believed to have the structure $\alpha_2\beta\beta'\sigma$. A second form of the holoenzyme was found in *E. coli*, which has a novel subunit designated as σ', with σ subunit-like activity, in place of the regular σ subunit (*3*). The core enzyme with the structure $\alpha_2\beta\beta'$, devoid of both σ and σ' subunits, possesses all the enzyme activities associated with the polymerase for carrying out RNA synthesis, except that it can scarcely initiate RNA synthesis from DNA duplexes (*4, 5*). The complex structure of the flexible multi-subunit RNA polymerase is believed to be required to match the complexity of the transcriptional controls. A number of attempts have been made to reveal the role each subunit plays in transcription, and the mechanisms underlying the

1

synthesis and assembly of the subunits of this multimeric enzyme. Detailed analyses of subunit interactions in RNA polymerase are required for a complete understanding of the mutual relationships between RNA polymerase subunit assembly and their significance in the control of transcription. The present article summarizes our studies on the subunit interactions during the assembly of *E. coli* RNA polymerase.

I. ROLE OF HOLOENZYME SUBUNITS

The roles of RNA polymerase subunits in transcription might be studied most directly by analysis of isolated subunits. However, many of the functions associated with each subunit are not seen in the isolated state; the subunit assembly is required for their intrinsic activities to be exposed. In addition, functions that are dependent on the preceding step reaction(s) carried out by other subunits cannot be studied in this way. Thus, a number of procedures have been developed to elucidate the subunit functions (*6, 7*).

The existence of multiple functional sites required for RNA synthesis was indicated by early studies by chemical modifications (*8*) and the results are summarized in Table I. Certain modifications with chemical

TABLE I

Alteration of RNA Polymerase Activity by Different Chemical Treatments

Treatment reagents	Primary modifica- tion	Enzyme activity			
		DNA- binding	NTP- binding	Initia- tion	Elonga- tion
I *p*-Chloromercuribenzoate HgCl$_2$, cyanuric fluoride N-Ethylmaleimide, iodoacetate Tetrathionate, isocyanate	SH	−	+	−	−
II β-Naphthoquinone-4-sulfonate Monochlorotrifluoro-*p*-benzoquinone	NH$_2$	+	−	−	−
III Diazo-1-H-tetrazole Rose bengal+phototoxidation	His	+	+	+	−

reagents inactivate RNA polymerase, yielding enzyme preparations which are still capable of catalyzing some of the discrete steps of RNA synthesis. A number of reagents which react with sulfhydryl groups render the enzyme inactive in binding to DNA, initiation of RNA synthesis and elongation of RNA chains, but have no effect on the binding of nucleoside triphosphates to the enzyme. Since initiation and elongation of

RNA chains depend on the binding of the enzyme to DNA, the sulfhydryl groups appear to be essential for this binding reaction. Likewise, amino groups play an essential role in the substrate-binding reaction. Modification of histidine residues of the enzyme results in the formation of an enzyme preparation able to bind DNA and nucleoside triphosphates and catalyze the initiation reaction but unable to catalyze the elongation reaction. These experiments suggested for the first time that multiple functional sites corresponding to at least three different regions on the enzyme are involved in the enzymatic synthesis of RNA.

The distribution of these sites among different subunits of the polymerase was then investigated by reversible dissociation of the enzyme into partially active subassemblies (*9–11*). Subassemblies with functions to carry out some of the discrete steps in RNA synthesis were also prepared by reassembly of isolated subunits in various combinations (*12–15*). The functional role of each subunit was deduced by testing whether such subassemblies sufficed for the performance of partial functions. Treatment of RNA polymerase with *p*-chloromercuribenzoate causes stepwise dissociation of the α subunit from the enzyme, resulting in producing $\alpha\beta\beta'$ and $\beta\beta'$ complexes; the dissociated α subunit is reassembled into the enzyme structure upon the reversal of inhibition by sulfhydryl reagents (Fig. 1). Since both the $\alpha\beta\beta'$ and $\beta\beta'$ complexes are fully active in binding substrates, β and/or β' subunits are required for substrate binding. Affinity labeling of RNA polymerase with uridine triphosphate analogues indeed demonstrated the location of the substrate binding site(s) on the β subunit (*16*). Modification experiments with *p*-chloromercuribenzoate also demonstrated that the binding site for the antibiotic rifampicin, a potent inhibitor of bacterial RNA polymerases, is located on the β and/or β' subunits. In fact, all mutations conferring resistance to rifampicin affect only the β subunit (*17–20*).

A second type of enzyme dissociation is afforded by low concentrations of urea, which leads to $\alpha\beta$ and $\alpha\beta'$ complexes (Fig. 1) (*9, 11*). Both combinations of subunits are capable of binding to DNA. It thus seems that the DNA binding capacity resides on the β and β' subunits, though the observed DNA binding does not imply the same specific interactions as the holoenzyme. Since the isolated β subunit, but not the β' one, lacks the ability to complex with DNA, subunit α seems to be required for the DNA-binding activity to be exposed (*12*). In agreement with this interpretation, we have recently found temperature-sensitive *E. coli* strains carrying mutations on the α–subunit gene (*rpoA*) and that RNA polymerase

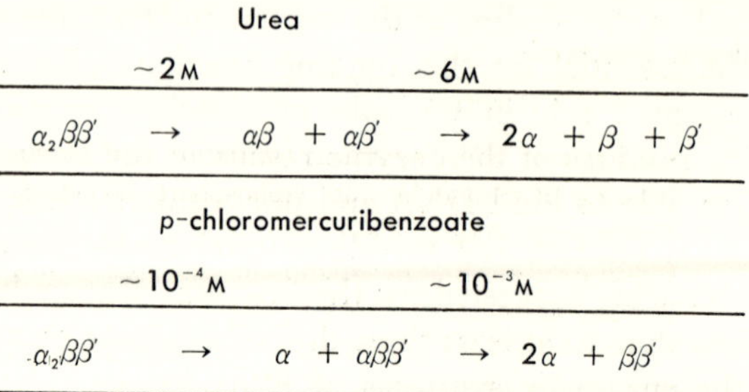

Fig. 1. Partial dissociation of RNA polymerase. RNA polymearse core enzyme was dissociated with 2 M urea into $\alpha\beta$ and $\alpha\beta'$ complexes which were further dissociated into individual subunits by treatment with 6 M urea. Treatment of core enzyme with 10^{-4} M p-chloromercuribenzoate (pCMB) yielded the $\alpha\beta\beta'$ complex which was further dissociated into free α and $\beta\beta'$ complexes upon increase of the pCMB concentration to 10^{-3} M. For details see Ref. *11*.

from these mutant cells exhibited considerably lower fidelity in transcription than wild-type RNA polymerases, implying that the α subunit is somehow involved in the recognition of the template sequence (*21*). The retention of DNA-binding activity appears to be essential for reassociation of these subassemblies to form the active enzyme, since reassociation is markedly enhanced by adding DNA (*12, 14*). Subassembly $\beta\beta'$ obtained by treatment with LiCl retains the ability to bind DNA (*9*). Poly-4-thio-thymidylic acid is cross-linked to the β' subunit (*16*) but photochemical cross-linking studies demonstrated the interaction of the β subunit with DNA (*22, 23*). Thus, the potential sites for DNA binding are supposed to reside on both the β and β' subunits. Studies with these partially dissociated subassemblies, however, cannot reveal the functions, when these depend on the preceding events in the sequence of RNA synthesis in which the deleted subunits participate. One possible way to circumvent this difficulty is to study the properties of enzymes reconstituted from one altered and other normal subunits. Establishment of a reconstitution system using isolated individual subunits is also needed to determine the mechanism underlying the formation of this multimeric enzyme and the control of subunit assembly. Affinity labeling with substrate and template analogues has also been employed for the identification of subunit functions. On the basis of our results, as well as those obtained in other laboratories by different procedures (*6, 7*), the distribution of the functional sites among different subunits is proposed to be as summarized in Table II.

TABLE II
RNA Polymerase Subunit Function

Subunit	Function
β	Catalysis of RNA synthesis (initiation and elongation)
	Binding of ribonucleoside triphosphate substrates
	Binding of product RNA
	Binding of antibiotics rifampicin and streptolydigin
β'	Binding of template DNA
α	Association of $\beta\beta'$ subunits
σ	Stimulation of initiation of RNA synthesis
	Stimulation of maturation of core enzyme

Subunit β is involved in binding both substrates and the antibiotics, rifampicin and streptolydigin, and in catalyzing phosphodiester bond formation during both initiation and elongation of RNA chains.

II. REVERSIBLE DISSOCIATION OF RNA POLYMERASE

Dissociation of *E. coli* RNA polymerase with high concentrations of urea leads to complete separation of all the subunits and, in appropriate conditions, removal of the denaturant leads to restoration of a major portion of the RNA polymerase activity (*18, 24, 25*). Reversible reactivation of the enzyme was first achieved by having glycerol present throughout the procedure (*24*), because in its absence the enzyme is irreversibly denatured by urea at concentrations above 2 M (*11*). Recovery of the enzyme activity depends on the ionic strength during both dissociation and reassociation and is remarkably enhanced upon exposure to high temperatures. It is well-known that subunit interactions in some enzymes are temperature dependent, where the hydrophobic interactions, which are markedly decreased at low temperatures, contribute as a major stabilizing force in subunit assembly (*26, 27*).

For the reconstitution of active polymerase from isolated individual subunits, the constituent subunits must first be separated and characterized. The initial reversible dissociation experiments indicated that subunits must be prepared in the continuous presence of glycerol and thus the first reconstitution was performed using subunits separated in the presence of glycerol by electrophoresis on cellulose acetate strips in buffer containing 6 M urea (*18*). The subunit isolation procedure is particularly useful if the amount of enzyme to be used is limited. For the preparation of large quantities of subunits, we employed the isolation procedure (Fig.

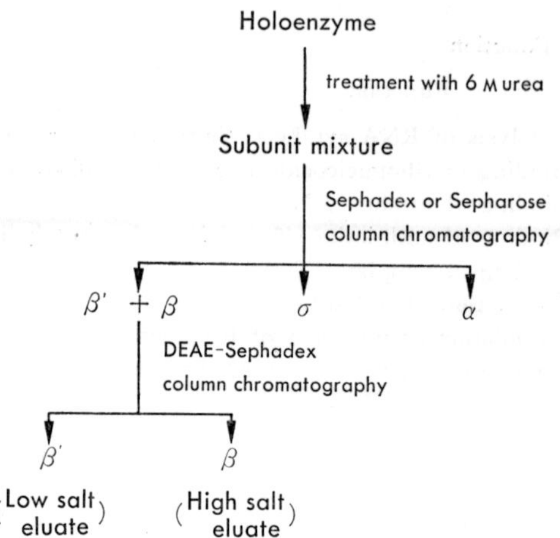

Fig. 2. Isolation of RNA polymerase subunits. RNA polymerase was completely dissociated into individual subunits by treatment with 6 M urea followed by fractionation of the resulting subunit mixture by two-step chromatography on Sepharose 6B (or Sephadex G200) and DEAE-Sephadex A50 columns. For details see Ref. *11*.

2), originally developed by Burgess (*1*), of a combination of sizing and ion-exchange chromatography and succeeded in the reconstitution of catalytically active enzymes from subunits thus isolated (*25*). Though the procedure of successive column chromatography allowed good separation of individual subunits and would be the first-choice method, a fairly satisfactory separation can be achieved by single-step chromatography on a DEAE-Sephadex or phosphocellulose column (*28, 29*). Subunits are also separated by chromatography on a blue-dextran-linked Sepharose column (*30*). Reconstitution of RNA polymerase can be achieved when the isolated subunits are mixed in the proper proportions. The best recovery of enzyme activity is obtained when the subunits are mixed in a molar ratio of $2:1:1$ for $\alpha:\beta:\beta'$, and excess amounts of the α subunit over the β and β' subunits are rather inhibitory to ordered assembly (*25*).

The addition of the fourth subunit, σ, during the core enzyme assembly significantly enhances the recovery of core enzyme activity (*31*). Likewise, the stimulation by DNA of the reactivation of RNA polymerase which has been dissociated with urea (*9, 24*) was observed when the enzyme reconstitution was performed with isolated subunits (*12*). The striking difference in the optimum salt concentrations for reconstitution in the presence and absence of these factors suggests that subunits are assembled in different ways depending on the presence or absence of these factors (*14, 32*). When the σ subunit or DNA is present during reconstitution, the factors may serve as nuclei for subunit assembly, leading to the formation of the holoenzyme or DNA-bound core enzyme, respectively.

The establishment of *in vitro* reconstitution from isolated subunits enabled the identification of altered subunits from mutant polymerases because hybrid enzymes in various combinations can be reconstituted. The technique of mixed reconstitution from subunits of wild-type and mutant polymerases indeed allowed the identification of the β subunit as the rifampicin- and streptolydigin-binding subunit (*17–20*). The identification of β′- (*33*) and α-subunit mutants (*21*) has also been carried out by the mixed reconstitution method.

III. PATHWAY OF SUBUNIT ASSEMBLY *IN VITRO*

The pathway of the assembly of RNA polymerase subunits has been extensively studied using the *in vitro* reconstitution system. In order to determine whether isolated subunits are assembled in a sequential process or in a concerted fashion, initial attempts were made to identify and characterize intermediate subassemblies. Formation of binary complexes was examined for all possible pairs, *i.e.*, α-β, α-β′, and β-β′. When the three combinations were dialyzed against the reconstitution buffer to remove urea and analyzed by glycerol gradient centrifugation, interaction was observed only between the α and β subunits but not between the α and β′ subunits (*25*). The α subunit that co-sedimented with the β subunit had a sedimentation coefficient of 9S, whereas the α subunit alone exhibited a sedimentation coefficient of about 4S. From the α subunit content and the known molecular weights of the two subunits, this α-β complex was found to have the structure $\alpha_2\beta$, not $\alpha\beta$. If subunit assembly is sequential, further addition of the third subunit β′ should convert this intermediate complex into active core enzyme. This was indeed shown to be the case only when the mixture was incubated at an elevated temperature. The active core enzyme thus formed had a sedimentation coefficient similar to that of native core enzyme. The recovery of enzyme activity is therefore accompanied by the recovery of a structural parameter. Detailed studies, however, revealed that subunit α associates to form a dimeric structure, which dissociates into a monomeric form at low protein concentrations, and the first reaction in the assembly sequence is the dimerization of the α subunit, because only the dimeric form of the α subunit is able to accept the β subunit (*14, 32*). Thus, the dimeric form of the α subunit is the immediate precursor of $\alpha_2\beta$ complex formation. Furthermore, the formation of active core enzyme from the $\alpha_2\beta$ complex and β' subunit was found to involve at least two step reactions (*12, 14*). At low temperatures, the $\alpha_2\beta$ complex and β' subunit associate to form the $\alpha_2\beta\beta'$ complex, which is identical to native core enzyme in subunit composition but is inactive in RNA synthesis; upon exposure to elevated temperatures, it regains the enzyme activity and, concomitantly, the conformation changes considerably (*34*). Thus, the pathway of the subunit assembly in the *in vitro* reconstitution system involves at least four step reactions (Fig. 3). As will be described later, evidence has accumulated which indicates that the subunit assembly *in vivo* proceeds in the same sequence as found *in vitro*.

Step	Reaction
I	$\alpha + \alpha \;\rightleftarrows\; \alpha_2$
II	$\alpha_2 + \beta \;\rightarrow\; \alpha_2\beta$
III	$\alpha_2\beta + \beta' \;\rightleftarrows\; \alpha_2\beta\beta'$
IV	$\alpha_2\beta\beta' \;\rightarrow\; E$
	$\alpha_2\beta\beta' + DNA \;\rightarrow\; E{\cdot}DNA$
	$\alpha_2\beta\beta' + Sigma \;\rightarrow\; E{\cdot}Sigma$

Fig. 3. Step-reactions of the *in vitro* assembly of RNA polymerase subunits. RNA polymerase core enzyme subunits were assembled by the four discrete step-reactions. The self-association of the α subunit is concentration dependent, and thus apparently reversible. The formation of premature core enzyme, *i.e.*, the inactive core complex, is also a concentration-dependent reversible reaction. The activation of the premature core enzyme, the fourth reaction, takes place *via* three different pathways, at least in our *in vitro* reconstitution system.

During the course of the reconstitution experiments, it was noticed that, in order to increase the recovery of enzyme activity, urea solutions must be freshly prepared prior to use (*28*). The reactivation of RNA polymerase, dissociated with urea, is markedly reduced when aged urea solutions are used without deionization. Prolonged treatment with high concentrations of urea inactivates enzyme subunits irreversibly, and upon reconstitution no appreciable reactivation can be achieved under the standard conditions. The presence of dithiothreitol and $MgCl_2$ in the dissociation buffer prevents this inhibition and, furthermore, reactivates denatured subunits. It is known that urea solutions develop considerable amounts of ammonium cyanate when stored (35), which reacts with the amino and sulfhydryl groups of proteins (35–37). The effect of potassium cyanate on the reactivation of RNA polymerase was then investigated (28). Upon exposure to cyanate, all the subunits are modified but it preferentially renders the β' subunit inactive, resulting in the accumulation of the $\alpha_2\beta$ complex and unassembled modified β' subunit. The treatment with cyanate has little effect on the formation of the $\alpha_2\beta$ complex, though both subunits are modified. The reaction of cyanate with amino groups is known to be slow and the product is stable over a rather wide range of pH (5 to 9), while its reaction with sulfhydryl groups is rapid and reversible at alkaline pH (35, 36). Since the inactivation of the β' subunit by cyanate is remarkable at acidic pH, the primary modification seems to be carbamylation of sulfhydryl groups, though modifications of imidazole, phenolic or carboxy groups (38) cannot be ruled out. This finding again supported the proposed sequence of the subunit assembly, in which the third step reaction is blocked by cyanate (Fig. 4). The important role of SH groups in the tertiary structure of polypeptides through hydrophobic interactions has already been suggested (39, 40). In fact, in the presence of p-chloromercuribenzoate, the assembly between $\alpha_2\beta$ and β' is blocked; but the modified $\alpha_2\beta$ complex is capable of associating with the native β' sub-

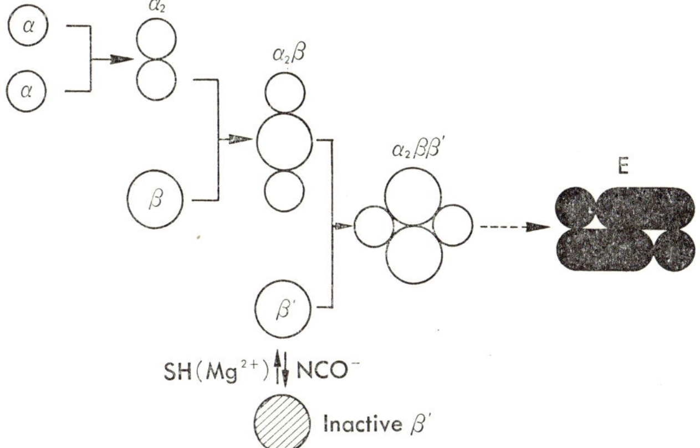

Fig. 4.   Inhibition of RNA polymerase assembly by cyanate. Cyanate developed in urea solutions binds primarily to sulfhydryl groups of RNA polymerase proteins. The modified β' subunit is unable to associate with the $\alpha_2\beta$ complex, while the modified $\alpha_2\beta$ complex is converted into active polymerase when the unmodified β' subunit is added. For details see Ref. 28.

unit (*28*). Modifications of SH groups render the polymerase inactive in binding to DNA (*8*). However, the DNA-binding activity of isolated β' subuint remains even after the modification with cyanate, implying that the DNA-binding site on the β' subunit does not overlap the site required for binding to the $\alpha_2\beta$ complex.

The kinetics of reconstitution were monitored by Palm *et al.* (*13*) after initiation by adding the β subunit to a solution containing the remaining subunits. Assembly of the $\alpha_2\beta$ complex is complete within 2 min; the $\alpha_2\beta$ complex, which reaches equilibrium rapidly with dissociated individual subunits, is stabilized by association with the β' subunit. We have found that the $\alpha_2\beta\beta'$ complex is also in rapid equilibrium with the dissociated $\alpha_2\beta$ complex and β' subunit (*14, 34*). Analogous to the nucleation in the folding of polypeptide chains (*40, 41*), sequential assembly seems to be required because the subunits are assembled in the proper sequence to yield intermediates that direct further assembly. The establishment of the sequential assembly in RNA polymerase allowed Friedman and Beychok (*42*) to propose that multi-subunit proteins with at least three nonidentical subunits assemble through multiple pathways.

IV. ENZYME MATURATION

The subunit σ is released from the holoenzyme after initiation of RNA synthesis (*43, 44*) and the resulting core enzyme with the structure $\alpha_2\beta\beta'$ is able to carry out the elongation of RNA chains. A model called the " sigma cycle " (*43*) was proposed in which the released σ subunit is reused for further initiation of transcription by binding to unused free core enzyme. Thus, the activity of core enzyme can be determined by measuring the rate of RNA chain elongation. DNA templates such as [d(A-T)] copolymer and denatured DNA lack the natural promoter structure and are therefore rather poor templates for holoenzyme; these artificial DNA's are, however, good templates for core enzyme (*3, 5, 45*). Since both holoenzyme and core enzyme exhibit similar activities with these templates, the activity of the core unit in enzymes reconstituted under various conditions and in the presence or absence of the σ subunit can be simply compared by measuring activities using these DNA templates. The model in which the assembly of core enzyme subunits is enhanced by the σ subunit (*14, 31, 32*) was first indicated by the reproducible findings that the RNA polymerase reconstituted in the presence of the σ subunit exhibited higher activity than an equivalent amount of core enzyme reconstituted in its absence, even if the σ subunit was added prior to the assay. The possibility was then raised that the core enzyme unit itself is functionally different, depending upon whether it is assembled in the presence of the σ subunit or not. Finally a quantitative relationship was obtained between the recovery of core enzyme activity and the amount of σ subunit added during the enzyme reconstitution.

On the other hand, the proposal that the reconstitution of RNA polymerase is enhanced by DNA was based on initial reversible dissociation experiments (*9, 11, 24*). Later, enhancement by DNA was observed in reconstitution from isolated individual subunits (*12, 14, 25*). When DNA is added, the active polymerase is formed from mixtures of isolated α, β, and β' subunits but lacking the σ subunit. Among the DNA samples tested, the order of enhancing activity was as follows: poly[d(A)]·poly [d(T)] > poly[d(A-T)]·poly[d(A-T)] > denatured DNA > native DNA > poly[d(G)]·poly[d(C)]. This order of reconstitution enhancement is identical to that of template activity for core enzyme. Moreover, the amounts of DNA required for enzyme formation are close to those needed for maximum activity with the same amounts of native core enzyme. These observations seem to suggest that only a limited number of unique sites serve as the site for construction of active polymerase. In the presence of DNA, active enzyme is indeed constructed on the DNA added as an enhancing factor, and initiates RNA synthesis without prior release from the DNA-enzyme complexes (*12*). For example, when the core enzyme was reconstituted in the presence of poly[d(G)]·poly[d(C)] and was assayed after the addition of poly[d(A)]·poly[d(T)], the reconstituted enzyme synthesized predominantly poly(G) or poly(C) but neither poly(A) nor poly(U). Likewise, when the enzyme reconstitution was carried out in the presence of poly[d(G)]·poly[d(C)], the incorporation was observed only for GTP or CTP, not for ATP and UTP. When the enzyme assay was performed at elevated KCl concentrations, however, significant incorporation of ATP or UTP was observed concomitantly with a decrease in the GTP and CTP incorporation. Thus, active enzymes are assembled on the DNA added during reconstitution and initiate RNA synthesis on the same DNA, but the enzymes remain active even after transferring to DNA added prior to the assay but after reconstitution.

Systematic attempts have been made to understand the molecular mechanism of the enhancement of enzyme assembly by either the σ subunit or DNA. Among the discrete steps of subunit assembly, the formation of the primary and secondary intermediates, *i.e.*, α_2 and $\alpha_2\beta$ complexes, is not affected by the presence or absence of the assembly-enhancing factors. The subsequent association of the $\alpha_2\beta$ complex with the β' subunit proceeds in the absence of the factors as well as in their presence. When the assembly reaction is carried out at low temperatures, however, this $\alpha_2\beta\beta'$ complex is totally inactive in catalyzing RNA synthesis. Thus, when the reconstitution is carried out at low temperatures, the subunit assembly stops until the inactive core complex is formed. This inactive $\alpha_2\beta\beta'$ complex was designated " premature core enzyme " (*12*). The formation of premature core enzyme proceeds rapidly under various conditions and even at low temperatures, but the activation of premature core enzyme takes place only under strict conditions and at high temperatures (*14, 32*). Incubation at elevated temperatures is essential for the premature core enzyme to be activated, but the temperature dependence of enzyme maturation is complex, exhibiting a nonlinear Arrhenius plot (Ishihama, unpublished observation). Differences between the premature and native core enzymes were further confirmed, as summarized below, by their different structural parameters, indicating that the temperature-dependent maturation is accompanied by the recovery of not only the activity but also of the structure. The rate-limiting step in the reconstitution of active polymerase is its temperature-dependent formation from the premature core enzyme and this step reaction is stimulated by either the σ subunit or DNA (*14, 32*).

In the course of activation of the premature core enzyme by the σ subunit, it may interact with both the $\alpha_2\beta$ complex and β' subunit. This concept is based on the following observations: subunit β'-DNA interaction is inhibited by addition of σ subunit and this inhibition of DNA-binding activity of the β' subunit is due to effective competition of the σ subunit with DNA for binding to the β' subunit (*12*), because the isolated σ subunit itself exhibits no binding to DNA (*12, 46*) (however, the σ subunits in DNA-RNA polymerase complexes can be photochemically cross-linked to DNA (*22, 23*)). This phenomenon is analogous to the observation that the addition of σ subunits to the core enzyme-DNA complexes leads to rapid dissociation of the core enzyme from the complex and as a result enables specific binding of holoenzyme to promoters (*47*). Taken together, one of the functions of the σ subunit is to lower the affinity of the β' subunit, and thus of the core enzyme, for its non-specific binding to DNA. Interaction of the σ subunit with the $\alpha_2\beta$ complex is evident because portions of holoenzyme can be dissociated into $\alpha_2\beta\sigma$ complexes, which can be isolated, and free β' subunits by passing it through phosphocellulose columns (*12, 46*). RNA polymerase from *Lactobacillus curvatus* is totally disassembled into the $\alpha_2\beta$ and $\beta'\sigma$ complexes by the same procedure (*48*). Such subassemblies can also be observed in the supernatant solution after enzyme preparations are mixed with excess DNA (*46*). During enzyme maturation, therefore, the σ subunit interacts with both $\alpha_2\beta$ and β', leading to tight assembly of the two constituents; in this model, the σ subunit is considered to be a macromolecular effector influencing the conformation of the flexible RNA polymerase core enzyme. Since the σ subunit is released from the holoenzyme for reuse once it has been used in the initiation of RNA synthesis (*43, 44*), the released σ subunit may function catalytically for the sequential maturation of core enzyme as well as for the integration of unused matured core enzyme into the transcription cycle (*31*) (Fig. 5).

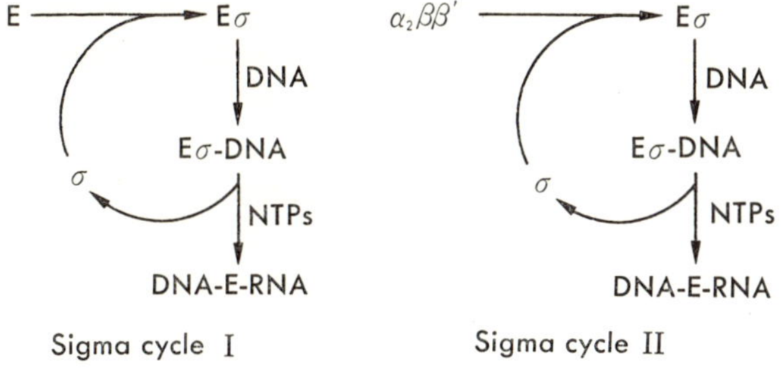

Fig. 5.   The "sigma cycle." Cyclic reuse of the σ subunit was proposed for transcription initiation (sigma cycle I) (*43*). In addition, the σ subunit was found to interact with the premature core enzyme and activated it (sigma cycle II) (*31*).

Another reconstitution-enhancing factor, DNA, acts in much the same way: among the discrete steps in the sequential assembly of RNA polymerase, the last step, *i.e.*, the activation of premature core enzyme, is enhanced by DNA (*12*). It is ruled out that the putative binary complexes, $\alpha_2\beta$-DNA or β'-DNA, are intermediates in DNA-dependent reconstitution. Instead, DNA may interact with both integral parts assembled into a premature core, and associate them to form conformationally tight complex with enzyme activity.

It has been proposed that the premature core enzyme can be activated in three different ways (Fig. 6): i) activation in the presence of σ (or σ') subunits leading to the formation of active holoenzyme (*3, 13, 14, 18, 31*); ii) activation in the presence of DNA leading to the formation of the

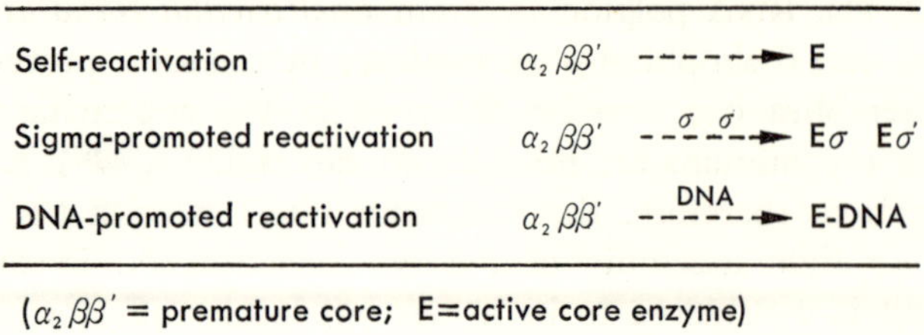

Self-reactivation	$\alpha_2\,\beta\beta'$ $- - - - - \rightarrow$ E	
Sigma-promoted reactivation	$\alpha_2\,\beta\beta'$ $- -\overset{\sigma}{-}\overset{\sigma}{-} -\rightarrow$ Eσ Eσ'	
DNA-promoted reactivation	$\alpha_2\,\beta\beta'$ $- -\overset{\text{DNA}}{- - -}\rightarrow$ E-DNA	

($\alpha_2\,\beta\beta'$ = premature core; E=active core enzyme)

Fig. 6.   Maturation *in vitro* of core enzyme. The premature core enzyme of inactive $\alpha_2\beta\beta'$ complex is converted into enzymatically active forms by three pathways. Among the three reactions, self-reactivation of the premature core takes place only in the presence of high concentrations of KCl or glycerol. The σ subunit-promoted reactivation can be observed in conditions supposedly close to those *in vivo*. See text for details.

active core enzyme-DNA complex (*12, 14*); and iii) activation in the absence of maturation-enhancing factors (*14, 29, 49, 50*). Considering the disagreement, it should be emphasized that several factors, including reaction temperature, protein concentration and medium composition, influence the rate and extent of the yield of enzyme activity, and that our reconstitution systems differ in these conditions from the other reconstitution systems so far reported. Systematic comparison of the yields of enzyme activity recovered under various conditions resolved this disagreement (*14, 32*): in the presence of high concentrations of salt or glycerol, the premature core alone is activated without the addition of either the σ subunit or DNA, but the premature core is activated even at salt concentrations as low as that *in vivo* and in the absence of glycerol if the incubation is carried out in the continuous presence of one of the factors. In fact, Yarbrough and Hurwitz (*29*) used a reconstitution buffer containing 0.5 M KCl and 20% glycerol, Harding and Beychok (*49*) used a buffer containing 0.45 M KCl and 10% glycerol, and Lill *et al.* (*50*) also used a buffer containing high concentrations of salt (0.4 to 0.5 M) and glycerol (44 to 55%) for enzyme reconstitution. In contrast all the experiments performed by us (*14, 31, 32*), and Zillig and his colleagues (*13, 18*), which indicated a positive effect of the σ subunit on enzyme reconstitution, were performed at lower concentrations of salt and glycerol. Since the optimum conditions for enzyme maturation are strict and specific for the three reactions, special care should be taken in choosing the concentrations of salt and glycerol for reconstitution to achieve complete recovery of enzyme activity. The optimum pH is approximately 9.0, 8.5, and 9.5 for self-, DNA-promoted, and σ subunit-promoted maturation, respectively. The concentration of premature core and the amounts of factors to be added should also be carefully determined for the best recovery of enzyme activity, because the premature core enzyme is in equilibrium with the mixture of the $\alpha_2\beta$ complex and β' subunit (*34*).

V. NATURE OF PREMATURE CORE ENZYME

One of the unique features in the assembly of RNA polymerase subunits is the formation of premature core enzyme. This proposal has been supported by others. Harding and Beychok (*49*) reported evidence indicating that denatured inactive polymerase, which was formed by the removal of guanidine at low temperature and thus might be identical to the premature core enzyme we proposed, was highly structured. Kinetic studies by Yarbrough and Hurwitz (*29*) also indicated that, when the denaturant was rapidly removed by dilution, denatured polymerase regains the physical characteristics, *e.g.*, protein fluorescence and circular dichroism (CD), of native enzyme before regaining the enzyme activity. In this respect, it might be worthwhile to recall the proposal of Teipel and Koshland (*51*) that differences in the rate of recovery of physical structure and of enzymatic activity indicate the presence of folded but inactive forms of enzymes, which are capable of undergoing further conformational changes to yield active enzymes. Thus, systematic comparison was made of the structure between the premature and native core enzymes (*34*). Although the inactive preparation of assembled polymerase may be a mixture of subassemblies, some evidence has accumulated which indicates that at least parts of the core subunits are indeed assembled into the $\alpha_2\beta\beta'$ complex, the premature core enzyme: all three core subunits are co-precipitated by specific antisera against each subunit (*52*); cross-liked aggregates of all possible subunit combinations are formed with bifunctional reagents (*34*); and all subunits are eluted simultaneously from a DEAE-Sephadex column (*34*). The gross structure is similar for the premature and native core enzymes, but minor differences exist, as summarized in Table III, which could be observed from absorption spectra, CD spectra, sedimentation behaviors, protease sensitivities, hydrogen-tritium exchange rates, reactions to bifunctional cross-linking reagents, and elution profiles from a phosphocellulose column (*34*).

Although the shape of absorption spectra is similar between the premature and native core enzymes, the degree of absorbance is higher for the premature core by 8.6% at 280 nm; the molecular extinction at

TABLE III

Differences in the Structure between Premature and Native Core Enzymes

	Premature	Native
Sedimentation coefficient	8.8–11.7S	12.8S
PC chromatography	$\alpha_2\beta+\beta'$	$\alpha_2\beta\beta'$
Hydrogen-tritium exchange rate	High	Low
Extent of $\beta\beta'$ cross-link	Low	High
Protease sensitivity	High	Low
Circular dichroism	(min. 208–209 nm, max. 193 nm)	
$[\theta]$ value 209 nm	0.80	1.0
222 nm	0.84	1.0
278 nm	0.67	1.0
269 nm/278 nm	0.95	0.85

280 nm is 2.863×10^5 for premature core and 2.636×10^5 for native core enzymes, respectively. The CD spectra in the far ultraviolet range are essentially similar for the premature, reconstituted and native core enzymes: minimum at 208–209 nm and maximum at 193 nm, and a less prominent minimum at near 220–222 nm. The molecular ellipticities at 222 nm, indicative of the α helix content, of premature and reconstituted core enzymes are 83.6 and 85.6%, respectively. The denatured core enzyme in 6 M urea buffer exhibits a completely different spectrum, but one essentially similar to that observed with guanidine-treated RNA polymerase (49). As summarized in Table IV, the amounts of α helix, β sheet and other structures determined from the CD spectra between 190 and 239 nm and according to the method of Chen et al. (53) are nearly equal among the three enzymes within experimental errors. Thus, the premature core enzyme has a secondary structure identical to or very similar to that of native core enzyme. However, the intensity of near-ultraviolet CD spectra, which is indicative of the optical activity of aromatic chromophores, is considerably smaller for premature core enzyme than native enzyme, the magnitude of the Cotton effect being 1.5-fold

TABLE IV

Helix and β Form Contents of RNA Polymerase

Enzyme	FH (%)	FB (%)	FR (%)	HNR (%)	HL (%)
Premature core	45.7	19.5	34.8	356.8	4.7
Reactivated core	46.0	21.2	32.7	372.3	4.5
Native core	44.3	18.3	37.3	303.0	5.3

FH, the fraction of helix; FB, the fraction of β form; FR, the fraction of unordered form; HNR, the number of helical segments; HL, the average size of helical segments.

higher for native core enzyme than premature core enzyme. In addition, the ratio of $[\theta]_{269nm}/[\theta]_{278nm}$ is lower for native (0.85) than for premature core (0.95) enzyme. The apparent difference in the shape of spectra in the near-ultraviolet region implies that the enzyme maturation may be accompanied by structural alteration at least in the environment of aromatic amino acid residues in core enzyme subunits. Since the intensity in the near-ultraviolet range does not increase to the native enzyme level if premature core enzyme is reactivated in the absence of maturation factors, *e.g.*, DNA or σ subunit, it is further shown that the maturation of RNA polymerase proceeds in the presence of factor(s).

The gross structural difference between premature and native core enzyme was also indicated by the difference in the rate of hydrogen-tritium exchange, which is known to be influenced by the state of exchangeable hydrogens in macromolecules. The tritium incorporated into the rapidly exchangeable hydrogen sites, which are located on the side chains or surfaces of subunit polypeptides, is exchanged-out within the time, approximately 10 min, required for separation of tritiated protein from tritium water by Sephadex column chromatography. The amount of stably bound tritium, which is retained on slowly exchangeable sites in buried portions of highly structured core enzyme, reaches a plateau level, about 20% of the total exchangeable hydrogens, after 7 days' exchange-in incubation at 4°C. The rate of exchange-out of the stably bound tritium is more than 10% higher after 1 hr of exchange-out for the premature core enzyme than the native core enzyme. However, the amount of most tightly bound tritium, which is not exchanged-out even after 5 hr, is virtually the same between the two enzyme proteins, *i.e.*, approximately 3–4% of the total exchangeable tritium. The backbone structure of RNA polymerase seems to be completely established when the premature core enzyme is assembled, and the temperature-dependent maturation of premature core enzyme should be accompanied by minor alterations in the structure, which produce labeled tritium shielded from contact with solvent or prevent water molecules from penetrating into some of the labeled sites buried due to the enzyme maturation. The determination of the tritium-exchange rate also demonstrated that the rate and extent of tritium exchange-out for isolated individual subunits are higher than those for both premature and native core enzymes (Aiba and Ishihama, unpublished observation).

The sensitivity of premature core to protease digestion is higher than native core enzyme. The two larger subunits, β and β', of premature core enzyme are degraded by trypsin into several fragments with molecular weights of 125,000 and 47,000 for the two major, and of 135,000, 115,000, 75,000, and 52,000 for the minor fragments. In addition, fragments of the α subunit with molecular weights smaller than 20,000 are produced in the presence of higher concentrations of trypsin. The origin of each polypeptide mentioned above can be determined by testing the cross-reaction to antisera against each subunit and this will be described elsewhere (Ishihama *et al.*, in preparation). In contrast, the native core enzyme is hardly digested by trypsin up to the concentration of 1 mg/ml in the presence of 10% glycerol, although the cleavage products obtained with higher concentrations of trypsin are essentially the same as those from premature core enzyme. The kinetics of digestion by α-chymotrypsin are much the same as found with trypsin. It has been demonstrated that assembly intermediates *in vivo* including premature core enzyme, accumulated in the temperature-sensitive assembly-defective *E. coli* mutants, are rapidly and preferentially degraded (*54–56*), though assembled RNA polymerase is one of the metabolically stable constituents in exponentially growing cells without detectable degradation (*57, 58*).

The sedimentation coefficients of enzymes reactivated by the three ways are essentially the same (12.5–13.4S) as those of native holo- and core enzymes if centrifugation was performed at an ionic strength above 0.25. In the absence of salt, however, both native and reconstituted core enzymes sediment faster, forming a peak around 20–25S, which indicates that the reconstituted core enzyme as well as native core enzyme is capable of associating to form a dimer or multimers. Both native and reconstituted core enzymes exhibit similar sedimentation behaviors to native holoenzyme when the σ subunit is added prior to centrifugation. In contrast, the sedimentation velocity of premature core enzyme is remarkably influenced by the protein concentration during centrifugation: the sedimentation coefficient at the lowest protein concentration examined was 8.9S, which is as low as that of the $\alpha_2\beta$ complex; and the value at the highest concentration was 11.7S, as high as that of native and reconstituted core enzymes. Nevertheless, all the subunits of premature core enzyme sediment together forming a single major peak, and neither the $\alpha_2\beta$ complex nor free β' subunit peak can be obtained through glycerol gradient. Moreover, when the premature core enzyme peak at the lowest protein concentration is treated with antibodies against individual subunits, at least half of all the core subunits are coprecipitated (*52*). Since subunit assemblies like native holoenzyme are dissociated, at least in part, by treatment with antibodies against individual subunits (Enami, unpublished observation), the observed values of co-precipitation are high enough to substantiate that major portions of the core subunits are assembled in premature core enzyme preparations. The concentration-dependent variation in the sedimentation coefficient might be best explained by saying that the premature core enzyme reaches equilibrium rapidly with a mixture of the $\alpha_2\beta$ complex and free β' subunit, in general agreement with the sedimentation behavior of complexes with instantaneous equilibration (59), whereas mature core enzyme is apparently not dissociated any further into subassemblies. Supporting this interpretation is the finding that the premature core enzyme is dissociated by passing it through a phosphocellulose column into the respective constituents. By passing premature core enzyme preparations through a DEAE-Sephadex column, however, most of the core proteins are recovered in a single peak. Since β' alone is a basic protein and tightly bound to DNA (12, 46), it is reasonable that polyanions such as phosphocellulose enhance the dissociation of premature core by removal of released β' into the polyanion-bound form. In accordance with this notion the $\alpha_2\beta$ complex is recovered in the supernatant when a mixture of premature core and DNA is centrifuged, while in the absence of DNA all core enzyme subunits are quantitatively recovered as complexes (Ishihama, unpublished observation). This polyanion-induced dissociation of premature core enzyme is no longer observed concomitant with the reactivation of premature core enzyme by exposure to elevated temperatures.

Finally, the difference in the quanternary structure between native and premature core enzymes was investigated by chemical cross-linking with bifunctional reagents, dimethylsuberimidate or dimethyladipimidate. The time course of the appearance of cross-linked products analyzed by electrophoresis under dissociating conditions indicated that three cross-linked bands were formed for native core enzyme, the molecular weights of each being 320,000, 210,000, and 190,000, with the estimation based on the mobility of non-cross-linked subunits as markers. These might represent the $\beta\beta'$, $\alpha\beta'$, and $\alpha\beta$ complexes, respectively, as identified by Hillel and Wu (60). The most extensive cross-linking involved the β and β' subunits for native core enzyme. Prolonged exposure to bifunctional reagents resulted in the formation of a small amount of the α dimer with a molecular weight of 80,000. The cross-linked products from premature core enzyme were essentially the same for those from native core enzyme. However, the rate and extent of cross-linking appeared to be different: the formation of the $\beta\beta'$ complex was slower and less for premature core enzyme, while that of the $\alpha\beta$ and $\alpha\beta'$ complexes was faster and more for premature core enzyme than native core enzyme. A difference in the relative yield of the cross-linked subunit complexes was observed when the mixtures of RNA polymerase core enzymes and various concentrations of bifunctional reagents were incubated for the same time period. These observations imply that the distance between the β and β' subunits is larger in premature core than in native core enzyme.

Although the premature core enzyme is incapable of catalyzing RNA synthesis, it is partly functional, retaining ribonucleoside triphosphates (*61*; Ishihama, unpublished observation), DNA (*12, 61*) and rifampicin binding activity (*61, 62*; Ishihama, unpublished observation). Thus, as summarized in Table V, most of the activities required for RNA synthesis

TABLE V

Activities of Assembly Intermediates

Activity	Subunit or subunit complex					
	α	β	β'	$\alpha_2\beta$	$\alpha_2\beta\beta'$	E
DNA binding	−	−	+	+	+	+
Rifampicin binding	−	−	−	+	+	+
Substrate binding	−	−	−	+	+	+
Sigma binding	−	−	+	+	+	+
RNA synthesis	−	−	−	−	−	+
Autogenous repression	−	−	−	+	?	+

are exposed until the premature core is formed except the catalytic activity of phosphodiester bond formation. Not only premature core enzyme but both the $\alpha_2\beta$ complex and β' subunit interact with the σ subunit (*12*), but the affinities are considerably smaller than the interaction between native core enzyme and the σ subunit. Thus, it is impossible to isolate the premature core-σ subunit complex by conventional glycerol or sucrose centrifugation. In contrast, the association constants between the σ subunit and native or reconstituted core enzymes are high enough to enable the isolation of the holoenzyme complex (*12, 63*). The catalytic activity is regained during the maturation reaction. However, detailed analyses indicated that the catalytic specificities may differ among enzymes reactivated in the three different ways. The evaluation of the structural and functional properties of reconstituted enzymes using the three different maturation methods relative to the initial and final states is in itself informative in the determination of the assembly pathway, and in particular of maturation. Core enzyme reactivated in the absence of maturation-enhancing factors, *i.e.*, self-reactivated enzyme, is as active in RNA synthesis as native core enzyme but differs in the following ways *(15)*: i) activity ratios of reactivated enzyme directed by varieties of DNA templates are not identical with those of native enzyme; ii) the K_m value of T7 DNA, but not of poly[d(A-T)]·poly[d(A-T)], for reactivated enzyme is higher than that for native enzyme; iii) the activation energy of RNA synthesis catalyzed by reactivated enzyme is lower than that of native enzyme; iv) the increase in RNA synthesis upon raising the temperature is smaller with reactivated enzyme than native enzyme; v) RNA synthesis with reactivated enzyme is stimulated by the σ subunit at lower concentrations than those needed for native core enzyme. Since these characteristics originate from the initiation among the discrete steps of RNA synthesis and since the rate of RNA chain elongation is indistinguishable between the reactions catalyzed by native and reactivated enzymes, it appears that self-reactivated core enzyme is different from the native core enzyme in the specificity of transcription initiation.

In concert with the difference in the transcription specificity, self-reactivated core enzyme gives a different shape of near-ultraviolet CD spectrum from that of native core enzyme *(34)*. But it remains to be determined that such a partially reactivated enzyme is an obligatory intermediate in the process of active enzyme formation, because structured species not on the renaturation pathway may be formed during the renaturation of some enzymes *(51, 64)*. These observations together with the finding that self-reactivation proceeds only in non-physiological conditions, *i.e.*, in the presence of high concentrations of salt or glycerol *(14)*, might suggest that an RNA polymerase whose structure and function are completely identical to those of native core enzyme can be formed *via* one or both of the other two maturation pathways. Studies are in progress on the properties of the enzymes formed in the presence of either DNA or the σ subunit, and should provide a critical test of the above hypothesis.

VI. ASSEMBLY INTERMEDIATES *IN VIVO*

On the basis of the understanding of the subunit assembly mechanism in the *in vitro* reconstitution system, our studies were then focussed on the *in vivo* assembly of RNA polymerase.

First, attempts were made to detect precursors or assembly intermediates in the intracellular pathway of RNA polymerase formation. For this purpose, the distribution of pulse-labeled polymerase proteins was examined (*65*): exponentially growing cells of wild-type *E. coli* were pulse-labeled with radioactive amino acids and the cell lysates were fractionated by glycerol gradient centrifugation; the distribution of pulse-labeled subunits was analyzed by the combination of precipitation with specific antibodies and polyacrylamide-gel electrophoresis of the antibody precipitates in the presence of sodium dodecyl sulfate. With cells labeled for 30 sec, major portions of the labeled subunits, *i.e.*, 80% of α, 40% of β, and 15% of β', are found in slowly sedimenting subassemblies, while the majority of polymerase proteins continuously labeled with radioactive amino acids are recovered in the RNA polymerase structure of 14–15S fractions. Both pulse- and continuously labeled α subunits exist in the unassembled free α peak of 4–5S, whereas only pulse-labeled α and β subunits can be recovered in the 8–10S $\alpha_2\beta$ complex peak, indicating the existence of an unassembled α subunit pool. The ratio of pulse-labeled subunits in 14–15S intact enzyme peak to slowly sedimenting fractions increased concomitantly with the chase and after a 15 min chase of a 30-sec labeled culture, the distribution of pulse-labeled subunits was identical to that of continuously labeled polymerase proteins.

In the pulse-labeled proteins which are precipitated by anti-holoenzyme serum as well as by anti–α subunit serum and which co-migrated with the α subunit upon electrophoresis, more than 80% is recovered in the 4–5S region after glycerol gradient centrifugation. Moreover, 15–20% of the continuously labeled α subunits is recovered in this unassembled fraction. Evidence has accumulated indicating that this 4–5S material indeed represents the unassembled free form of the α subunit (*65, 66*): i) it is precipitated by the specific antiserum against the α subunit; ii) the electrophoretic mobility is identical to the authentic α subunit; iii) it can be assembled into the polymerase structure by the addition of native β and β' subunits isolated from intact polymerase; iv) upon chasing the pulse-labeled culture, the labeled α subunit in unassembled fractions is incorporated, at least in part, into the polymerase structure; and v) the tryptic peptide patterns of the putative free α subunit pulse- or continuously labeled with radioactive lysine or arginine are identical to those of the α subunit integrated into the intact RNA polymerase. The amount of free forms of the α subunit is considerably large, approximately 80% of the pulse-labeled α and, even after a prolonged chase, about 10% of the α subunits remains unassembled. This is consistent with the observations that the synthesis ratio of subunit polypeptides is 2–3:1:1 for $\alpha : \beta : \beta'$ (55, 57, 58, 67) and thus significantly more of the α subunit than the amount needed to match the β and β' subunits is produced. In contrast, the pool size of the $\alpha_2\beta$ complex is very small; less than 1% of the pulse-labeled α subunit is recovered in this complex, which is rapidly incorporated into the enzyme structure upon chasing. Since the sedimentation velocity of premature core enzyme varies and is dependent on the concentration between 8–15S, the fractionation method is not suitable for separation of premature core enzyme from native enzyme and the $\alpha_2\beta$ complex. Thus, it has been impossible to detect the premature core, if any, in extracts from wild-type *E. coli* cells. However, as described later, the *in vivo* existence of premature core enzyme was determined with use of assembly-defective mutants (54, 56). The identification of assembly

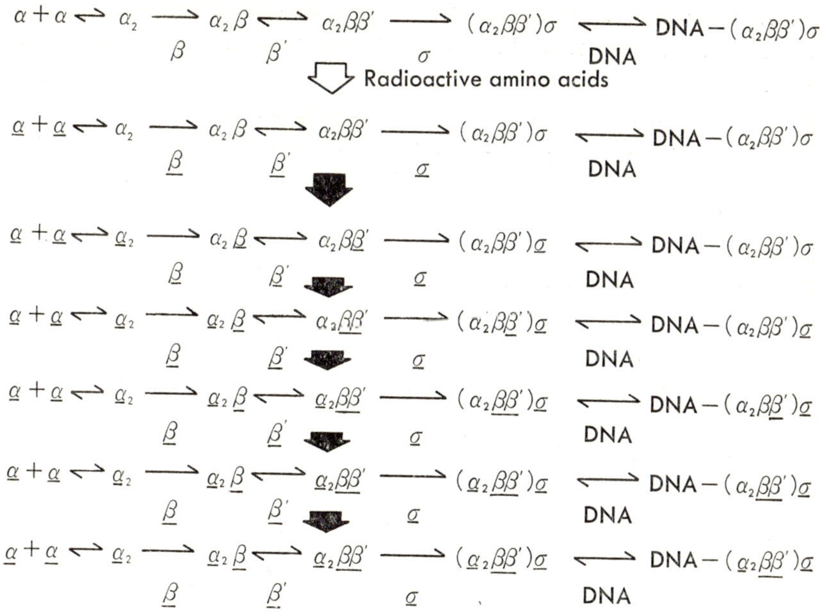

$$\alpha + \alpha \rightleftharpoons \alpha_2 \xrightarrow[\beta]{} \alpha_2\beta \rightleftharpoons \alpha_2\beta\beta' \xrightarrow[\sigma]{} (\alpha_2\beta\beta')\sigma \xleftrightarrow[DNA]{} DNA-(\alpha_2\beta\beta')\sigma$$

$$\Downarrow \text{ Radioactive amino acids}$$

$$\underline{\alpha} + \underline{\alpha} \rightleftharpoons \alpha_2 \xrightarrow[\underline{\beta}]{} \alpha_2\beta \rightleftharpoons \alpha_2\beta\beta' \xrightarrow[\underline{\beta'}]{} (\alpha_2\beta\beta')\sigma \xleftrightarrow[DNA]{} DNA-(\alpha_2\beta\beta')\sigma$$
$$\underline{\sigma}$$

$$\Downarrow$$

$$\underline{\alpha} + \underline{\alpha} \rightleftharpoons \underline{\alpha}_2 \xrightarrow[\underline{\beta}]{} \alpha_2\underline{\beta} \rightleftharpoons \alpha_2\beta\underline{\beta'} \xrightarrow[\underline{\sigma}]{} (\alpha_2\beta\beta')\underline{\sigma} \xleftrightarrow[DNA]{} DNA-(\alpha_2\beta\beta')\sigma$$

$$\alpha + \alpha \rightleftharpoons \alpha_2 \xrightarrow[\beta]{} \alpha_2\underline{\beta} \rightleftharpoons \alpha_2\underline{\beta\beta'} \xrightarrow[\underline{\sigma}]{} (\alpha_2\beta\beta')\underline{\sigma} \xleftrightarrow[DNA]{} DNA-(\alpha_2\beta\beta')\underline{\sigma}$$

$$\alpha + \alpha \rightleftharpoons \alpha_2 \xrightarrow[\beta]{} \alpha_2\beta \rightleftharpoons \underline{\alpha}_2\beta\beta' \xrightarrow[\underline{\sigma}]{} (\alpha_2\underline{\beta\beta'})\underline{\sigma} \xleftrightarrow[DNA]{} DNA-(\alpha_2\beta\underline{\beta'})\underline{\sigma}$$

$$\alpha + \alpha \rightleftharpoons \alpha_2 \xrightarrow[\beta]{} \alpha_2\underline{\beta} \rightleftharpoons \underline{\alpha}_2\beta\beta' \xrightarrow[\underline{\sigma}]{} (\underline{\alpha}_2\beta\beta')\underline{\sigma} \xleftrightarrow[DNA]{} DNA-(\alpha_2\underline{\beta\beta'})\underline{\sigma}$$

$$\underline{\alpha} + \underline{\alpha} \rightleftharpoons \alpha_2 \xrightarrow[\underline{\beta}]{} \alpha_2\beta \rightleftharpoons \alpha_2\beta\beta' \xrightarrow[\underline{\sigma}]{} (\underline{\alpha}_2\beta\beta')\underline{\sigma} \xleftrightarrow[DNA]{} DNA-(\underline{\alpha}_2\beta\beta')\underline{\sigma}$$

Fig. 7. Kinetics of subunit assembly *in vivo*. To exponentially growing wild-type *E. coli* cells, radioactive amino acids were added. From the experimental results (28), the incorporation of radioactive amino acids into RNA polymerase subunits, subassemblies and the matured enzyme structure is illustrated. The underlined subunits represent radioactively labeled molecules, which appear in the enzyme structure in the following order: $\beta' \rightarrow \beta \rightarrow \alpha$.

intermediates suggests, but is not evidence for, a particular pathway of assembly. The simplest interpretation is that it lies directly on the pathway from the initial to final state, but it may also be an intermediate not directly on the assembly pathway. These two alternatives can be kinetically distinguished from one another. The pulse-chase experiments revealed, as illustrated in Fig. 7, that the pulse-labeled enzyme subunits appear in the enzyme structure in the order of $\beta' \rightarrow \beta \rightarrow \alpha$ *(65)*, which is in good agreement with the order expected from the known sequence of subunit assembly *in vitro*.

VII.   ASSEMBLY-DEFECTIVE MUTANTS

To characterize further the nature of intermediate subassemblies *in vivo*, efforts have been made to find mutants defective in the assembly of RNA polymerase leading to accumulation of unassembled subunits or subassemblies. A number of temperature-sensitive *E. coli* strains carrying mutations in the genes encoding RNA polymerase subunits have been examined for the assembly of newly synthesized subunits in these cells at nonpermissive temperatures. Since the intracellular concentration of RNA polymerase is maintained at levels characteristic of the rate of cell growth *(32, 57, 67)*, it is expected that cells which exhibit abnormal rates of subunit synthesis might be defective in the assembly of enzyme subunits. In accordance with this expectation, most of the assembly-defective mutants so far identified exhibited abnormal features of subunit synthesis *(55, 68)*.

First, two β'-subunit mutations were found to cause a defect in the assembly reactions where altered β' subunits were involved *(54)*. The strain Ts4 carries a temperature-sensitive mutation, *rpoC*4, in the gene coding for the β' subunit *(69)*. When log-phase cultures of Ts4 grown at 30°C are transferred to 42°C, the majority of pulse-labeled α and β subunits exist in the unassembled $\alpha_2\beta$ complex with a sedimentation coefficient of 8–9S, and in addition the radioactive α subunit alone can be identified in the slowly sedimenting fraction of 4–5S. The unassembled free β' subunit is rapidly degraded. On the contrary, under the same experimental conditions, most of the pulse-labeled subunits are recovered in the enzyme structure for the wild-type parental strain X240. Thus, it was concluded that the Ts4 mutant has a defect in the RNA polymerase assembly at a step later than the formation of the $\alpha_2\beta$ complex, resulting in the accumulation of this intermediate complex.

A temperature-sensitive mutant carrying another β'-subunit mutation, *rpoC*1, was found to be an assembly-defective mutant but the defect was determined to be at the activation step of premature core enzyme. The mutation *rpoC*1, originally called *ts*X (*70, 71*), is a genetic marker of the β' subunit of RNA polymerase; enzymological comparison of the mutant enzyme with wild-type polymerase by the technique of mixed reconstitution revealed that the unusual temperature sensitivity of the mutant enzyme is due to the alteration of β' subunit protein (*33*). Analysis of the pulse-labeled polymerase proteins with this mutant indicated that not only the α and β subunits, but also the β' subunit, were found in the 9–11S fractions (*54*). Thus, when grown at non-permissive temperatures, this strain seems to accumulate the premature core enzyme containing all the core enzyme subunits. It has been reported that the *ts*X mutation weakens protein-protein interactions between RNA polymerase subunits, thus leading to easy disintegration of the enzyme structure (*70*). This effect seems to be strengthened by some mutations in the structural gene for the β subunit, probably due to a further decrease in the affinity between the altered β subunit and the thermosensitive β' subunit. These observations seem to be in good agreement with the finding that the RNA polymerase from this mutant was thermosensitive, unlike that from the Ts4 mutant strain (Taketo and Ishihama, unpublished observation).

A conditionally assembly-defective mutant was also identified for a temperature-sensitive β-subunit mutant, strain A2R7, which was originally isolated by Kirschbaum *et al.* (*69*) and is known to carry the mutations *rpoB*2 and *rpoB*7 in the structural gene for the β subunit. Unusual accumulation of both the $\alpha_2\beta$ complex and premature core is observed even at 30°C and in addition unassembled α increases at 42°C, indicating that the assembly is inefficient at all the steps where altered β is involved and thus the mutant β is incapable of associating with the α subunit at non-permissive temperature (*56*).

Finally, in the two temperature-sensitive α-subunit mutants we identified recently (*21*), the abortive initiation of subunit assembly was found with a temperature-sensitive mutant HN316*ts*112 carrying the mutation in the α-subunit gene (*rpoA*112) (*68*). This mutant has peculiar properties (*21*): considerable thermolability of purified RNA polymerase; low fidelity of *in vitro* transcription catalyzed by the purified polymerase; cessation of RNA synthesis after a considerable lag in the temperature rise, which immediately suggested a defect in the assembly of RNA polymerase, resulting ultimately in the cessation of RNA synthesis. Accordingly, it was found that the initiation, *i.e.*, either the dimerization of newly synthesized α subunits or the subsequent association of altered α dimers with β subunits, is blocked in this particular mutant, in particular at non-permissive temperatures. Thus, a large amount of unassembled α subunits accumulates in this mutant. The degradation fragments of β and β' subunits identified in this mutant are identical at least in size to those formed by limited digestion *in vitro* of unassembled β and β' subunits by proteases but differed from those derived from assembled RNA polymerase (Ishihama *et al.*, in preparation).

TABLE VI

Mutations Affecting the Assembly of RNA Polymerase Subunits

Mutation	Step-reaction of subunit assembly[a]				Accumulation of assembly intermediate	Synthesis rate[b]			
	$2\alpha \rightleftarrows \alpha_2 \rightarrow \alpha_2\beta \rightleftarrows \alpha_2\beta\beta' \rightarrow E$					α	β	β'	σ
*rpoA*112	$\pm$	$-$	$-$	$-$	Free α	$\pm$	$+$	$+$	$\pm$
*rpoB*2·*rpoB*7	$+$	$\pm$	$-$	$-$	Free α	$\pm$	$-$	$-$	$\pm$
*rpoC*4	$+$	$+$	$-$	$-$	$\alpha_2\beta$	$\pm$	$+$	$+$	$\pm$
*rpoC*1	$+$	$+$	$+$	$-$	$\alpha_2\beta\beta'$	$\pm$	$+$	$+$	$\pm$

[a] $+$, unaffected; $-$, inhibited at non-permissive temperatures.
[b] $+$, increased after temperature up-shift; $-$, decreased after temperature up-shift.

As summarized in Table VI, we have so far identified all possible mutants which have defects in the subunit assembly of RNA polymerase at the steps where mutant subunits participate. Since all the subassemblies accumulated in the assembly-defective mutants are recovered in cytoplasm (*72*), it appears that the subunit assembly takes place in cytoplasm until the premature core is formed. The subassemblies accumulated at non-permissive temperatures are assembled at least in part into the enzyme structure following a drop in the culture temperature. The identification of assembly-defective mutants therefore provided the conclusive evidence indicating that the intracellular pathway of subunit assembly is identical with that established in the *in vitro* reconstitution system. But no assembly-defective mutations have been found that can be mapped on the *E. coli* chromosome outside the structural genes for core enzyme subunits. Along this line, it is interesting to determine whether assembly-defective mutants exist carrying the mutations in the σ-subunit gene, because the σ subunit has been implied to be a factor involved in the activation of premature core enzyme at least in the *in vitro* reconstitution system (*14, 31*).

VIII. INTERACTION BETWEEN SUBUNIT ASSEMBLY AND SUBUNIT SYNTHESIS

Studies on the synthesis of individual subunits have been performed under various growth conditions and it is well-established that the synthesis of core enzyme subunits, α, β, and β', is coordinated under both balanced and unbalanced growth of wild-type *E. coli* (*57*, *67*), although the genes for the α and $\beta\beta'$ subunits are not linked in the *E. coli* chromosome (*7*, *73*). The rate of core subunit synthesis, and thus the amount of these subunits, is maintained at certain levels characteristic of the rate of cell growth. However, mutants have been found which produce subunits at either increased or decreased rates (*55*, *68*, *74*). Further analyses of these mutants indicated that these abnormalities in the subunit synthesis are often accompanied by defects in the assembly of RNA polymerase (see Table VI). Thus, it was proposed that cells somehow recognize the intracellular concentration of functional RNA polymerase but not of total amounts of subunit proteins (*7*, *52*, *75*, *76*). Thus, the synthesis and assembly of RNA polymerase are closely coupled and disorder of one of the successive processes affects to a greater or lesser extent another process in enzyme formation.

The simplest mechanism to explain the above results for RNA polymerase synthesis is so-called autogenous regulation (*7*, *75*, *77*). The frequent occurrence of *rpoBC* mutations differentially affecting the catalytic activities and subunit synthesis, as well as the specific effect of rifampicin on subunit synthesis and the absence of gene dosage effects, is best explained if the regulatory molecule itself consists of one or more enzyme subunits. In fact, Fukuda *et al.* (*76*) found that the synthesis of the β subunit *in vitro* directed by DNA from transducing phages carrying the *rpoB* gene is specifically inhibited by the addition of exogeneous RNA polymerase holoenzyme or the $\alpha_2\beta$ complex, but not core enzyme or individual subunits. Thus, the regulatory site for the autogenous regulation (repression) may be located in the β subunit. The finding of repressor activity for the $\alpha_2\beta$ complex raised the novel concept in biological regulation that some intermediate subassemblies in the pathway of the formation of molecular assemblies, *e.g.*, enzyme complexes, themselves play some biological functions.

The determination of mRNA coded for by the β and β' subunit genes (*rpoBC*) in the *in vitro* system demonstrated that the amount of *rpoBC* mRNA in the repressed conditions is not different from that in the absence

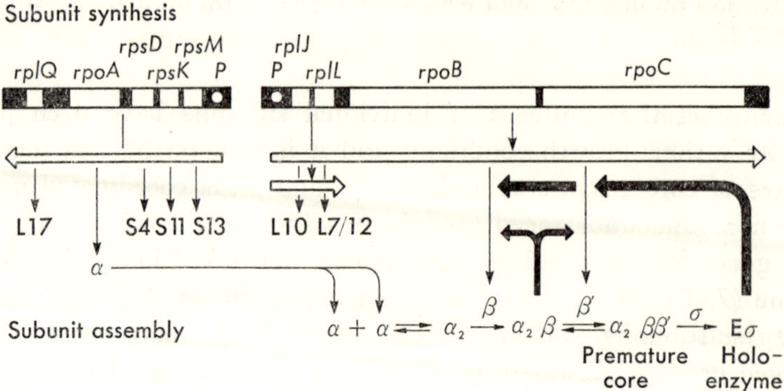

Fig. 8.　Autogenous regulation of RNA polymerase synthesis. In *E. coli* the RNA polymerase α subunit gene and the β and β' subunit genes are organized in two separate transcription units, each including some ribosomal protein genes (*73*). The genetic organization of the $\beta\beta'$ operon located near 89 min on the chromosome is $rpoP_\beta$ (promoter) → *rp1J* (ribosomal protein L10) → *rp1L* (ribosomal protein L7/12) → *rpoB* (β subunit) → *rpoC* (β' subunit). The majority of transcription initiated from $rpoP_\beta$ terminates before the *rpoB* gene and thus the transcription of *rpoBC* genes is regulated not only by transcription initiation at $rpoP_\beta$ but also by termination at the attenuation site (*78*). Translation of *rpoBC* mRNA is also regulated whereby RNA polymerase holoenzyme and the $\alpha_2\beta$ complex function as regulatory molecules with repressor activity (*76, 78, 97*). For details see Ref. *79*.

of excess repressors, either holoenzyme or the $\alpha_2\beta$ complex (*78*). This observation raised the possibility that autogenous regulation (repression) operates post-transcriptionally, presumably at the level of translation. The model shown in Fig. 8 is described in detail in a previous review article from this laboratory (*79*). This novel regulation of prokaryotes has also been suggested for the synthesis of some ribosomal proteins (*80, 81*). The molecular mechanism of the translational regulation is being studied.

IX.　INTERPLAY OF RNA POLYMERASE WITH TRANSCRIPTION FACTORS

The transcription specificity of assembled RNA polymerase is thought to be controlled through interactions with a number of proteins, generally called transcription factors (*82–85*), and nucleotide effectors such as ppGpp and tRNA (*82–84*). However, details of the structure and function have been reported for none of the proposed regulatory proteins.

In order to identify transcription factors with regulatory functions, we have started systematic analyses of proteins co-purified with RNA polymerase for their function and structure, including the σ' subunit (3), groE protein (86, 87) with ATPase activity (87, 88), stringent starvation protein (SSP) (89) which is a predominant protein synthesized under extreme amino acid starvation (90), and τ and ω proteins (1). An *E. coli* protein called L factor, which is necessary for β-galactosidase synthesis *in vitro* (91) and for the functioning of the N gene product in phage λ-infected cells (92), also forms a stable complex with RNA polymerase (A. Ishihama, unpublished observation). The identification of transcription factors was based on the following examinations: i) cross-reaction of enzyme-associated proteins to antibodies against enzyme subunits; ii) precipitation of proteins in cell extracts by subunit-specific antibodies and comparison with enzyme-associated proteins; iii) identification of degradation fragments of enzyme subunits by treatment with proteases and comparison with enzyme-associated proteins; iv) comparison of peptide maps between putative factors and enzyme subunits. Proteins that were identified not to be degradation fragments of the enzyme subunits have been studied extensively for their function and structure. To identify these proteins as transcription factors, it is further required that transcription of certain genes or gene groups is specifically affected when RNA polymerase holoenzyme but not core enzyme complexes with the proteins. But for conclusive evidence of the *in vivo* function it is necessary to identify the genes coding for the proteins and to isolate mutants in the genes which affect transcription of other gene(s).

Travers (84) hypothesized that components of the translation machinery such as $tRNA_f^{Met}$, initiation factor IF2 and elongation factor EF-Tu may interact with holoenzyme and alter its transcription specificity. In fact, charged but not uncharged $tRNA_f^{Met}$ specifically binds to holoenzyme and stimulates *lac* mRNA synthesis (93). Nucleotide effectors such as ppGpp and ppApp are also believed to interact with RNA polymerase, thereby regulating its transcription specificity (82–84). Several mutations have been identified which phenotypically suppress certain *rif(rpoB)* mutations and are mapped outside of the known *rpo* genes (7). Proteins encoded by these genes might interact with RNA polymerase β subunit. All this current evidence supports the idea that transcription is regulated at least in part through the interaction of RNA polymerase holoenzyme with proteins that we propose to call transcription factors

(85). However, virtually nothing is known about the molecular basis of the assembly between RNA polymerase and transcription factors.

X. ACTIVE AND INACTIVE RNA POLYMERASE

Previous examinations of the amount of RNA polymerase in *E. coli* cells growing at various rates indicated that, although it is maintained at certain levels characteristic of the rate of cell growth *(57, 67)*, there is a surplus of nonfunctioning polymerase, which may be grouped into two classes: active (functional) free RNA polymerase and inactive RNA polymerase *(94, 95)*. It is indeed known that a considerable amount of RNA

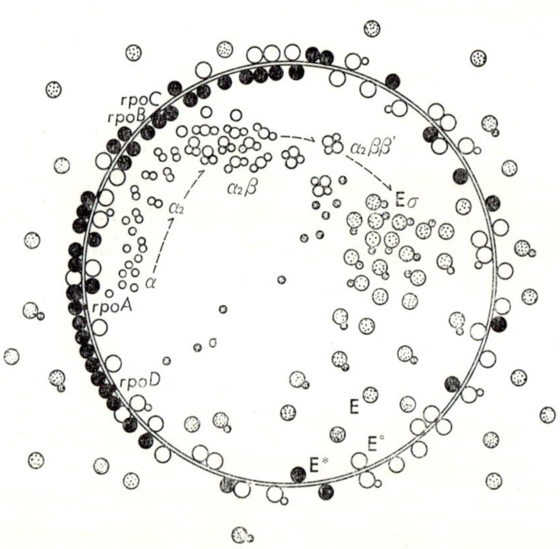

Fig. 9.   Intracellular states of RNA polymerase proteins. The genes for RNA polymerase σ *(rpoD)*, α *(rpoA)*, and $\beta\beta'$ *(rpoBC)* subunits are located at 67, 72, and 89 min, respectively, on the *E. coli* chromosome *(7, 73)*. The newly synthesized subunits are assembled sequentially in the order *(14, 32)*: $2\alpha \rightarrow \alpha_2 \rightarrow \alpha_2\beta \rightarrow \alpha_2\beta\beta'$ (premature core enzyme) $\rightarrow$ Eσ (active holoenzyme). The subunit assembly takes place in cytoplasm until the premature core enzyme is formed *(72)*. In exponentially growing wild-type *E. coli* cells, about two-thirds of the RNA polymerase are associated with nucleoids *(58)*, of which approximately half the molecules are synthesizing RNA at the time of cell lysis *(72)*. E, E°, and E* represent cytoplasmic active core, nucleoid-bound core, and transcribing core enzymes, respectively, while Eσ represents σ subunit-bound core or holoenzyme.

polymerase with virtually no enzyme activity exists in minicells free from chromosomes (*96*). Comparison of the structure and function between the premature core and the unused inactive polymerase remains to be made, but it is reasonable to consider that RNA polymerase is stored in inactive forms through interaction with transcription factor(s) which repress binding of the polymerase to the DNA template and keep it in an unused but metabolically stable state(s). In connection with this phenomenon, we have obtained evidence indicating that the partition control of RNA polymerase operates in *E. coli*, by which the amount of functioning RNA polymerase recovered on nucleoids is regulated depending upon the cell growth (*58*): about two-thirds of the core enzyme is located on folded chromosomes during the exponentially growing phase, but the level decreases to one-third at the beginning of the stationary phase. All the known states of RNA polymerase proteins are illustrated together in Fig. 9.

During the studies on the metabolism of RNA polymerase during the cell cycle of wild-type *E. coli* cells, we noticed that, even in wild-type cells, newly synthesized subunits are degraded without being assembled into the enzyme structure only at the beginning of the stationary phase (*58*). Thus, it should be considered that the subunit assembly itself is the reaction which is subject to metabolic control.

SUMMARY

The isolated subunits of *Escherichia coli* DNA-dependent RNA polymerase are reassembled in a stepwise manner in the following sequence: $2\alpha \rightarrow \alpha_2 \rightarrow \alpha_2\beta \rightarrow \alpha_2\beta\beta'$ (premeature core enzyme) $\rightarrow$ E (active core enzyme). When the *in vitro* reconstitution is performed at low temperature, the subunit assembly is prevented until the assembled but inactive premature core enzyme is formed, which is similar to native core enzyme in many parameters of gross conformation but differs from it in several minor and local conformations. The temperature-dependent activation of premature core enzyme at a salt concentration as low as that *in vivo* takes place only in the continuous presence of either the σ subunit or DNA. The σ subunit is therefore proposed to be a regulatory protein which influences the conformation of core subunit assembly in multiple ways from the initial enzyme maturation to the final initiation of transcription.

Evidence has accumulated which indicates that the subunit assembly *in vivo* proceeds *via* the same pathway as that identified *in vitro*, including the identification of all species of the assembly intermediates in cell extracts, the identification of all possible types of assembly-defective mutants among temperature-sensitive α-, β-, and β'-subunit mutants, the kinetics of the appearance of pulse-labeled subunits in the enzyme structure as expected from the assembly sequence and the integration of labeled subassemblies into the enzyme structure upon chasing.

The functional complexity of RNA polymerase coupled with transcriptional control appears to depend on its structural flexibility which fluctuates through the assembly with various transcription factors. This type of transcriptional control is being thoroughly considered but a final conclusion awaits further examinations.

Acknowledgments
The author wishes to acknowledge the pleasant and successful collaboration with Ryuji Fukuda, Koreaki Ito, Yoichiro Iwakura, Shigetaka Naito, Makoto Taketo, Tsunao Saitoh, Kiyoshi Kawakami, Masayuki Kajitani, Haruko Nagasawa-Fujimori, and Masayoshi Enami on many aspects of the work described in the present report. He also thanks Ryukichi Takeda, Masaaki Yamamoto, and Masuo Higashii for their technical assistance and Junko Asano, Akiko Katayama, and Michiko Tsutsumi for their secretarial help. The work was supported by grants from the Ministry of Education, Science and Culture, of Japan, the Matsunaga Science Foundation and the Asahi Press.

REFERENCES

1 R. R. Burgess, *Annu. Rev. Biochem.*, **40**, 711 (1971).
2 R. R. Burgess, "RNA Polymerase," ed. by R. Losick and M. Chamberlin, Cold Spring Harbor Laboratory, New York, p. 69 (1976).
3 R. Fukuda, Y. Iwakura, and A. Ishihama, *J. Mol. Biol.*, **83**, 353 (1974).
4 M. Chamberlin, *Annu. Rev. Biochem.*, **43**, 721 (1974).
5 M. Chamberlin, "RNA Polymearse," ed. by R. Losick and M. Chamberlin, Cold Spring Harbor Laboratory, New York, p. 159 (1976).
6 W. Zillig, P. Palm, and A. Heil, "RNA Polymerase," ed. by R. Losick and M. Chamberlin, Cold Spring Harbor Laboratory, New York, p. 101 (1976).
7 T. Yura and A. Ishihama, *Annu. Rev. Genet.*, **13**, 59 (1979).
8 A. Ishihama and J. Hurwitz, *J. Biol. Chem.*, **244**, 6680 (1969).
9 A. Ishihama, *J. Cell. Physiol.*, **74** (Suppl. 1), 223 (1969).
10 V. S. Sethi, W. Zillig, and H. Bauer, *FEBS Lett.*, **6**, 339 (1970).
11 A. Ishihama, *Biochemistry*, **11**, 1250 (1972).

12 R. Fukuda and A. Ishihama, *J. Mol. Biol.*, **87**, 523 (1974).
13 P. Palm, A. Heil, D. Boyd, B. Grampp, and W. Zillig, *Eur. J. Biochem.*, **53**, 283 (1975).
14 T. Saitoh and A. Ishihama, *J. Mol. Biol.*, **104**, 621 (1976).
15 T. Saitoh and A. Ishihama, *Biochemistry*, **18**, 979 (1979).
16 A. M. Frischauf and K. H. Scheit, *Biochem. Biophys. Res. Commun.*, **53**, 1227 (1973).
17 D. Rabussay and W. Zillig, *FEBS Lett.*, **5**, 104 (1969).
18 A. Heil and W. Zillig, *FEBS Lett.*, **11**, 165 (1970).
19 Y. Iwakura, A. Ishihama, and T. Yura, *Mol. Gen. Genet.*, **121**, 181 (1973).
20 M. Kawai, A. Ishihama, and T. Yura, *Mol. Gen. Genet.*, **143**, 223 (1976).
21 A. Ishihama, N. Shimamoto, H. Aiba, K. Kawakami, H. Nashimoto, A. Tsugawa, and H. Uchida, *J. Mol. Biol.*, **137**, 137 (1980).
22 Z. Hillel and C.-W. Wu, *Biochemistry*, **17**, 2954 (1978).
23 R. B. Simpson, *Cell*, **18**, 277 (1979).
24 U. I. Lill and G. R. Hartmann, *Biochem. Biophys. Res. Commun.*, **39**, 930 (1970).
25 A. Ishihama and K. Ito, *J. Mol. Biol.*, **72**, 111 (1972).
26 C. Frieden, *Annu. Rev. Biochem.*, **50**, 653 (1971).
27 C. Chothia and J. Janin, *Nature*, **256**, 705 (1975).
28 K. Ito and A. Ishihama, *J. Mol. Biol.*, **79**, 115 (1973).
29 L. R. Yarbrough and J. Hurwitz, *J. Biol. Chem.*, **249**, 5400 (1974).
30 C. M. Halling, K. C. Burtis, and R. H. Doi, *J. Biol. Chem.*, **252**, 9024 (1977).
31 A. Ishihama, R. Fukuda, and K. Ito, *J. Mol. Biol.*, **79**, 127 (1973).
32 A. Ishihama, M. Taketo, T. Saitoh, and R. Fukuda, "RNA Polymerase," ed. by R. Losick and M. Chamberlin, Cold Spring Harbor Laboratory, New York, p. 485 (1976).
33 S. R. Panny, A. Heil, B. Mazus, P. Palm, W. Zillig, S. Z. Mindlin, S. Ilyina, and R. B. Khesin, *FEBS Lett.*, **48**, 241 (1974).
34 A. Ishihama, H. Aiba, T. Saitoh, and S. Takahashi, *Biochemistry*, **18**, 972 (1979).
35 G. R. Stark, W. H. Stein, and S. Moore, *J. Biol. Chem.*, **235** 3177 (1960).
36 G. R. Stark, *J. Biol. Chem.*, **239**, 1411 (1964).
37 A. Cerami and J. M. Manning, *Proc. Natl. Acad. Sci. U.S.*, **68**, 1180 (1971).
38 G. R. Stark, "Methods in Enzymology," ed. by C. H. W. Hirs and S. N. Timasheff, Academic Press, New York and London, Vol. 25, p. 579 (1972).
39 R. Cecil and M. A. W. Thomas, *Nature*, **206**, 1317 (1965).
40 D. B. Wetlaufer, *Proc. Natl. Acad. Sci. U.S.*, **70**, 697 (1973).
41 D. B. Wetlaufer and S. Ristow, *Annu. Rev. Biochem.*, **42**, 135 (1973).
42 F. K. Freidman and S. Beychok, *Annu. Rev. Biochem.*, **48**, 217 (1979).
43 A. A. Travers and R. R. Burgess, *Nature*, **222**, 537 (1969).
44 J. S. Krakow, K. Daley, and M. Karstadt, *Proc. Natl. Acad. Sci. U.S.*, **62**, 432 (1969).
45 D. C. Hinkle and M. Chamberlin, *J. Mol. Biol.*, **70**, 157 (1972).
46 W. Zillig, K. Zechel, D. Rabussay, M. Schachner, V. S. Sethi, P. Palm, A. Heil, and W. Seifert, *Cold Spring Harbor Symp. Quant. Biol.*, **35**, 47 (1971).
47 M. J. Chamberlin, W. Mangel, G. Rhodes, and S. Stahl, "Control of Ribosome Synthesis," ed. by N. O. Kjeldgaard and O. Maaløe, Munksgaard, Copenhagen, p. 22 (1976).
48 K. O. Stetter and W. Zillig, *Eur. J. Biochem.*, **48**, 527 (1974).

49 J. D. Harding and S. Beychok, *Proc. Natl. Acad. Sci. U.S.*, **71**, 3395 (1974).
50 U. I. Lill, E. M. Behrendt, and G. R. Hartmann, *Eur. J. Biochem.*, **52**, 411 (1975).
51 J. W. Teipel and D. E. Koshland, Jr., *Biochemistry*, **10**, 798 (1971).
52 M. Taketo, R. Fukuda, and A. Ishihama, *Mol. Gen. Genet.*, **165**, 7 (1978).
53 Y.-H. Chen, J. T. Yang, and K. H. Chau, *Biochemistry*, **13**, 3350 (1974).
54 M. Taketo and A. Ishihama, *J. Mol. Biol.*, **102**, 297 (1976).
55 M. Taketo, A. Ishihama, and J. Kirschbaum, *Mol. Gen. Genet.*, **147**, 139 (1976).
56 M. Taketo and A. Ishihama, *J. Mol. Biol.*, **112**, 65 (1977).
57 Y. Iwakura, K. Ito, and A. Ishihama, *Mol. Gen. Genet.*, **133**, 1 (1974).
58 K. Kawakami, T. Saitoh, and A. Ishihama, *Mol. Gen. Genet.*, **174** 107 (1979).
59 G. Kegeles and J. R. Cann, " Methods in Enzymology," ed. by C. H. W. Hirs and
    S. N. Timasheff, Academic Press, New York and London, Vol. 48, p. 258 (1978).
60 Z. Hillel and C.-W. Wu, *Biochemistry*, **16**, 3334 (1977).
61 J. D. Harding and S. Beychok, " RNA Polymerase," ed. by R. Losik and M. Cham-
    berlin, Cold Spring Harbor Laboratory, New York, p. 355 (1976).
62 P. A. Lowe and A. D. B. Malcolm, *Eur. J. Biochem.*, **64**, 577 (1976).
63 T. S. Campbell and P. A. Lowe, *Biochem. J.*, **163**, 177 (1977).
64 A. Ikai and C. Tanford, *Nature*, **230**, 100 (1971).
65 K. Ito, Y. Iwakura, and A. Ishihama, *J. Mol. Biol.*, **96**, 257 (1975).
66 M. Taketo, A. Ishihama, A. Muto, and S. Osawa, *Mol. Gen. Genet.*, **145**, 311 (1976).
67 Y. Iwakura and A. Ishihama, *Mol. Gen. Genet.*, **142**, 67 (1975).
68 K. Kawakami and A. Ishihama, *Biochemistry*, **19**, 3491 (1980).
69 J. B. Kirschbaum, I. V. Claeys, S. Nasi, B. Molholt, and J. G. Miller, *Proc. Natl.
    Acad. Sci. U.S.*, **72**, 2375 (1975).
70 R. B. Khesin, S. Z. Mindlin, Zh. M. Gorlenko, and T. S. Ilyina, *Mol. Gen. Genet.*,
    103, 194 (1968).
71 T. S. Ilyina, M. I. Ovadis, S. Z. Minklin, Zh. M. Gorlenko, and R. B. Khesin, *Mol.
    Gen. Genet.*, **110**, 118 (1971).
72 T. Saitoh and A. Ishihama, *J. Mol. Biol*, **115**, 403 (1977).
73 M. Nomura, E. A. Morgan, and R. Jaskunas, *Annu. Rev. Genet.*, **11**, 297 (1977).
74 M. Taketo and A. Ishihama, " Control of Ribosome Synthesis," ed. by O. N. Kjeld-
    gaard and O. Maaløe, Munksgaard, Copenhagen, p. 79 (1976).
75 J. Scaife, " RNA Polymerase," ed. by R. Losick and M. Chamberlin, Cold Spring
    Harbor Laboratory, New York, p. 207 (1976).
76 R. Fukuda, M. Taketo, and A. Ishihama, *J. Biol. Chem.*, **253**, 4501 (1978).
77 R. F. Goldberger, *Science*, **183**, 810 (1974).
78 M. Kajitani, R. Fukuda, and A. Ishihama, *Mol. Gen. Genet.*, **179**, 489 (1980).
79 A. Ishihama and R. Fukuda, *Mol. Cell. Biochem.*, **31**, 177 (1980).
80 R. Fukuda, *Mol. Gen. Genet.*, **178**, 483 (1980).
81 J. L. Yates, A. E. Arfsten, and M. Nomura, *Proc. Natl. Acad. Sci. U.S.*, **77**, 1837
    (1980).
82 R. Losick and J. Pero, " RNA Polymerase," ed. by R. Losick and M. Chamberlin,
    Cold Spring Harbor Laboratory, New York, p. 227 (1976).
83 R. H. Doi, *Bacteriol. Rev.*, **41**, 568 (1977).
84 A. Travers, *Nature*, **263**, 641 (1976).
85 A. Ishihama, R. Fukuda, K. Kawakami, M. Kajitani, and M. Enami, " Genetics and

Evolution of RNA Polymerase, tRNA and Ribosomes," ed. by S. Osawa, H. Ozeki, H. Uchida, and T. Yura, Univ. Tokyo Press, Tokyo and Elsevier/North-Holland, Amsterdam-New York-Oxford, p. 105 (1980).

86 T. Hohn, B. Hohn, A. Engel, M. Wurtz, and P. R. Smith, *J. Mol. Biol.*, **129**, 359 (1979).

87 A. Ishihama, T. Ikeuchi, and T. Yura, *J. Biochem.*, **79**, 917 (1976).

88 A. Ishihama, T. Ikeuchi, A. Matsumoto, and S. Yamamoto, *J. Biochem.*, **79**, 927 (1976).

89 A. Ishihama and T. Saitoh, *J. Mol. Biol.*, **129**, 517 (1979).

90 S. Reeh, S. Pedersen, and J. D. Freisen, *Mol. Gen. Genet.*, **149**, 279 (1976).

91 H. F. Kung, C. Spears, and H. Weissbach, *J. Biol. Chem.*, **250**, 1556 (1975).

92 J. Greenblatt, J. Li, S. Adhya, D. I. Friedman, L. S. Baron, B. Redfield, H.-F. Kung, and H. Weissbach, *Proc. Natl. Acad. Sci. U.S.*, **77**, 1991 (1980).

93 O. Pong and N. Ulbrich, *Proc. Natl. Acad. Sci. U.S.*, **73**, 3064 (1976).

94 H. Bremer and D. G. Dalbow, *Biochem., J.* **150**, 9 (1975).

95 D. P. Nierlich, *Annu. Rev. Microbiol.*, **32**, 393 (1978).

96 A. C. Rogerson, *Biochem. Biophys. Res. Commun.*, **66**, 665 (1975).

97 M. Taketo, R. Fukuda, and A. Ishihama, *Mol. Gen. Genet.*, **165**, 7 (1978).

Received for publication December 11, 1979.

Adv. Biophys., Vol. 14, pp. 37–138 (1981)

DYNAMIC ANALYSIS OF HIGHER ORDER BIOLOGICAL SYSTEMS

KENSUKE SATO

Research Foundation on Traffic Medicine, Tokyo, Japan

I. EXCITABILITY AND ACTIVITY OF BIOLOGICAL SYSTEMS

Biological systems (bio-systems), such as an individual, system, organ, tissue, cell, cell membrane, *etc.*, of humans and animals and/or their subsystems exhibit various biological phenomena (bio-phenomena) as responses, provided they were stimulated by certain physical, chemical, and physicochemical states in their external and internal environments (*1*), each of which is nothing but natural phenomenon on the Earth on which they are living. These stimulations can be considered to be transformed into responses by the systems through their activities, such as excitation, inhibition, fascilitation, occlusion, *etc.*

Excitation is one of the most fundamental biological activities (bio-activity) and can be defined quantitatively by the well known *excitability* given by the inverse of the *threshold stimulus* of the bio-system in a physiological state. Here, therefore, let the amount of the threshold stimulus and excitability of the bio-system be X_u and E, respectively, and then the equation of the definition is easily given as

$$X_u \cdot E = 1. \tag{1.1}$$

Here, the " 1 " on the right side has no biological significance, but both X_u and E on the left side have an important one. It would be natural, therefore, to give it some biological significance. Hence, let it indicate the amount of the response with unity, *i.e.*, *unit response*, then this equation gives such a *stimulus-system-response* relation (2) that *delivery of a threshold stimulus amounting X_u to a bio-system having excitability amounting to E will cause it to elicit its response with unity (law of excitability)*, as illustrated by the top block diagram in Fig. 1, so that the excitability E gives a part of the characteristics of the bio-system quantitatively.

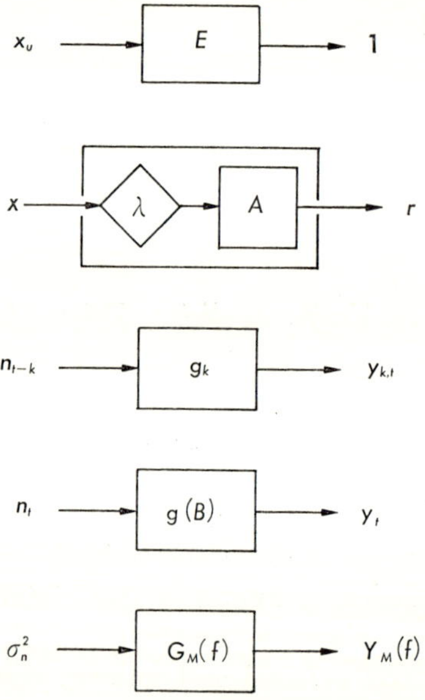

Fig. 1. Block diagrams of various *stimulus(input)-system-response(output)* relations in biological systems (*18*).

There are three well known essential factors for an effective stimulus to elicit a response from a bio-system endorsing generation of its excitation, which are termed the *strength*, *slope*, and *time* factors, respectively. E in the above Eq. 1. 1 is the excitability concerning the strength factor, whereas the *minimal gradient* and *chronaxy* (*3*) are those concerning the slope and time factors, respectively. Consequently, the inverses of the minimal gradient and chronaxy of a bio-system indicate, respectively, the excitability concerning the slope and time factors of the bio-system (*4, 5*). In the same way, it was also elucidated that *Weber's law*, well known in psychophysics and sensory physiology, is equivalent to the law of excitability concerning the *relative noticeable threshold* (4, 5).

As a generalization of Eq. 1.1, the following *law of activity (4–7)* is given as

$$X \cdot \lambda A = r, \qquad (1.2)$$

which indicates such a stimulus-system-response relation that *delivery of a stimulus having strength X to a bio-system and/or bio-subsystem having activity (6, 7) amounting to A evokes a response r times larger in size than that due to the threshold stimulus X_u,* as demonstrated by the second block diagram in Fig. 1, where

$$\lambda = 1, \quad \text{when} \quad X \geqq X_u,$$
$$\lambda = 0, \quad \text{when} \quad 0 < X < X_u. \qquad (1.3)$$

That is, the bio-system sorts out $\lambda = 0$ to reject the stimulus X, when the stimulus delivered from the external and/or internal environments is subthreshold $(X < X_u)$; thus $r = 0$, so that the system exhibits no response, whereas the system sorts out $\lambda = 1$ to accept the stimulus delivered, when it is not subthreshold $(X \geqq X_u)$, and the system transforms the accepted stimulus through A into its response amounting to $r(>0)$ in size. Therefore, λ and A respectively can be considered as *all-or-nothing judging and transforming activities (5, 6)* characterizing the bio-system.

Not only have the well known *Fechner's law* and *Plateau-Stevens power function (8)* in psychophysics and sensory physiology been proved to be equivalent to the laws of activity concerning the strength factor of an effective stimulus in the semi- and double logarithmic domains, respectively, but also the laws concerning the slope, time and strength-time factors have also given in each of the linear, semi- and double logarithmic domains, respectively (5).

II. HIGHER ORDER ACTIVITY IN BIO-SYSTEMS

1. *Biological Sway in Bio-phenomena*

Humans and animals exhibit various bio-phenomena over the course of time, from the past to the present and into future, up to just before their death. These phenomena are enormous in number and each state at any time depends in stochastic fashion not only on its past history, but those of various natural phenomena, *i.e.,* those in the external and internal environments. Some of these natural phenomena will also fluctuate severely to cause such marked changes in the state of a bio-phenomenon that endanger the life of humans and animals, provided they do not have the well known indispensable function termed *homeostasis* (1, 9), by which the changes will be reduced enough to preserve their normal life. Though Cannon (9) advocated *homeostasis* in 1932 as a general principle in regulating bio-phenomena concerning only the autonomic nervous system, various other bio-phenomena would also be regulated by this principle, since each state of numerous bio-phenomena varies more or less with sways in regard to their respective average values, which suggest the level of *homeostasis* (10, 11). Consequently, some essential mechanisms of a bio-system would be hidden in each sway of the bio-phenomena exhibited by the system. If the sway around the average state of a bio-phenomenon is measured at equally spaced time intervals, say τ, chosen appropriately for the phenomenon under study, then a random discrete time series with average zero will be obtained. Let the measured values of this time series at an arbitrary time $t-k\tau$, $k=0, 1, 2, \cdots, N-1$, be y_{t-k}, then the discrete time series given as

$$y_{t-N+1}, \cdots\cdots, \quad y_{t-2}, \quad y_{t-1}, \quad y_t,$$
$$(N: \text{large engugh positive integer}) \tag{2.1}$$

would be able to be considered, at least for the first step, as random variables of a (weakly) stationary normal stochastic process. In addition, each value in this time series will depend in stochastic fashion on previous values, as noted above. If each value is dependent on M previous values, this time series is capable of being described by an autoregressive (AR-) process of order M (11–19), where $M=1, 2, 3, \cdots$. Hence, y_t at an arbitrary time t will be given as a weighted sum of M previous values and a random residual independent of all values in the time series. Let this residual be n_t at time t, then y_t will be given as

$$y_t = a_1 y_{t-1} + a_2 y_{t-2} + \cdots + a_M y_{t-M} + n_t,$$
$$(t = M+1, M+2, \cdots), \tag{2.2}$$

where a_m, $m=1, 2, 3, \cdots, M$ are the weights termed *autoregressive* (*AR-*) *coefficients* and the residual n_t can be considered as the weighted sum of unmeasured values of enormous other bio- and natural phenomena (12, 13). This sum n_t is, therefore, able to be considered by the *law of large number* as a purely random time series with average zero, for the first step, such that

$$E[n_t] = 0,$$
$$E[n_{t-k} \cdot n_t] = 0, \quad \text{when} \quad k \neq 0,$$
$$= \sigma_n^2, \quad \text{when} \quad k = 0, \quad (\sigma_n^2 : \text{variance of } n_t) \qquad (2.3)$$

and

$$E[n_t \cdot y_{t-k}] = 0, \quad (k = 0, 1, 2, \cdots), \qquad (2.3\text{-}1)$$

where [ ] indicates an average of the values in the brackets. In other words, the amount of the sway of a bio-phenomenon y_t at an arbitrary time t is composed of two parts, one determined by its past history in stochastic fashion and the other by the external and internal environments of the bio-system. And in the former will be hidden an essential characteristic of the bio-system which will be maifested by the latter.

Though it is necessary to determine the value of the order M, it can be practically estimated by obtaining the minimum value of the *final prediction error FPE* of Akaike (*20, 21*) or that of the *amount of information criterion AIC* estimate (*22*) such that

$$FPE(M) = \sigma_n^2(1+(M-1)/N)/(1-(M+1)), \qquad (2.4)$$

and

$$AIC(M) = N \ln \sigma_n^2 + 2(M+1) \qquad (2.5)$$

respectively, where N is the length of the time series y_t in Eq. 2.1.

2. Higher Order Activity of Bio-systems

The circular rotation of the earth, causes day and night to occur alternatively and the revolution of the earth around the sun causes changes in the four seasons during the year. Both circular motions of the earth divide time into days and years, respectively. Based on these phenomena, man invented the clock to make more detailed divisions of time into hours, minutes, and seconds through a circular motion of hands on a clock. Consider a point that moves counterclockwise $1/\tau$ times in a unit time around the circumference of a circle with unit radius, then an arbitrary time point t will be shifted backward to $t-k\tau$, $k=1, 2, 3, \cdots$, provided this point moved counterclockwise k times around the circumference, so that the values of the time series y_t and n_t at time t in Eq. 2.2 will be shifted to y_{t-k} and n_{t-k}, respectively. This movement causinng a shift in time can be replaced by the *backward shift operator* (*23*), say B, such that

$$B^k y_t = y_{t-k}, \quad B^k n_t = n_{t-k}, \quad |B| \leq 1,$$
$$(k = 1, 2, 3, \cdots). \qquad (2.6)$$

Substituting this into Eq. 2. 2, we obtain the following stimulus-system-response relation

$$n_t \cdot g_M(B) = y_t \qquad (2.7)$$

that can be illustrated by the fourth block diagram in Fig. 1, where $g_M(B)$ is termed the *transfer function* (*23*) of the AR-process y_t such that

$$g_M(B) = 1/a_M(B), \qquad (2.8)$$

where $a_M(B)$ is temed the *characteristic function* (*23*) of the AR-process given as

$$a_M(B) = 1 - a_1 B - a_2 B^2 - \cdots - a_M B^M. \qquad (2.9)$$

It can be recognized from a comparison between Eqs. 1. 2 and 2. 7 and between the second and fourth block diagrams in Fig. 1 that the random natural stimulation n_t delivered from the external and internal environments to the bio-system, the time series y_t and the transfer function $g_M(B)$, respectively, are extensions of the stimulus X, the response r and the judging and transforming activities λA of the bio-system. Consequently, the transfer function $g_M(B)$ can be termed the *higher* (*M-th*) *order activity* (*12–19*) to transform the natural stimulation n_t into the bio-phenomenon y_t as the response of the bio-system. In addition, λA in Eq. 1. 2 can be regarded as a zero-order activity, since $\lambda A = \lambda A B^0$ (*14, 15*).

The characteristic function $a_M(B)$ is less than 1, provided the time series y_t is of stationary stochastic AR-process (*23*). Hence, the higher order activity $g_M(B)$ of the bio-system:

$$\begin{aligned}
g_M(B) &= 1/(1 - a_1 B - a_2 B^2 - \cdots - a_M B^M) \\
&= g_0 + g_1 B + g_2 B^2 + \cdots, \qquad (g_0 = 1)
\end{aligned} \qquad (2.10)$$

will converge on the right side. Consequently, Eq. 2. 7 can be rewritten as

$$\begin{aligned}
n_t g_0 + n_{t-1} g_1 + n_{t-2} g_2 + \cdots &= \textstyle\sum_{k=0}^{\infty} n_{t-k} g_k \\
&= \textstyle\sum_{k=0}^{\infty} y_{kt} \\
&= y_t, \qquad (2.11)
\end{aligned}$$

where

$$n_{t-k} g_k = y_{kt}, \qquad (k = 0, 1, 2, \cdots) \qquad (2.12)$$

This gives also a stimulus-system-response relation as illustrated by the third block diagram in Fig. 1 (*12–14, 19*).

Assume here a case such that the natural stimulation n_u at time u is

$$n_u = 1, \quad \text{when} \quad u = t-k$$
$$n_u = 0, \quad \text{otherwise}, \quad (k = 0, 1, 2 \cdots) \qquad (2.13)$$

then the natural stimulation n_u indicates a unit impulse at time u in this situation, and thus y_{kt} in Eq. 2. 12 becomes a *unit impulse response* of the bio-system elicited by a *unit impulse stimulus* delivered to the bio-system at time $t-k\tau$ and

$$g_k = y_{kt}, \quad (n_{t-k} = 1). \qquad (2.12\text{-}1)$$

It is obvious, therefore, that g_k, $k=1, 2, 3, \cdots$, in Eq. 2. 10 is the *unit impulse response series* (*12–15, 17, 19, 23*) of the bio-system.

Equation 2. 12 and the third block diagram in Fig. 1 show that *a stimulus delivered to a bio-system in the past at time $t-k\tau$ from the external and internal environments will be transformed by activity g_k of the system to elicit its response amounting to y_{kt} at the present time t.* In addition, Eq. 2. 11 indicates that the amount of the response y_t at time t will be composed of the total sum of the responses y_{kt} at this time, each of which was caused by stimuli n_{t-k} delivered at each time point $t-k$ prior to the present time t.

3.   Bio-informing Activity of Bio-systems

Assume an exceptional situation such that the bio-system exerts no transforming activity on the natural stimulation n_t in Eq. 2. 7, which will be delivered from the external and internal environments. Then y_t exhibited by the bio-system in this situation does not differ from the purely random stimulation n_t, *i.e.*,

$$y_t = n_t, \qquad (2.14)$$

so that y_t in this situation does not depend upon any preceding values. Hence, all of the AR-coefficients in Eq. 2. 2 will vanish, *i.e.*,

$$a_1 = a_2 = \cdots = a_M = 0. \qquad (2.15)$$

Therefore,

$$a_M(B) = g_M(B) = 1. \qquad (2.8\text{-}1)$$

Let the *entropy* of y_t in this exceptional case be H_0, then it will be given as that of a normal Gaussian and purely random process with average zero and variance σ_y^2 such that

$$H_0 = 1/2 \cdot \mathrm{lb}(2\pi e \sigma_y^2) \quad \text{bit}, \quad (\mathrm{lb} = \log_2). \tag{2.16}$$

On the other hand, a bio-system in general exerts its higer order transforming activity on the natural stimulation n_t to manifest its response that obeys an AR-process of order M given by Eq. 2.2. Let the entropy of the response y_t in this case be H_M, then

$$H_M = 1/2 \cdot \mathrm{lb}(2\pi e \sigma_y^2) + 1/2 \cdot \mathrm{lb}\, K_M \quad \text{bit}, \quad (K_M = a_M(B)) \tag{2.16-1}$$

(*12–15, 18, 19*). Consequently, the *amount of bio-informing activity* exerted by the bio-system will be given as the following *mutual information amount*, say I_M, i.e.,

$$I_M = H_0 - H_M = -1/2 \cdot \mathrm{lb}\, K_M = 1/2 \cdot \mathrm{lb}\, g_M(B) \quad \text{bit} \tag{2.17}$$

(*12, 18*). Let the autocovariance and autocorrelation functions with time lag m of the time series y_t be R_m and ρ_m, respectively, then

$$R_m = E[y_{t-m} \cdot y_t] = E[y_t \cdot y_{t-m}] = R_{-m}, \tag{2.18}$$
$$\rho_m = R_m/R_0 = \rho_{-m}, \quad (m = 0, 1, 2, 3, \cdots) \tag{2.19}$$
$$R_0 = E[y_t^2] = \sigma_y^2, \quad (\text{variance of } y_t). \tag{2.18-1}$$

It is obvious that

$$\rho_0 = 1,$$

therefore,

$$B^m = B^m \rho_0 = \rho_{-m} = \rho_m. \tag{2.19-1}$$

Hence, the value of the characteristic function $a_M(B)$ will be given as the following K_M such that

$$K_M = a_M(B) = 1 - \sum_{m=1}^{M} a_m B^m = 1 - \sum_{m=1}^{M} a_m \rho_m. \tag{2.20}$$

Thus, the value of the higher order activity $g_M(B)$ of order M is given as the following G_M such that

$$G_M = 1/K_M. \tag{2.21}$$

Consequently, the bio-informing activity amount I_M is given also as

$$I_M = 1/2 \cdot \mathrm{lb}\, G_M \quad \text{bit}. \tag{2.17-1}$$

Though it is impossible to measure the values of the random time series n_t, its variance can be obtained by the following procedures. From Eq. 2.7 and 2.8, we obtain

$$n_t = a_M(B) \cdot y_t.$$

Multiplying y_t on the both sides of the above equation and quoting Eq. 2.20, we obtain

$$n_t y_t = K_M y_t^2.$$

Substituting Eq. 2.2 into the left side of this, we obtain

$$n_t(\sum_{m=1}^{M} a_m y_{t-m} + n_t) = K_M y_t^2.$$

Taking here averages of both sides of the above, then we obtain

$$\sum_{m=1}^{M} a_m E[n_t y_{t-m}] + E[n_t^2] = K_M E[y_t^2].$$

The first term on the left side of the above vanishes, since n_t does not correlate to y_{t-m}, and the second term gives the variance of n_t, say σ_n^2, then we obtain

$$\sigma_n^2 = K_M \sigma_y^2. \tag{2.22}$$

4. *Power Spectral Density of Higher Order Bio-phenomena*
The backward shift operator B can be considered as

$$B^k = \exp(-i2\pi mf), \quad (k = 1, 2, 3, \cdots) \tag{2.6-1}$$

since B indicates a counterclockwise circular motion, where $i^2 = -1$ and f is circular frequency. Then the power spectral density of the M-th order AR-process y_t is given as the following $Y_M(f)$:

$$Y_M(f) = 2\sigma_n^2 / |1 - \sum_{m=1}^{M} a_m \exp(-i2\pi mf)|^2$$

$$= 2\sigma_n^2 / \{(\sum_{m=0}^{M} a_m \cos 2\pi mf)^2 + (\sum_{m=0}^{M} a_m \sin 2\pi mf)^2\},$$
$$(a_0 = -1, \quad 0 \leq f \leq 1/2\tau) \tag{2.23}$$

(23). Let here the power spectral density of the activity $g_M(B)$ of order M be $G_M(f)$, then

$$G_M(f) = 2 / |1 - \sum_{m=1}^{M} a_m \exp(-i2\pi mf)|^2. \tag{2.23-1}$$

Therefore, Eq. 2.23 is rewritten easily as follows

$$\sigma_n^2 \cdot G_M(f) = Y_M(f). \tag{2.24}$$

The variance of the natural stimulation delivered to the bio-system σ_n^2 is none other than the power spectral density of the stimulation, since the stimulation is a normal purely random stochastic process, as given by Eq. 2. 3. Consequently, $\sigma_n{}^2$ in the above Eq. 2. 24 can be considered the power spectral density of the purely random natural stimulation n_t. Thus the above relation can be regarded as the stimulus-system-response relation in the frequency domain (*12, 13, 25*) of the bio-system having an activity of order M, as illustrated by the fifth block diagram in Fig. 1.

5. *Higher Order Activities in Electroencephalograms (EEGs)*

An EEG potential can be considered as a normal stationary, at least weakly stationary, stochastic process in most situations (*26–29*) and as an autoregressive-moving average (ARMA) process (*30–32*) as well as an AR-process (*33–38*). Each AR-coefficient a_m, $m=1, 2, 3, \cdots, M$, in Eq. 2. 2 of EEG potentials are expected to show some fluctuations obeying a normal Gaussian distribution, since the time series of the EEG potential described by Eq. 2. 2 is given by a linear combination of AR-coefficients a_m, $m=1, 2, \cdots\cdots, M$.

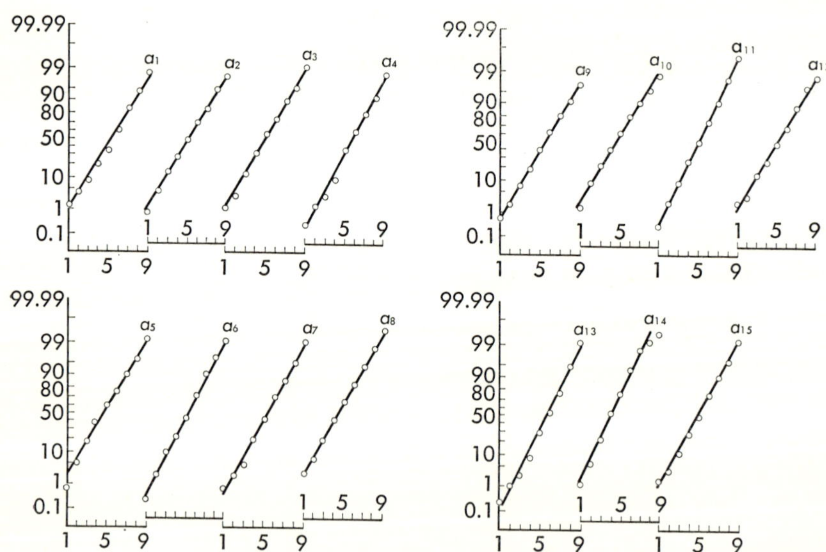

Fig. 2.　Percentage cumulative frequency distributions of AR-coefficients $a_1, a_2, \cdots, a_{15}$ of each EEG led bipolarly from the Fz-Pz regions of 366 normal adults plotted on normal probability paper (*40*). Unit amount of abscissa was taken as one-tenth of the difference between the maximal and minimal values of each AR-coefficient. Each distribution displays an oblique straight line, suggesting a normal distribution.

Figure 2 demonstrates percentage cumulative distributions of from the first, a_1, to the fifteenth, a_{15}, AR-coefficients of each EEG led from Fz-Pz regions of 366 normal adult males plotted on the normal probability papers. All displayed an oblique straight line to depict normal distribution (*39, 40*). The average of the first coefficient, say $\bar{a}_1$, showed a positive maximal value, whereas that of the second one, say $\bar{a}_2$, took a negative minimal value, as seen in Fig. 3. The third and higher order average coefficients also deviated from zero, showing some sway as a whole, which decayed to zero with increasing the order m and it is suggested that the coefficients higher than about the thirteenth order of EEGs vanish in the averages.

An example of an adult EEG led from the left occiput is illustrated on the upper left in Fig. 4. The former half tracing (A) of this EEG dis-

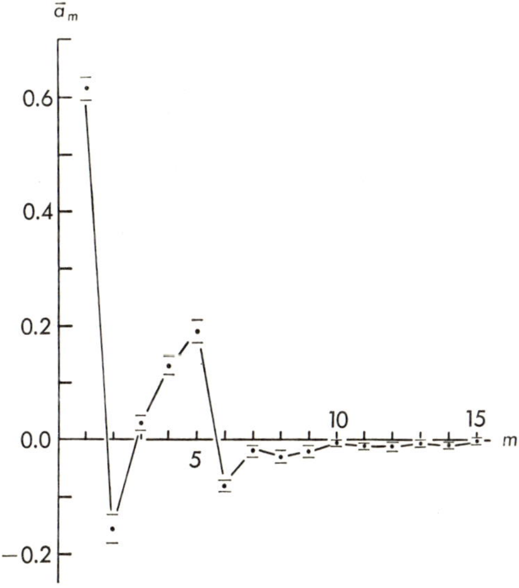

Fig. 3.   Averages of AR-coefficients and their fiducial intervals of 366 normal adult males EEGs (*40*). Abscissa (*m*) and ordinate: the order ($m=1, 2, 3, \cdots, 15$) and amount of average AR-coefficients $\bar{a}_1$, $\bar{a}_2$, $\cdots$, $\bar{a}_{15}$. The vertical lines around each average AR-coefficients (•) show the fiducial interval at the significant level of 0.99. Average AR-coefficients of higher order than the twelfth, $\bar{a}_{13}$, $\bar{a}_{14}$, and $\bar{a}_{15}$, can be considered as zero, so that they are negligible. AR-coefficients were estimated from the time series of each EEG, which was led from the Fz-Pz regions bipolarly, measured at every 20 msec interval in the length of 20 sec.

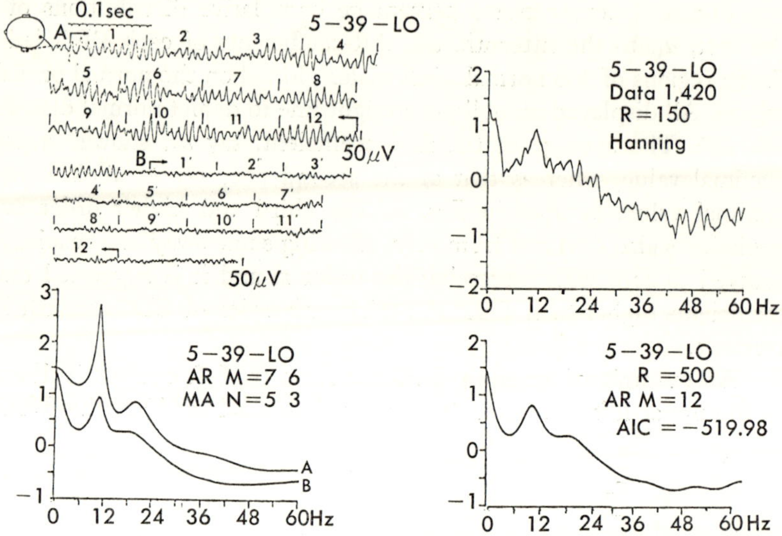

Fig. 4.   EEG of normal adult male and power spectral densities (*42*). Left top: EEG led monopolarly between the left occiput and earlobe (LO). A (upper half EEG tracing), high amplitude alpha wave dominant pattern; B (lower half tracing), low voltage "alpha blocking" pattern. Left bottom: autoregressive moving average (ARMA) power spectral densities of alpha dominant (A) and low voltage (B) EEGs. Right top: fourier power spectral density of the low voltage EEG (B) with the Hanning window function applied for smoothing. Right bottom: AR-power spectral density of the low voltage EEG (B). Each of the discrete digital time series of alpha dominant (A) and low voltage (B) EEGs were obtained by sampling data at every 1/120 sec in each length of 12 sec (1,440 data in number). Though the Hanning window function was applied for smoothing in the Fourier power spectral density, many irregular and nonsignificant peaks were displayed to mask significant peaks, whereas smooth, regular, and significant peaks appeared in the AR-power spectral density.

plays a dominant pattern of high amplitude alpha waves with frequencies of about 11 Hz, whereas the latter half (B) indicates a low voltage *alpha-blocking* pattern. Each percentage cumulative distribution of the time series of both the EEGs plotted on normal probability paper also displayed an oblique straight line, as depicted in Fig. 5 (*28*). On the top right in Fig. 4 is illustrated the Fourier power spectral density (Fourier transform of the autocovariance) of the latter half EEG (B), and on the bottom right the AR-power spectral density $Y_M(f)$ given by Eq. 2. 23 is shown. Though the Hanning *window function* (*41*) for smoothing was applied in the former,

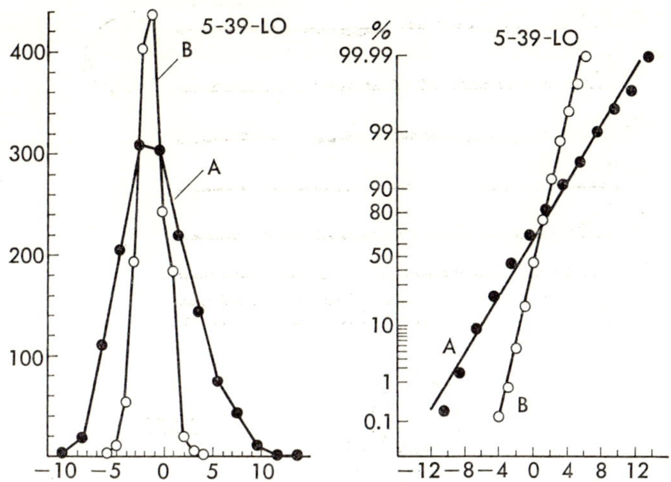

Fig. 5. Frequency polygons (left) and percentage cumulative frequency distributions (right) of sways in EEG (*28*). Two discrete digital time series of EEG (left upper in Fig. 4) were obtained from the former half (A) and latter half (B), respectively, for each length of 12 sec, wherein data were sampled at 1/120 sec (8.3 msec $= \tau$ and $N = 1{,}440$ in Eq. 2.1). Oblique straight line distributions on normal probability paper depicted in the right suggest that not only the alpha wave dominant (A) EEG, but also the low voltage alpha blocking (B) one sways, at least for the first approximation, obeying a normal Gaussian distribution.

a far more irregular spectral pattern was observed than that in the latter (*42*). Not only did delta and alpha wave peaks occur at 0 and about 11 Hz, respectively, but also beta wave peaks were easily identifiable around 20, 40, and 65 Hz, respectively, in the AR-power spectral density, whereas only delta and alpha peaks were suggested in the Fourier spectral density and those of other waves were masked by many irregular nonsignificant peaks. The left bottom is the ARMA spectral densities of the alpha dominant (A) and alpha-blocking low voltage (B) EEG tracings of the top left. The former spectral density was of the seventh order AR and fifth order MA processes, in which a sharp alpha wave peak was pierced high up at about 11 Hz and lower delta and beta wave peaks were also observed at 0 and around 20 Hz, respectively. In the spectral density of the low voltage alpha-blocking EEG (B), on the other hand, the sharpest and highest peak appeared at 0 Hz, which corresponds to irregular slow sways in EEG tracing B. Though the EEG configuration of the alpha-blocking state is usually called the low voltage fast pattern, the peak at around

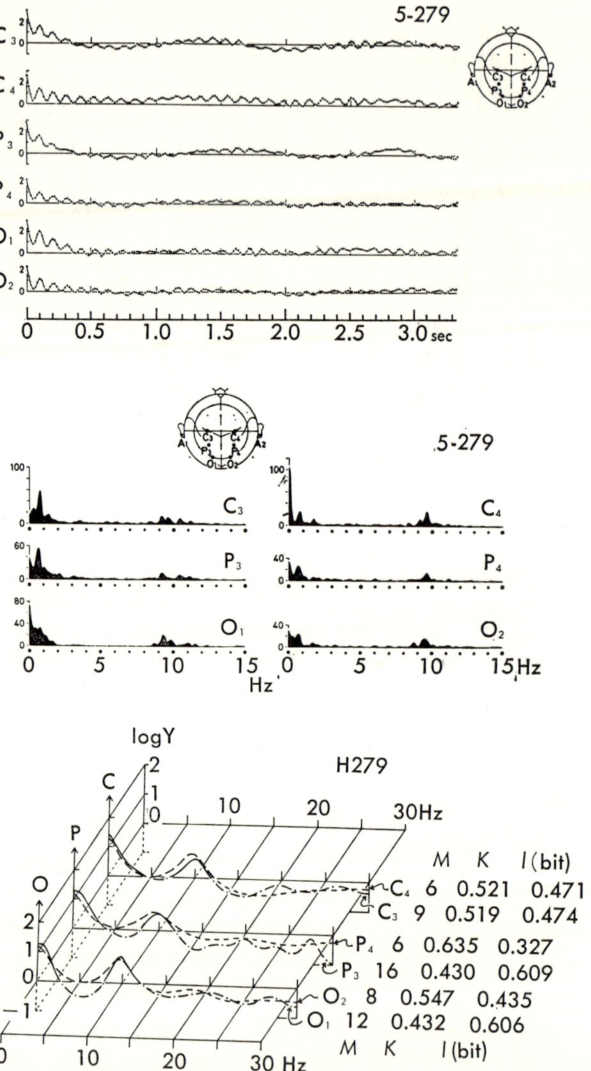

Fig. 6. Autocovariances (top), Fourier (middle), and AR-power spectral densities (bottom) of normal adult male EEGs led monopolarly from the bilateral central (C_3 and C_4), occipitoparietal (P_3 and P_4), and occipital regions (O_1 and O_2) (*18*). Each EEG was measured at every 1/60 sec (τ) for 10 sec length ($N=600$). Ordinate of the Fourier and AR-power spectral densities were taken, respectively, in linear and logarithmic scales. M and K, order and value of the " characteristic function " of the AR-process, respectively; I, " bio-informing activity amount."

20 Hz was lower than that in the spectral density of alpha dominant EEG tracing A. This obviously suggests, therefore, that not only alpha waves, but beta waves would also be depressed in the alpha-blocking state, as already depicted in the same state elicited by photic stimulation (43).

Another example of the autocovariances and AR-power spectral densities of EEGs of a normal adult male are shown by the top and bottom in Fig. 6, respectively. The Fourier power spectral densities obtained by the computer program of Southworth (44) are also demonstrated in the middle, in which the window function for smoothing was also applied. Some irregular peaks in each frequency range of delta, theta and alpha waves were also observed in these Fourier power spectral densities. It is impossible, however, to identify respective bio-activities with each of the irregular peaks. The pattern in the AR-power spectral densities, on the contrary, was far more smooth and regular. There seems to be some differences by side in the alpha waves between the left (O_1) and right (O_2) in the occipital region and between the left and right occipito-parietal regions (P_3 and P_4), since the peak at about 10 Hz on the right side (P_4 and O_2) is broader than that on the left side (P_3 and O_1). A peak at a frequency except 0 Hz in the power spectral density in general indi-cates a damped oscillation in the time domain, and the broader the peak is, the greater is the decay of the oscillation and *vice versa*. Oscillation of the autocovariances of about 10 Hz in the left occipital O_1 and occipito-parietal regions P_3, respectively, in Fig. 6 are slightly greater than those of the respective regions on the right side O_2 and P_4. This evidence would coincide with the above difference in the power spectral densities (18). The average amount of bio-informing activity given by Eq. 2. 17-1 was less in the above two regions on the right side than those on the left, as seen on the bottom right in Fig. 6. That is, those in the left occipito-parietal (P_3) and occipital regions (O_1) are 0.609 and 0.606 bit, respectively, but those in the respective right regions P_4 and O_2 are 0.327 and 0.435, respectively. In the central regions C_3 and C_4, however, no difference was observed, since these amounts were 0.474 and 0.471 bit, respectively. Oscillation in the autocovariance would be more regular with less decay and peaks in the AR-power spectral density would be more regular and narrower when a greater amount of bio-informing activity is exerted by the bio-system. This evidence is depicted in the top and bottom in Fig.6.

6. Higher Order Activities in Human Postural Sways
Erect standing posture is peculiar to humans, since no animal can

maintain a posture supporting the body with the hind limbs only. However, this posture is very unstable, because the human body is not only slender, but movable around the neck, waist, knee, and ankle joints, *etc.* The constant pull of the earth's gravity upon the human body serves to disturb standing posture. When the body sways in one direction the muscle spindles in the muscles on the other side will be stretched to cause reflexogenic proprioceptive muscle contraction, through which the sway of the body will be corrected. In addition, various postural reflexes such as righting and attitudinal reflexes will also be exerted to restore the inclination of the body. The former righting reflex is caused by stimulations delivered to the receptors on the labyrinth, photoreceptors in the retina and various proprio- and exteroceptors, whereas the latter attitudinal reflex is caused by the position of head in relation to the earth, trunk and four limbs. Both reflexes are composed of extensor, crossed extensor, supporting, tonic labyrinthine, tonic neck, tonic lumber reflexes, *etc.* Therefore, gravity and stimulations to the various receptors which cause the above postural reflexes while standing posture is being maintained can be regarded on the whole as random natural stimulation, n_t in Eq. 2. 7, which exhibits postural sway as the response y_t of a *posture holding (control) system* in the human body.

Figure 7 illustrates an example of minute sways of the center of gravity and ankle joints during human erect posture. Each of the percentage cumulative distributions of both the sways also traced an oblique straight line on normal probability paper (*18, 45–47*) as did the head sway recorded by a *goniometer* (*48*) placed on the neck. Therefore, human sway during

→ Fig. 7. Static sensograph for recording sways of human erect posture (top), antero-posterior sway of body gravity and sway of dorsi-plantar flexion of ankle joint (middle)(*18*), their frequency polygones and cumulative frequency distributions (bottom left), and percentage cumulative frequency distributions plotted on normal probability paper (bottom right). The subject stands on the "detective (triangular) plate" of the "Static Sensograph" (Sanei Sokki, Tokyo) with his eyes open and looks at a signal mark on the wall ahead adjusted to eye height. The antero-posterior and lateral sways are transformed into changes in electrical currents by transducers (TRANSD) 1 and 2 underneath the detective plate, respectively. Changes in the currents are amplified and recorded magnetically by an analogue data recorder and ink-written simultaneously by a pen recorder and the dorsi-plantar flexion of the ankle joint are also recorded simultaneously by an "electrogoniometer" (*48*). Each of both the percentage cumulative frequency distributions of the sway of body gravity (A) and that of ankle joint (B) displayed an oblique straight line to suggest a normal distribution, at least for the first step.

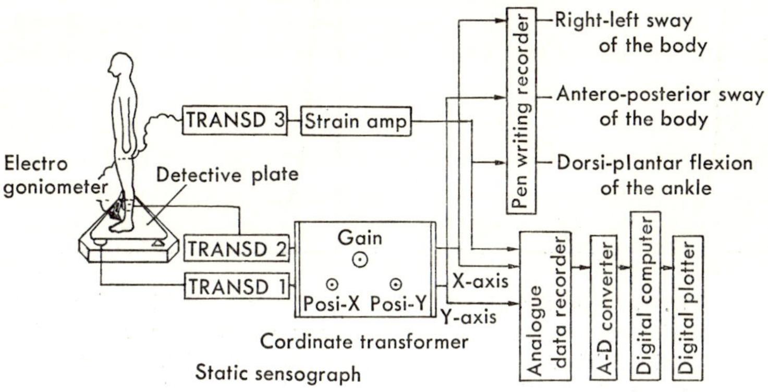
Block diagram of experimental apparatus
Right-left sway
of the body
Antero-posterior sway
of the body
Dorsi-plantar flexion
of the ankle
TRANSD 3
Strain amp
Pen writing recorder
Electro
goniometer
Detective plate
TRANSD 2
TRANSD 1
Gain
Posi-X Posi-Y
X-axis
Y-axis
Cordinate transformer
Static sensograph
Analogue data recorder
A-D converter
Digital computer
Digital plotter

Antero-posterior sway of body
A10
5sec
P10

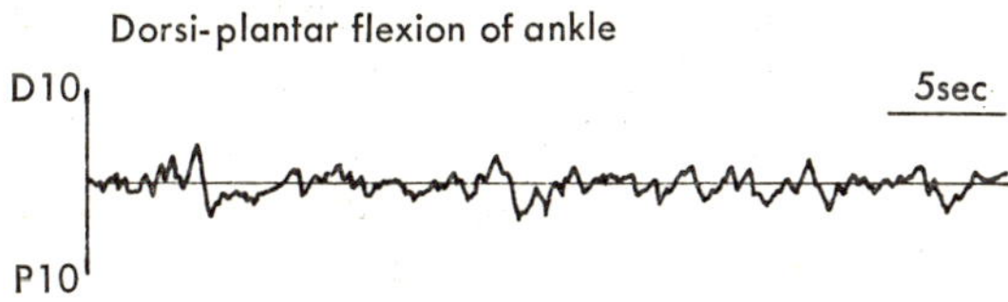
Dorsi-plantar flexion of ankle
D10
5sec
P10

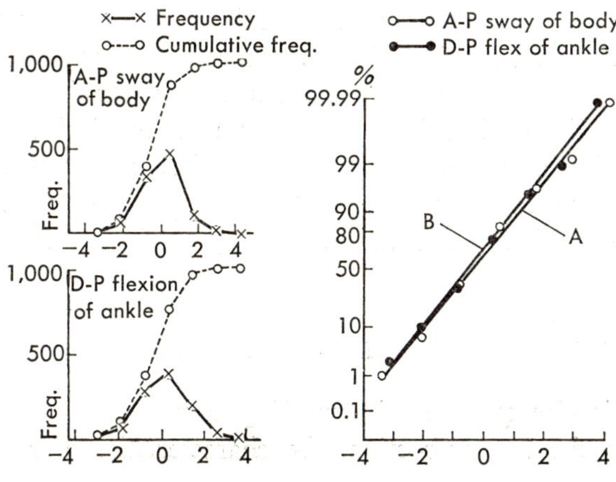
Frequency
Cumulative freq.
A-P sway of body
D-P flex of ankle
1,000
A-P sway
of body
500
Freq.
-4 -2 0 2 4
1,000
D-P flexion
of ankle
500
Freq.
-4 -2 0 2 4
%
99.99
99
90
80
50
10
1
0.1
B
A
-4 -2 0 2 4

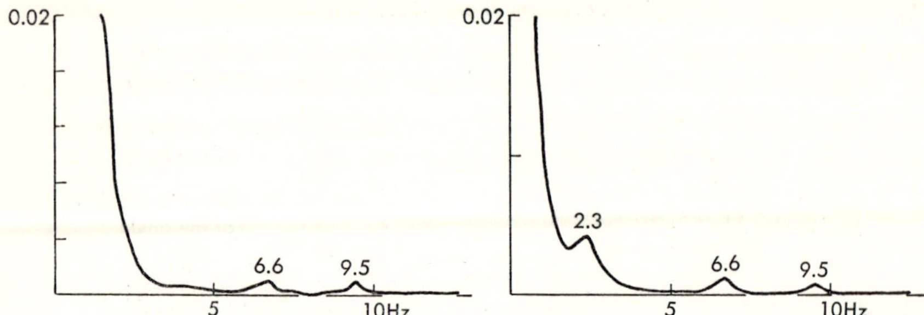

Fig. 8. Autoregressive (AR-) power spectral densities of postural sways during erect standing posture *(18)*. Left: antero-posterior sway of the body gravity. Right: dorsi-plantar flexion of the ankle joint. The order of the AR-process of both the sways was $M=11$. (H. Noguchi).

erect standing posture can also be regarded, at least for the first step, as an example of a normal AR-process. Consequently, the AR-power spectral density of a human postural sway given by Eq. 2. 23 will manifest a higher order activity of a "posture holding (control) system." In Fig. 8 are depicted AR-power spectral densities of the antero-posterior sway of the body (center of gravity) and of dorsi-plantar flexion of the ankle joint. A most powerful and nonoscillatory peak scaled out at around 0 Hz was observed along with lower oscillatory peaks at 6.6 and 9.5 Hz in both the spectra. An additional peak at 2.3 Hz which was higher than the two former peaks was seen in the ankle joint flexion, whereas a lower and broader peak at around 4 Hz and a hump suggesting a peak around about 2 Hz at the base of the highest peak at 0 Hz were observed in the spectral density of the body sway. The orders of both the posture holding activities were estimated to be 11. It may be possible, therefore, that a very slow oscillatory peak will be hidden in the highest peak around 0 Hz, which was scaled out. The sway of around 9 Hz during standing posture was also demonstrated by Mori *(49)*, which was related intimately to rhythmic grouping discharges of antigravity muscles, *e.g.*, the triceps surae, soleus and gastrocnemius muscles, *etc.*, and suggested that the sway around 9 Hz will be the "posture control (holding) activity" through the stretch reflx loop.

III. PATTERN DISCRIMINATION OF BIO-PHENOMENA

1. Pattern Discrimination of an Arbitrary Bio-phenomenon
As already noted in the former section, the characteristics of a higher

(M-th) order normal stationary stochastic bio-phenomenon exhibited by a bio-system can be given by a set of AR-coefficients a_m, $m = 1, 2, \cdots$, M. Consider here a group of standard bio-phenomena of sufficiently large number N, none of them with any significant difference from each other. And let a $K(\geq M)$ dimensional column vector of the AR-coefficients of the j-th phenomenon in this group be A_j such that

$$A_j = (a_{1j}, a_{2j}, \cdots, a_{Kj})', \qquad (K \geq M), \tag{3.1}$$

where $(\ )'$ indicates transposition of the vector and

$$a_{mj} = 0, \quad \text{when} \quad M < m \leq K \tag{3.1-1}$$

Let further the average vector of A_j be $\bar{A}$ and the average of the k-th AR-coefficient a_{kj} be $\bar{a}_k$ such that

$$\bar{a}_k = (1/N)\sum_{j=1}^{N} a_{kj}, \qquad (k = 1, 2, 3, \cdots, K) \tag{3.1-2}$$

then

$$\bar{A} = (\bar{a}_1, \bar{a}_2, \cdots, \bar{a}_K)'. \tag{3.2}$$

Consequently, the variance-covariance matrix of this group, say S, is given as

$$S = (1/N)\sum_{j=1}^{N}(A_j - \bar{A})' \cdot (A_j - \bar{A})$$

$$= \begin{bmatrix} S_{11} & S_{12} & \cdots & S_{1k} & \cdots & S_{1K} \\ S_{21} & S_{22} & \cdots & S_{2k} & \cdots & S_{2K} \\ & & \cdots\cdots\cdots \\ S_{k1} & S_{k2} & \cdots & S_{kk} & \cdots & S_{kK} \\ & & \cdots\cdots\cdots \\ S_{K1} & S_{K2} & \cdots & S_{Kk} & \cdots & S_{KK} \end{bmatrix}, \tag{3.3}$$

where S_{kk} is the variance and S_{jk}, $j \neq k$, is the covariance of the standard group and

$$S_{jk} = (1/N)\sum_{n=1}^{N}(a_{jn} - \bar{a}_j)(a_{kn} - \bar{a}_k),$$
$$(j, k = 1, 2, 3, \cdots, K). \tag{3.4}$$

It is obvious, therefore, that the AR-coefficient vector of the above standard group of bio-phenomena will fluctuate obeying the following K-variates normal distribution, say $f(A_j)$, such that

$$f(A_j) = \{1/(2\pi)^{K/2}|S|^{1/2}\} \cdot \exp\{-1/2 \cdot (A_j - \bar{A}) \cdot S^{-1} \cdot (A_j - \bar{A})\} \tag{3.5}$$

where $|S|$ and S^{-1}, respectively, are the determinant and inverse matrix of the variance-covariance matrix S. Consider here an arbitrary bio-phenomenon to be tested, regardless of whether it belongs to the standard group or not, and let its AR-coefficients be a_{xk}, $k=1, 2, \cdots, K$ and let further the coefficient column vector be

$$A_x = (a_{x1}, a_{x2}, \cdots, a_{xK})'. \tag{3.6}$$

There is a useful measure termed *Mahalanobis' generalized distance* to give the distance of the vector A_x from the average vector $\bar{A}$. Let it be $D^2(A_x, \bar{A})$, then

$$D^2(A_x, \bar{A}) = (A_x - \bar{A})' \cdot S^{-1} \cdot (A_x - \bar{A}). \tag{3.7}$$

Provided here the following *null* and its *alternative hypotheses*, say H_0 and H_1, be such that

H_0:   *The bio-phenomenon to be tested does not differ from the standard group, so this phenomenon belongs to the standard group,*   (3.8)

and

H_1:   *This bio-phenomenon differs in its pattern from that of the standard group, i.e., this does not belong to the standard group.*   (3.8-1)

Then the following distance F_x of this vector A_x from the average vector $\bar{A}$ of the standard group

$$F_x = \{N(N-K)/K(N+1)\} \cdot D^2(A_x, \bar{A}) \tag{3.9}$$

will follow the well known Fisher's F-distribution with degrees of freedom

$$n_1 = K \quad \text{and} \quad n_2 = N-K, \tag{3.9-1}$$

provided the above " null hypothesis " H_0 is accepted. If the *level* of the *significance* or *risk* is taken to have a sufficiently small α, such as 0.05 or 0.01, the amount of the distance at this level can be obtained by referring to the " Table of F-distribution." Let it be $F(n_1, n_2, \alpha)$, then the " null hypothesis " H_0 cannot be rejected, when

$$F_x \leq F(n_1, n_2, \alpha), \tag{3.10}$$

but can be rejected and the " alternative hypothesis " H_1 cna be accepted, when

$$F_x > F(n_1, n_2, \alpha). \tag{3.10-1}$$

Hence, $F(n_1, n_2, \alpha)$ would be the *critical limit* for the pattern discrimination of the bio-phenomenon (*39, 40*).

Often it would be more useful to consider two levels of significance α_1 and $\alpha_2(<\alpha_1)$, for instance $\alpha_1=0.05$ and $\alpha_2=0.01$, respectively. In this case, the " null hypothesis " H_0 cannot be rejected, but can be accepted, when

$$F_x \leqq F(n_1, n_2, \alpha_1), \tag{3.11}$$

whereas the " null hypothesis " H_0 can be rejected and the " alternative hypothesis " H_1 can be accepted, when

$$F_x > F(n_1, n_2, \alpha_2). \tag{3.11-1}$$

However, not only H_0 cannot be rejected, but H_1 also cannot be accepted, when

$$F(n_1, n_2, \alpha_1) < F_x \leqq F(n_1, n_2, \alpha_2). \tag{3.11-2}$$

Figure 9 depicts the histograms of the distances F_x given by Eq. 3.9 of each EEG in a group of EEGs led from the Fz-Pz regions of 90 normal adult males during the resting relaxed state (CONT), during hyperventi-

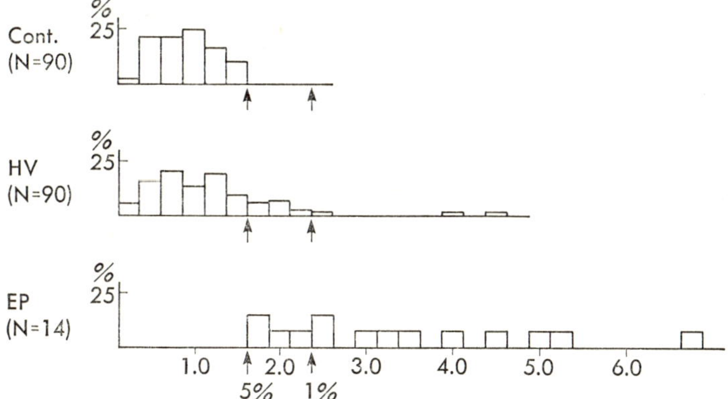

Fig. 9. Histograms of the distance of EEG pattern from the average pattern of the standard group (*39*). Abscissa, amount of the distance F_x given by Eq. 3.9. Vertical arrows beneath the abscissa indicate the " critical limits " of the distance at the levels of 0.05 (left) and 0.01 (right), respectively. Ordinate, percentage frequency. N, total amount of EEGs in number; Cont. and HV, EEGs of 90 normal adults during resting relaxed condition and hyperventilation, respectively; EP, EEGs of 14 epileptic patients in the Center for Epilepsy, Shizuoka.

lation (HV) in the same group and EEGs of 14 epileptic patients (EP) led from the F_3-C_3 regions. All distances F_x in the 90 EEGs during the resting relaxed state were within the " critical limit " $F(15, 75, 0.05)$, so that this group composed of 90 EEGs during resting relaxed state can be considered as a standard group of resting adult male EEGs led from the Fz-Pz regions. On the other hand, during hyperventilation the distances of seven and three EEGs were greater than the critical limits, $F(15, 75, 0.05)$ and $F(15, 75, 0.01)$, respectively. Hence, moderate and distinct effects due to hyperventilation were observed in four and three cases, respectively, among the former seven EEGs. An EEG from this group is shown in Fig. 10. Its power spectral density displayed a similar pattern to that of the average one, which was given by substituting average AR-coefficients, $\bar{a}_1, \bar{a}_2, \cdots, \bar{a}_{15}$, into Eq. 2. 23, $i.e.$,

$$Y_{15}(f) = 2\sigma_n{}^2/|1 - \textstyle\sum_{k=1}^{15}\bar{a}_k \exp(-i2\pi kf)|^2,$$
$$(0 \leqq f \leqq 1/2\tau). \tag{3. 12}$$

The highest sharp peak at 0 Hz in the spectral density indicates sway of the base line in the EEG tracing, which is a first-order aperiodic damped exponential component activity in the autocovariance of the EEG given by Eq. 4. 13-1 in Sect. IV. 2. The second dominant peak was enhanced at about 11.5 Hz indicating the alpha wave component given as the second-order damped oscillatory activity given by Eq. 4. 25 in Sect. IV. 3. In addition, more two low humps were observed at about 5.5 and 19 Hz, respectively,each of which shows theta and beta wave component waves, respectively, in the EEG also given as the second-order damped oscillatory activities described by the autocovariance in Eq. 4. 25.

An example of another normal adult during hyperventilation which showed a distance larger than the critical limit at the level of 0.01 is shown in the right top of Fig. 10. Some marked differences were observed not only in the EEG tracing, but also in its AR-spectral density during hyperventilation. In agreement with high amplitude alpha wave pattern in the EEG, a higher and sharper alpha wave peak than that during the resting condition appeared at about 9 Hz in the AR spectral density accompanying another lower alpha wave peak at about 11.5 Hz. Further, not only delta and theta wave peaks at 1 and 5 Hz, respectively, but also those of two beta waves at about 17 and 23 Hz, respectively, were observed in the spectral density during hyperventilation, suggesting a significant difference from that of the resting state.

In the 14 epileptic EEGs, all of the distances were greater than the

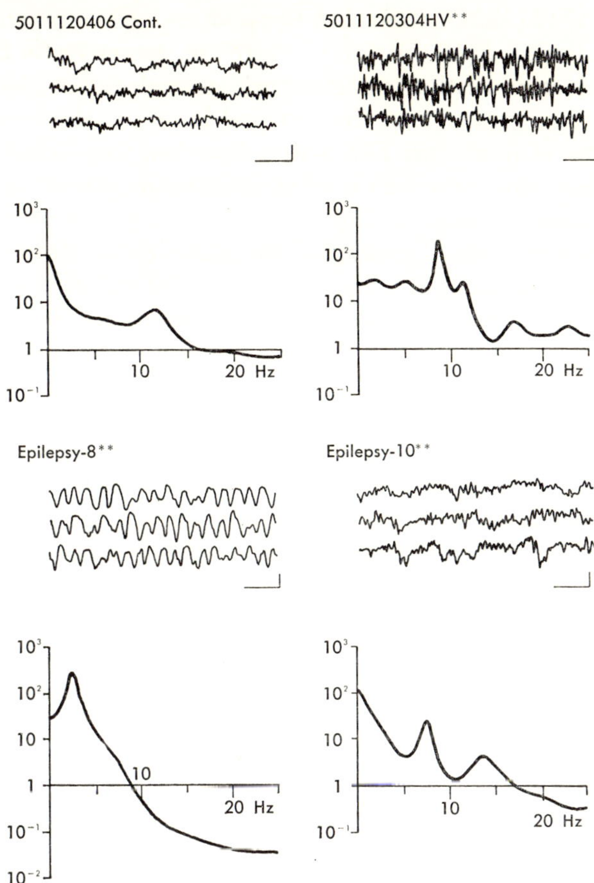

Fig. 10. Examples of EEG and AR-power spectral densities (*39*). Top left: an EEG (Fz-Pz bipolar lead) in the standard group having a spectral pattern similar to that of the average one of the standard group with constituent EEGs of 90 normal adult males in the resting condition. Top right: an example of an EEG during hyperventilation in the standard group, which depicted a marked difference from the standard group. The difference F_x was the largest in the histogram in the middle of Fig. 9. Bottom: two examples of epileptic EEGs (F_3-C_3 bipolar lead) with a slow wave pattern having a high peak at 2 Hz and a hump at about 6 Hz in the AR-spectral density (left) and relatively fast theta wave of about 7 Hz and slow beta wave of about 14 Hz in its AR-spectral pattern (right). Both were markedly different from the standard group in their patterns. Each distance F_x was larger than the "critical limit" at the level 0.01. Scale underneath the EEG tracings is 1 sec and 50 μV calibration.

" critical limit " $F(15, 75, 0.05)$, but those of four EEGs were less than the " critical limit " $F(15, 75, 0.01)$. Therefore, ten epileptic EEGs can be recognized as indicating an abnormal spectral pattern, since their distances F_x were greater than the " critical limit " at the level of 0.01. However, those of four other EEGs were less than this " critical limit," but greater than that at the 0.05 level to indicate the case of Eq. 3. 11-2, which can be classified as border line or " $\pm$ " in being between the normal " $-$ " and abnormal " $+$ " patterns. Two examples of these are demonstrated by the lower in Fig. 10. The one on the bottom left is a slow wave EEG pattern with a sharp delta wave peak at about 2.5 Hz followed by a low theta wave hump at about 7 Hz in its spectral density. The other one on the bottom right shows three peaks of delta, theta and beta waves at 0, about 7.5 and 14 Hz, respectively, in its spectral density. Both EEGs were of course different from the standard group in their patterns at the level of 0.01. Though the EEGs of epileptic patients were led from regions different from those of the standard group, the differences in the spectral pattern can be recognized as significant, since the EEGs of both groups were recorded bipolarly.

2. *Pattern Discrimination between Two Groups of Bio-phenomena*

There are some such cases that need to be tested to determine whether two groups of bio-phenemena are different or not, each of which can be considered, at least for the first approximation, as a normal weakly stationary stochastic process. Let here the number of bio-phenomena in the two groups be N_1 and N_2 and the AR-coefficient vector of each bio-phenomenon in the two groups be A_{1i} and A_{2j} such that

$$A_{1i} = (a_{11i}, a_{12i}, \cdots, a_{1Ki}),$$
$$(i = 1, 2, 3, \cdots, N_1) \tag{3.13}$$

and

$$A_{2j} = (a_{21j}, a_{22j}, \cdots, a_{2Kj}),$$
$$(j = 1, 2, 3, \cdots, N_2) \tag{3.13-1}$$

respectively. Then the average vectors of these groups are given as the following $\bar{A}_1$ and $\bar{A}_2$:

$$\bar{A}_1 = (\bar{a}_{11}, \bar{a}_{12}, \cdots \bar{a}_{1K})', \tag{3.14}$$

$$N_1\bar{a}_{1k} = \sum_{i=1}^{N_1} a_{1ki}, \quad (k = 1, 2, \cdots, K) \tag{3.14-1}$$

and

$$\bar{A}_2 = (\bar{a}_{21}, \bar{a}_{22}, \cdots, \bar{a}_{2K})', \tag{3.15}$$

$$N_2\bar{a}_{2k} = \sum_{j=1}^{N_2} a_{2kj}, \quad (k = 1, 2, \cdots, K), \tag{3.15-1}$$

respectively. Let further their variance-covariance vector be V_1 and V_2 such that

$$N_1 V_1 = \sum_{i=1}^{N_1} (A_{1i} - \bar{A}_1)' \cdot (A_{1i} - \bar{A}_1), \tag{3.16}$$

$$N_2 V_2 = \sum_{j=1}^{N_2} (A_{2j} - \bar{A}_2)' \cdot (A_{2j} - \bar{A}_2). \tag{3.16-1}$$

Then the difference, say D, between the two average vectors, which shows the difference between the two groups, is given as

$$D = \bar{A}_1 - \bar{A}_2,$$

which belongs to a K-variates normal distribution with average zero, provided such " null hypothesis " H_0 is accepted so that both groups do not differ in their patterns. Hence, the following $F(n_1, n_2)$:

$$F(n_1, n_2) = \{(N-K-1)N_1 N_2/KN\} D' V^{-1} D, \tag{3.17}$$

where

$$N = N_1 + N_2, \quad V = N_1 V_1 + N_2 V_2, \tag{3.17-1}$$
$$n_1 = K, \quad n_2 = N - K - 1 \tag{3.17-2}$$

follows an F-distribution with degrees of freedom $n_1 = K$ and $n_2 = N - K - 1$. Therefore, the difference between the above two average vectors, each of which characterizes the pattern of the two groups of bio-phenomena, can be tested. An example will be shown in a later section.

IV. DECOMPOSITION OF HIGHER ORDER ACTIVITY OF BIO-SYSTEMS

1. Component Activity of Bio-systems

The *characteristic equation* $a_M(B)=0$ *of M degree* with respect to the *backward shift operator* B given by Eq. 2.6 can be factorized as follows:

$$a_M(B) = (1 - S_1 B)(1 - S_2 B)\cdots(1 - S_M B) = 0, \tag{4.1}$$

where $\{S_m^{-1}\}$, $(m=1, 2, \cdots, M)$ is the root of the " characteristic equation," each of which can be obtained by the well known Bairstow's method, *etc.*, and in practice all roots can be as simple. When one of these roots is imaginary, its complex conjugate is also a root of this equation. Let here, therefore, real roots be V in number, say ϕ_v^{-1}, $(v=1, 2, \cdots, V)$ and W pairs be complex conjugate roots, say $\{\eta_w^{-1}, \eta_w^{*-1}\}$, $(w=1, 2, \cdots,$

W). Then the characteristic equation $a_M(B)$ can be rewritten by products of the first- and second-order characteristic functions, say $a_{1v}(B)$ and $a_{2w}(B)$, respectively (12, 13), such that

$$a_M(B) = \{\Pi_{v=1}^{V} a_{1v}(B)\} \cdot \{\Pi_{w=1}^{W} a_{2w}(B)\}, \qquad (V+2W = M), \qquad (4.2)$$

where

$$a_{1v}(B) = 1-\phi_v B \qquad (4.3)$$
$$a_{2w}(B) = 1-\varphi_{1w}B-\varphi_{2w}B^2 \qquad (4.4)$$
$$\varphi_{1w} = \eta_w+\eta_w^* \qquad (4.5)$$
$$\varphi_{2w} = -\eta_w \cdot \eta_w^*. \qquad (\eta_w, \eta_w^*:\text{ complex conjugate}) \qquad (4.5\text{-}1)$$

The values of the characteristic function of the first- and second-order AR-processes $a_{1v}(B)$ and $a_{2w}(B)$ are easy to obtain by letting $M=1$ and 2 in Eq. 2.20, respectively. Let them be K_{1v} and K_{2w}, respectively, then

$$K_{1v} = 1-\phi_v^2, \qquad (4.6)$$
$$K_{2w} = 1-\varphi_{2w}^2-(1+\varphi_{2w})\varphi_{1w}^2/(1-\varphi_{2w}). \qquad (4.6\text{-}1)$$

Consequently, the time series of the M-th order bio-phenomenon y_t can be decomposed into the first- and second-order processes of V and W in number, respectively (50). Therefore, the higher order activity $g_M(B)=1/a_M(B)$ of a bio-system given by Eq. 2.8 and the fourth block diagram in Fig. 1 can be decomposed into the first- and second-order activities of V and W in number, respectively, say $g_{1v}(B)$ and $g_{2w}(B)$, respectively, such that

$$g_M(B) = \{\Pi_{v=1}^{V} g_{1v}(B)\} \cdot \{\Pi_{w=1}^{W} g_{2w}(B)\}, \qquad (V+W = M), \qquad (4.7)$$

where

$$g_{1v}(B) = 1/a_{1v}(B), \qquad (4.8)$$

and

$$g_{2w}(B) = 1/a_{2w}(B). \qquad (4.8\text{-}1)$$

Consequently, the bio-system having M-th order activity $g_M(B)$ can be considered as being constituted of the first- and second-order bio-sub-systems of V and W in number, respectively, having respective activities $g_{1v}(B)$, $v=1, 2, \cdots, V$ and $g_{2w}(B)$, $w=1, 2, \cdots, W$, respectively.

The values of the first- and second-order activities, say G_{1v} and G_{2w}, respectively, are easy to obtain as inverses of values of the first- and second-order characteristic functions K_{1v} and K_{2w} given by Eq. 4.6 and

4. 6-1, respectively, such that

$$G_{1v} = 1/K_{1v}, \qquad (v = 1, 2, \cdots, V) \qquad (4.9)$$
$$G_{2v} = 1/K_{2w}, \qquad (w = 1, 2, \cdots, W). \qquad (4.9\text{-}1)$$

The autocovariance function of the AR-process y_t of order M, which can be considered as being exhibited by the bio-system of order M as its response caused by the natural stimulation n_t (see Fig. 1, fourth block diagram) will be given by the following R_k (23) with time lag k such that

$$R_k = E[y_{t-k}y_t] = H_1 S_1{}^k + H_1 S_2{}^k + \cdots + H_M S_M{}^k,$$
$$(k = 0, 1, 2, \cdots) \qquad (4.10)$$

where $S_m{}^{-1}$ is the root of the characteristic equation $a_M(B)=0$, and $H_m, m = 1, 2, \cdots, M$ is given as

$$H_m = \sigma_n{}^2/a_M(S_m)\cdot a3(S_m{}^{-1}), \qquad (m = 1, 2, \cdots, M) \qquad (4.11)$$

where $\sigma_n{}^2$ is the variance of the random natural stimulation n_t in Eq. 2. 3 already given as

$$\sigma_n{}^2 = K_M \sigma_y{}^2 \qquad (2.22)$$

and $a3(S_m{}^{-1})$ is given by substituting $B = S_m{}^{-1}$ into the following equation (30)

$$a3(B) = M - (M-1)a_1 B - (M-2)a_2 B^2 - \cdots - a_{M-1}B^{M-1}. \qquad (4.12)$$

As the real and imaginary roots were assumed to be V in number and W pairs, respectively, and all of them are simple in practice, the above autocovariance function with time lag k is rewritten as

$$R_k = \sum_{v=1}^{V} H_v \phi_v{}^k + \sum_{w=1}^{W}(H_w \eta_w{}^k + H_w{}^* \eta_w{}^{*k})$$

$$= \sum_{v=1}^{V} R_{1vk} + \sum_{w=1}^{W} R_{2wk},$$

$$(k = 0, 1, 2, \cdots ; M = V + 2W) \qquad (4.10\text{-}1)$$

where R_{1vk} and R_{2wk} are the autocovariance functions of the first- and second-order AR-processes, respectively, while not only η_w and $\eta_w{}^*$, but also H_w and $H_w{}^*$ are complex conjugates.

2. The First-order Activity of a Bio-system
The autocovariance function of the first-order AR-process exhibited by a first-order bio-system is given by the first term on the right side of

Eq. 4. 10-1 such that

$$R_{1vk} = H_v\phi_v{}^{k\tau}, \qquad (\tau: \text{sampling interval})$$
$$(v = 1, 2, \cdots, V; \; k = 0, \pm 1, \pm 2, \cdots) \qquad (4.13)$$

where ϕ_v^{-1} is one of the real roots, V in number of the charcteristic equation $a_M(B)=0$ and H_v is given as the case of $M=1$ and S_1^{-1} in Eq. 4. 11 and $B=\phi_v$ in Eq. 4. 3 such that

$$H_v = \sigma_n{}^2/a_{1v}(\phi_v)\cdot a_{1v}3(\phi_v{}^{-1})$$
$$= \sigma_n{}^2/(1-\phi_v{}^2) = \sigma_n{}^2/K_{1v} = \sigma_n{}^2 G_{1v}. \qquad (4.14)$$

Let here

$$\phi_v = d_{1v}\exp(-1/T_{1v}), \qquad (4.15)$$

where

$$d_{1v} = 1, \qquad \text{when} \quad \phi_v > 0,$$
$$= -1, \qquad \text{when} \quad \phi_v < 0. \qquad (4.15\text{-}1)$$

Then the autocovariance function R_{1vk} given by Eq. 4. 13 is rewritten as the following damped exponential form

$$R_{1vk} = d_{1v}{}^k\sigma_n{}^2 G_{1v}\exp(-|k|\tau/T_{1v}), \qquad (k = 0, \pm 1, \pm 2, \cdots) \qquad (4.13\text{-}1)$$

and this decays exponentially to zero with increasing absolute value of time lag k, as depicted by the left top in Fig. 11. The time constant of the decay is given by T_{1v} such that

$$T_{1v} = \tau/1\text{n}|\phi_v| \qquad (4.16)$$

when ϕ_v is positive, whereas the decay oscillates in sign when ϕ_v is negative (*23*).

The first-order AR-process will be given as the case of $M=1$ in Eq. 2. 2 such that

$$y_{1vt} = \phi_v y_{1v,t-1}+n_{1vt}, \qquad (v = 1, 2, \cdots, V) \qquad (4.17)$$

then the power spectral density of this process, say $Y_{1v}(f)$, will be given as the case of $M=1$ in Eq. 2. 23 such that

$$Y_{1v}(f) = 2\sigma_{n1v}{}^2/|1-\phi_v\exp(-i2\pi f)|^2$$
$$(0 \leqq f \leqq 1/2\tau, \qquad v = 1, 2, \cdots, V), \qquad (4.18)$$

where $\sigma_{n1v}{}^2$ is the variance of the random process n_{1vt}, which indicates the random natural stimulation to the first-order bio-subsystem exhibiting y_{1vt} at an arbitrary time t in Eq. 4. 17.

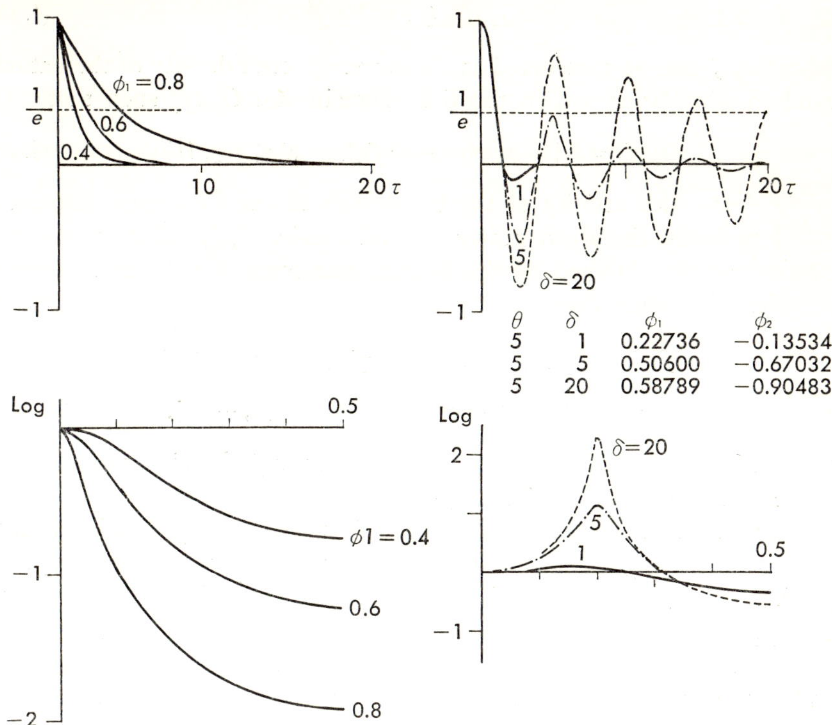

Fig. 11. Three autocorrelation curves (top) and the power spectral densities (bottom) of the first- (left) and second-order (right) autoregressive (AR-) processes. The first AR-coefficients of three first-order AR-processes ϕ_1 are 0.4, 0.6, and 0.8, respectively, each autocorrelation curve displays a damped exponential with its "time constant" (T) 1.09, 1.96, and 4.48, respectively, where τ is the sampling time interval of the time series of the AR-process given by the case of $M=1$ in Eq. 2.2. Their power spectral densities (left bottom) show a peak having a summit at 0 Hz. The larger the AR-coefficient ϕ_1, the longer the time constant of the autocorrelation curve and the narrower the peak in the spectral density and *vice versa*. Each autocorrelation curve of three second-order AR-processes (top right) displays a damped oscillatory curve with its "damping time" $\delta=1\,\tau$, $5\,\tau$, and $20\,\tau$, respectively, and a peak is enhanced with its summit at a frequency higher than 0 Hz and lower than $1/2\,\tau$ Hz in the power spectral density. The longer the damping time of the autocorrelation curve, the narrower the peak in the spectral density and *vice versa*.

The "stimulus-system-response" relation of the first-order bio-subsystem in the frequency domain will be given as the case of $M=1$ in Eq. 2.23 such that

$$\sigma_{n1v}{}^2 \cdot G_{1v}(f) = Y_{1v}(f) \tag{4.19}$$

where $\sigma_{n1v}{}^2$ can be considered as the power spectral density of the natural random stimulation n_{1vt} in the time domain and $G_{1v}(f)$ such that

$$G_{1v}(f) = 2/|1-\phi_v \exp(-i2\pi f)|^2, \qquad (0 \leqq f \leqq 1/2\,\tau) \tag{4.20}$$

is the first-order activity of the bio-system in the frequency domain.

Practically, the above power spectral density $Y_{1v}(f)$ of the first-order process can be obtained by the Fourier transform of the autocovariance function R_{1vk} such that

$$Y_{1v}(f) = 2\tau \sum_{k=-K}^{K} R_{1vk} \exp(-i2\pi fk), \qquad (0 \leqq f \leqq 1/2\,\tau),$$

$$= 2\tau \sum_{k=-K}^{K} R_{1vk} \cos 2\pi fk, \qquad (v = 1, 2, \cdots, V), \tag{4.18-1}$$

where K is taken as a positive integer large enough for elimination of the autocovariance function, *i.e.*, $R_{1vK} \fallingdotseq 0$.

The spectral density $Y_{1v}(f)$ and the activity $G_{1v}(f)$ display a maximum value at $f=0$ and decay monotonously with increasing frequency f, when ϕ_v is positive, as depicted by the left bottom in Fig. 11, whereas the opposite frequency pattern will be displayed when ϕ_v is negative, that is, the maximal value is displayed at the highest frequency $f_{max}=1/2\tau$ and monotonous decay occurs with decreasing frequency. The process can, therefore, be termed in this latter case the *high frequency active (high pass) type* of bio-system.

The narrower this peak at $f=0$ or $f=f_{max}$ in the power spectral density, the longer is the time constant T_{1v} of the autocovariance or autocorrelation function of the first-order bio-phenomenon and *vice versa*. It is obvious that the power of the first-order bio-system will be given by

$$R_{1v0} = H_v. \tag{4.21}$$

The " amount of bio-informing activity " of the first-order bio-system is given as the case of $M=1$ in Eq. 2.14-1. Let it be I_{1v}, then

$$I_{1v} = -0.5 \,\mathrm{lb}\, K_{1v} = -0.5 \,\mathrm{lb}(1-\phi_v{}^2)$$
$$= 0.5 \,\mathrm{lb}\, G_{1v} \quad \mathrm{bit}, \qquad (\mathrm{lb} = \log_2). \tag{4.22}$$

As $G_{1v}=1/K_{1v}$ is the first-order " activity amount," the first-order " bio-informing activity amount " is one half bit of the logarithm of the " activity amount " G_{1v} to the base two.

3. The Second-order Activity of a Bio-system

The time series of the second-order AR-process will be given by the case of $M=2$ in Eq. 2. 2. Let it be y_{2wt} at an arbitrary time t, then

$$y_{2wt} = \varphi_{1w}y_{2w,t-1}+\varphi_{2w}y_{2w,t-2}+n_{2wt}.$$
$$(w = 1, 2, \cdots, W) \qquad (4.23)$$

Let the complex conjugate roots η_w and $\eta_w{}^*$ of the characteristic equation $a_M(M)=0$ of the M-th order AR-process given by Eq. 4. 1 be

$$\eta_w = d_{2w} \exp\left(-\tau/T_{2w}-i\omega_w\right), \qquad (w = 1, 2, \cdots, W) \qquad (4.24)$$
$$\eta_w{}^* = d_{2w} \exp\left(-\tau/T_{2w}+i\omega_w\right), \qquad\qquad (4.24\text{-}1)$$

where $\omega_w=2\pi f_{Dw}$, and ω_w and f_{Dw} are angular and circular frequencies, respectively. Then the autocovariance function of the second-order AR-process R_{2wk} with respect to time lag $k\tau$ in Eq. 4. 10-1 is given as the following damped oscillatory form, as depicted by the right top in Fig. 11,

$$R_{2wk} = d_{2w}{}^k \exp\left(-|k|\tau/T_{2w}\right)\cdot(P_{2w}\cos 2\pi f_{Dw}k\tau+Q_w \sin 2\pi f_{Dw}k\tau)$$
$$= d_{2w}{}^k \exp\left(-|k|\tau/T_{2w}\right)\Omega_w \cos\left(2\pi f_{Dw}k\tau-\lambda_w\right),$$
$$(w = 1, 2, \cdots, W) \qquad (4.25)$$

where

$$P_{2w} = H_w+H_w{}^* = 2\mathrm{Re}[H_w], \qquad (4.26)$$
$$iQ_w = H_w-H_w{}^* = i2\mathrm{Im}[H_w], \qquad (4.26\text{-}1)$$

and

$$d_{2w} = 1, \qquad \text{when} \quad \varphi_{1w} > 0,$$
$$= -1, \qquad \text{when} \quad \varphi_{1w} < 0, \qquad (4.27)$$

and

$$\Omega_w = P_{2w}/\cos \lambda_w. \qquad (4.28)$$

Here Re[ ] and Im[ ] indicate, respectively, the real and imaginary parts of the value in the brackets. This autocovariance function R_{2wk} displays a damped oscillation with a *damping frequency* f_{Dw} such that

$$f_{Dw} = \tan^{-1}\left(\mathrm{Im}[\eta_w]/\mathrm{Re}[\eta_w]\right) = \tan^{-1}\left(h_2/h_1\right)$$
$$= \theta_w/2\pi\tau, \qquad (4.29)$$
$$h_1 = \mathrm{Re}[\eta_w], \qquad (4.30)$$
$$h_2 = \mathrm{Im}[\eta_w], \qquad (4.30\text{-}1)$$

when the real part $\mathrm{Re}[\eta_w]$ of the complex root is positive, where θ_w is given as

$$\theta_w = \tan^{-1}\left((-\varphi_{1w}^2 - 4\varphi_{2w})^{1/2}|\varphi_{1w}|\right)$$
$$= 2\pi\tau \tan^{-1}(h_2/h_1). \tag{4.31}$$

The damped oscillation of this autocovariance R_{2wk} has a *phase* λ_w

$$\lambda_w = \tan^{-1}(Q_w/R_{2w}), \tag{4.32}$$

where Q_w is given as

$$Q_w = 2\sigma_n^2(h_1 h_4 + h_2 h_3)/h, \tag{4.33}$$

where

$$h_3 = \mathrm{Re}[a3(\eta_w^{-1})] \tag{4.34}$$
$$h_4 = \mathrm{Im}[a3(\eta_w^{-1})] \tag{4.34-1}$$

and

$$h = (h_1^2 + h_2^2)(h_3^2 + h_4^2).$$

Here Q_w and $\tan \lambda_w = Q_w/P_w$ indicate, respectively, the *skewness* of the autocovariance R_{2wk} and the peak at the *resonance frequency* f_{Rw} given later by Eqs. 4.44 and 4.44-1 in the power spectral desity $Y_{2w}(f)$ of the second-order process.

When the real part $\mathrm{Re}[\eta_w]$ or φ_{1w} is negative, the autocovariance R_{2wk} changes alternatively in sign with respect to time lag $k\tau$ (23) to indicate the " high frequency active type " in the second-order bio-subsystem of the M-th order bio-system. The decay in the amplitude of the damped oscillation is given by the *damping coefficient*. Let it be ζ_w, *then*

$$\zeta_w = \delta_w(\delta_w^2 + \theta_w^2)^{-1/2} \tag{4.35}$$

where

$$\delta_w = -0.5 \ln(\varphi_{2w}) \tag{4.36}$$

or by the *damping time*, say T_{2w}, which gives the time constant of the envelope of the damped oscillation, given as $\exp(-k\tau/T_{2w})$ such that

$$T_{2w} = -\tau/\ln|\eta_w|. \tag{4.37}$$

It is obvious here that the *power* of the second-order AR-process exhibited by the second-order bio-subsystem is P_{2w} in Eq. 4.25, since

$$R_{2w0} = P_{2w} \tag{4.38}$$

given by the case of $k=0$ in Eq. 4.25 and this will be given also as

$$P_{2w} = 2\sigma_n^2(h_1 h_3 - h_2 h_4)/h. \tag{4.38-1}$$

The power spectral density of the second-order AR-process is given by the case of $M=2$ in Eq. 2.23. Let it be $Y_{2w}(f)$, then

$$Y_{2w}(f) = 2\sigma_{2w}^2/|1-\varphi_{1w}\exp(-i2\pi f)-\varphi_{2w}\exp(-i4\pi f)|^2$$
$$= 2\sigma_{2w}^2/(1+\varphi_{1w}^2+\varphi_{2w}^2-2\varphi_{1w}(1-\varphi_{2w})\cos 2\pi f-2\varphi_{2w}\cos 4\pi f),$$
$$(0 \leq f \leq 1/2\tau, \quad w=1,2,\cdots,W), \tag{4.39}$$

where σ_{2w}^2 is the variance of the random natural stimulation n_{2wt} to the second-order bio-subsystem and is also the power spectral density of the natural stimulation. Then, the " stimulus-system-response " relation in the frequency domain is given as

$$\sigma_{2w}^2 \cdot G_{2w}(f) = Y_{2w}(f), \tag{4.40}$$

where $G_{2w}(f)$ is the power spectral density of the second-order activity in the frequency domain such that

$$G_{2w}(f) = 2/(1+\varphi_{1w}^2+\varphi_{2w}^2-2\varphi_{1w}(1-\varphi_{2w})\cos 2\pi f-2\varphi_{2w}\cos 4\pi f). \tag{4.41}$$

Practically, the power spectral density $Y_{2w}(f)$ of the second-order AR-process can also be obtained as the Fourier transform of the autocovariance function R_{2wk} such that

$$Y_{2w}(f) = 2\tau \sum_{k=-K}^{K} R_{2wk}\exp(-i2\pi fk\tau)$$
$$= 2\tau \sum_{k=-K}^{K} R_{2wk}\cos 2\pi fk\tau$$
$$(0 \leq f \leq 1/2\tau, \quad w=1,2,\cdots,W), \tag{4.39-1}$$

where K is taken as a positive integer lage enough for the R_{2wk} to disappear (46), i.e., $R_{2wK} \fallingdotseq 0$.

A second-order AR-process displaying a damped oscillation in average in the time-domain can be considered in general as being composed of two processes. The one is an oscillatory process without decay in its amplitude and the other is a *decrementing process* that causes amplitude decay. The rhythmicity of the former process will be characterized by the *natural frequency*. Let this frequency of the w-th second-order process described by Eq. 4.25 be f_{Nw}, then

$$f_{Nw} = (\delta_w^2+\theta^2)^{1/2}/2\pi\tau, \tag{4.42}$$

where τ is the sampling interval of the time series y_{2wt} of the second-order AR-process given by Eq. 4.23.

The " damping time " T_{2w} given by Eq. 4.37 characterizes the latter process in the second-order process and it is related to the above " natural frequency " and the " damping coefficient " ζ_w given by Eq. 4. 35 such that

$$T_{2w} = (2\pi f_{Nw}\zeta_w)^{-1}. \tag{4.43}$$

The frequncy at the summit of the peak of a second-order process in the spectral density $Y_{2w}(f)$ is the *resonance frequency*. Let it be f_{Rw}, then

$$f_{Rw} = f_{Nw}(1-2\zeta_w{}^2)^{1/2}, \tag{4.44}$$

where

$$\zeta_w < 1/\sqrt{2} = 0.7071. \tag{4.45}$$

A similar expression of the " damping frequency " f_{Dw} given already by Eq. 4. 29 is well known as

$$f_{Dw} = f_{Nw}(1-\zeta_w{}^2)^{1/2}. \tag{4.29-1}$$

Therefore, the relationship between the " resonance " and " damping frequency " is given as

$$\begin{aligned} f_{Rw} &= f_{Dw}(1-\zeta_w{}^2/(1-\zeta_w{}^2))^{1/2} \\ &= f_{Dw}(1-\zeta^2-\zeta^4-\cdots)^{1/2}. \end{aligned} \tag{4.44-1}$$

Equation 4. 44 denotes that the smaller the " damping coefficient " ζ_w, the longer the oscillation of the second-order process will be sustained on average and *vice versa*. On the other hand, Eq. 4. 44 discloses that the rhythmicity of the oscillation will disappear and no peak will be enhanced in the spectral density $Y_{2w}(f)$, when the " damping coefficient " exceeds 0.7071, which can be, therefore, termed the *critical damping coefficient*.

In addition to the " damping coefficient " and " damping time," it would be useful to consider how many swells of oscillation in a second-order process will be displayed on average over the duration of the " damping time." The larger they are in amount, the longer the oscillation is sustained on average. Therefore, this can be termed the *continuity* (46) of the oscillation of the second-order process. Let it be C_w, then

$$C_w = f_{Dw} \cdot T_{2w}. \tag{4.46}$$

The second-oreder AR-process having negative first AR-coefficient ($\varphi_{1w}<0$) is a process of " high frequency active (high pass) type," so that the " natural," " damping " and " resonance frequencies " in this process can be given as the following f_{Nw}', f_{Dw}', and f_{Rw}', respectively, such that

$$f_{Nw}' = f_{max} - f_{Nw}, \qquad (f_{max} = 1/2\,\tau) \tag{4.47}$$
$$f_{Dw}' = f_{max} - f_{Dw}, \tag{4.48}$$

and

$$f_{Rw}' = f_{max} - f_{Rw}. \tag{4.49}$$

The "bio-informing activity amount" of the second-order bio-subsystem of the M-th order bio-system exhibiting a second-order AR-process is given by the case of $M=2$ in Eq. 2.14-1. Let it be I_{2w}, then

$$I_{2w} = -0.5\,\text{lb}\,K_{2w} = 0.5\,\text{lb}\,G_{2w} \quad \text{bit}, \qquad (\text{lb} = \log_2) \tag{4.50}$$

where G_{2w} is given by Eq. 4.9-1 as the "activity amount" of the bio-subsystem.

V.   COMPONENT ACTIVITIES IN EEGs

1. Classification of EEG Component Waves

An EEG is an example of higher order AR-bio-phenomena exhibited by a brain site neighboring the lead electrode(s). An example of a normal adult EEG is demonstrated in Fig. 12. Alpha waves of about 10 Hz frequency are contaminated by irregular, slow fluctuations. A sample of 1,024 (N) digital discrete time series y_t in Eq. 2.1 was obtained by AD transformation at $20(\tau)$ msec sampling intervals from the MT recording of the EEG. The order M of the AR-process of this EEG that gave the minimum "final prediction error" FPE was 11. The autocovariance of the above EEG and the summation of the autocovariances of the first- and second-order processes of the EEG components given as the left and right sides of Eq. 4.10-1, respectively, are displayed in Fig. 13. The difference between both the curves was so small that it was nonsignificant.

The AR-power spectral densities on logarithmic and linear scales are displayed, respectively, as B and C in Fig. 12 under the EEC tracing (A), and D depicts the AR-power spectral density of the EEG components on a linear scale. In the power spectral densities displayed as B and C, two dominant peaks are enhanced at 0 and about 9 Hz, respectively. The former is the peak of the first-order process indicating the aperiodic transient delta wave termed δ_0 wave and the latter is that of the second-order process depicting the alpha wave oscillation. The former is not necessarily easy to recognize in the EEG tracing (A), but the latter is seen as a domi-

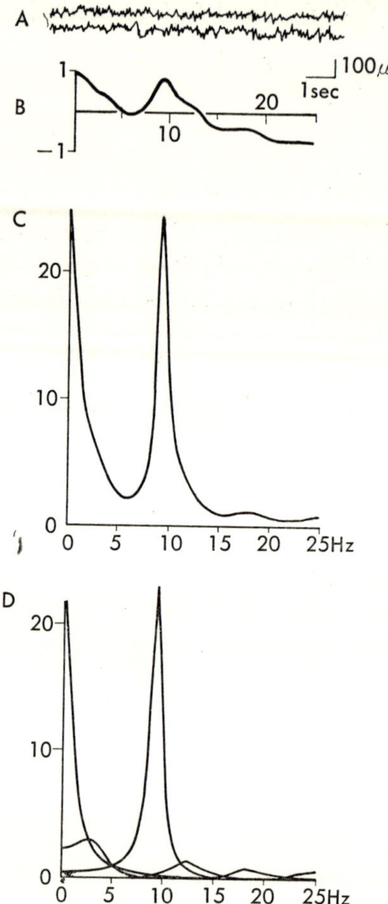

Fig. 12.   An example of a normal adult male EEG, its AR-power spectral density and components (50). A : EEG led from the Fz-Pz regions bipolarly during resting relaxed supine position with eyes closed. B and C: AR-power spectral densities on logarithmic and linear scales, respectively. D : AR-power spectral density of the EEG components on linear scale.

nant oscillation. Each of the two peaks in C is accompanied by a low hump on the right tail indicating the second-order processes of theta and fast alpha wave components in D, respectively. In addition, the second-order processes of the low beta wave peaks are also observed at 18 and 25 Hz in C and D, respectively. That is, this eleventh-order EEG process was decomposed into one first- and five second-order processes, as depicted

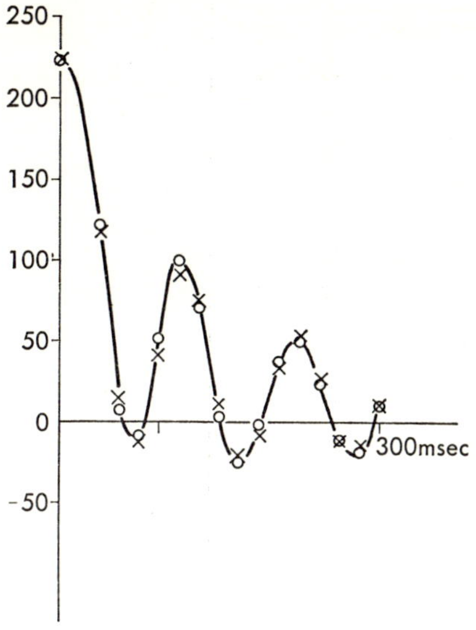

Fig. 13.   Autocovariance curve of EEG and sum of autocovariance of the EEG components (50). Empty circles (○), autocovariance R_k of the EEG in Fig. 12A given as the left side term of Eq. 4. 10-1; crosses (×), sum of autocovariances of EEG components given by the right terms of Eq. 4. 10-1. Explanation in the text.

in the power spectral density D at the bottom of Fig. 12, as the case of $V=1$ in Eq. 4. 18-1 and $W=5$ in Eq. 4. 39-1. The first-order component is certainly an aperiodic transient delta wave of 0 Hz, while the five second-order components are damped oscillatory waves showing a low theta wave, the most powerful alpha and the very low fast alpha waves, and two very low beta waves, respectively.

Component waves in normal adult EEGs are classified by each frequency of oscillation as delta (less than 3.4 Hz), theta (3.5–7.4 Hz), alpha (7.5–13.4 Hz), and beta waves (13.5–30.4 Hz), respectively (51). As has already been demonstrated, two or three alpha wave components with slightly different frequencies from each other (19, 40, 50, 52–54) can be classified, for convenience's sake, as " slow alpha " waves (α_1) of 7.5–9.4 Hz, " typical or second alpha " waves (α_2) of 9.5–11.4 Hz and " fast alpha " waves (α_3) of 11.5–13.4 Hz " damping frequencies ", respectively. Usually, about three second-order components of beta waves with different frequencies also appear in relatively many EEGs, each of which can be termed, for convenience's sake, waves of β_1 (" damping frequency " f_D:

TABLE I
Characteristic Values of EEG Component Waves in Fig. 12 ($M=11$) (50)

	f_N	f_D	f_R	ζ	T_1 / T_2 (sec)	C	P_1, P_2	Q	λ (deg)	I (bit)	K	G
δ_0	—	—	—	—	0.150	—	71.54 32.04%	—	—	1.049	0.234	4.273
θ	3.90	3.34	2.65	0.520	0.078	0.26	28.02 12.55%	−0.105	−0.215	1.422	0.139	7.194
α α_1	9.21	9.19	9.16	0.074	0.233	2.14	95.60 42.81%	0.042	0.025	1.019	0.244	4.098
α_2	12.48	12.33	12.17	0.156	0.082	1.01	15.02 6.73%	−0.064	−0.244	0.341	0.623	1.605
β β_1	17.64	17.88	18.13	0.252	0.086	1.54	8.83 3.95%	−0.005	−0.032	0.695	0.381	2.625
β_2	21.51	22.95	—	0.809	0.056	1.28	4.29 1.92%	0.002	−0.027	1.464	0.131	7.634
Total							223.30			5.99		

f_N, f_D, and f_R, natural, damping, and resonance frequencies, respectively; ζ, damping coefficient; T_1 and T_2, time constant of the first-order component and damping time of the second-order component, respectively; C, continuity ($f_D \times T_2$); P_1 and P_2, power (μV^2) and percent power; Q, skewness; λ, phase angle; I, bio-informing activity amount; K, value of characteristic function; G, activity amount.

13.5–17.4 Hz), β_2 (17.5–21.4 Hz), and β_3 (21.5–25 Hz), respectively (40).

Various characteristics of each component EEG wave in Fig. 12 are illustrated in Table I. The highest peak in Fig. 12D is the "slow alpha" (α_1) waves at 9.2 Hz showing the highest power (95.60 μV^2, 42.81%), which has the least "damping coefficient" (0.074), the longest "damping time" (0.233 sec), the best "continuity" (2.14), and the least "phase angle" (0.025 degrees) of the autocorrelation curve among those of the component waves. However, the "activity amount" (4.098) and "bio-informing activity amount" (1.019 bit) of this wave is not the highest, but those of the "second beta" (β_2) waves (7.634 and 1.464 bit, respectively) are the highest and those of the "theta" (θ) waves (7.194 and 1.422 bit, respectively) are also higher than those of the slow alpha waves. Such evidence that the most powerful component wave does not necessarily have the highest "activity amount" and "bio-informing activity amount" would suggest some unknown biological significance.

Another example of the EEG led monopolarly from the left occiput O_1 of an eight-year-old child is dispalyed in Fig. 14A. Its basic rhythm is the "second alpha" wave (α_2) of about 10 Hz damping freqeuncy. The AR-power spectral density on a logarithmic scale is depicted at the top of the lower (Fig. 14P). This was obtained from a sample of 2,000(N) time series values of Eq. 2. 1 taken at 20-msec (τ) intervals from the MT recording of the above EEG. The order M of the AR-process in Eq. 2.2 that indicated the "minimal final prediction error" FPE was 13 and the "activity" $g_{13}(B)=1/a_{13}(B)$ given by Eq. 2. 6 is shown underneath the spectral density (Fig. 14G). This thirteenth-order activity was decomposed into six ($W=6$) second-order and one ($V=1$) first-order component activities, each of which is illustrated in the lower of Fig. 14 and in Table II.

In the power spectral density of this thirteenth-order activity, two dominant peaks appear at 0 and 10 Hz "resonance frequencies" f_R, respectively, each of which indicates delta and alpha waves, respectively. The former is separated into a peak at 0 Hz in the power spectral density and autocovariance of the first-order damped exponential component activity, as depicted at the bottom (δ_0) of Fig. 14, and the latter was separated into those of the second-order damped oscillatory component activity of alpha waves, as depicted in the third line (α) in the lower. The alpha wave peak is the sharpest among the the six second-order activities (δ_1, θ, α, β_1, β_2, and β_3) and its "damping frequency" (f_D) is 10.03 Hz in the autocorrelation curve. The duration of oscillation in this curve is also the longest,

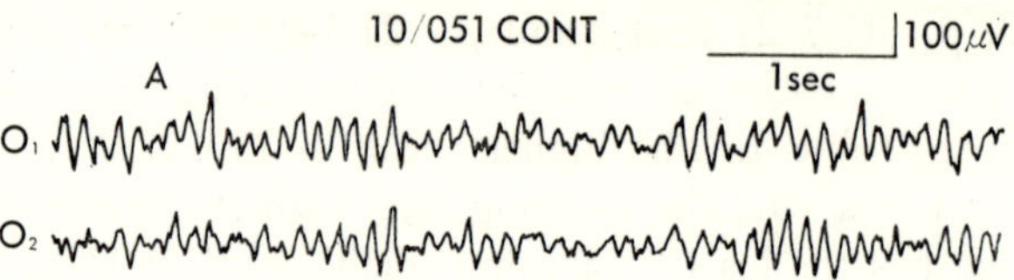

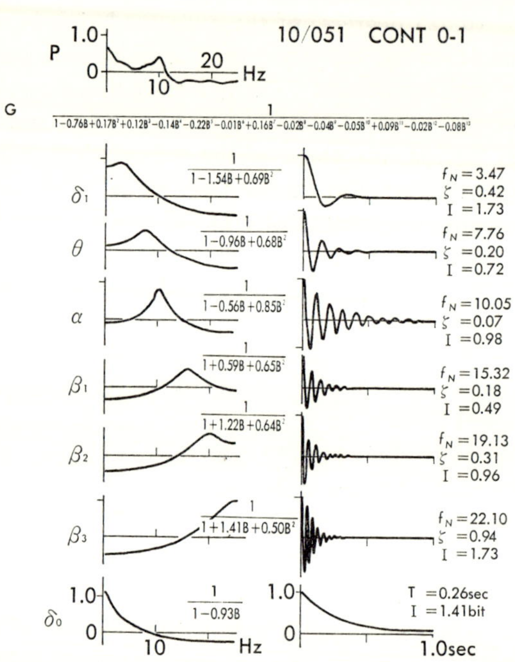

Fig. 14. An example of a child EEG, its AR-power spectral density, formula of thirteenth-order activity and the first- and second-order EEG components (50). A, EEG led monopolarly from the left occiput (O1-A1) (A1, left ear lobe) of an 8-year-old child. From the top downward, P, AR-power spectral density of the EEG on logarithmic scale; G, thirteenth-order activity $(g_{13}(B)=1/a_{13}(B))$, δ_1, θ, α, β_1, β_2, and β_3; (oscillatory second-order components); δ_0, nonoscillatory first-order component activity. Power spectral densities on logarithmic scale (left column), formulae of activities (middle column) and autocorrelation curves of components (right column); F_N, natural frequency; ζ, damping coefficient; I, bioinforming activity amount.

lasting for about 1 sec, which is supported by the least " damping coefficient " (0.07), the longest " damping time " (0.24 sec) and the best " continuity " (2.41) among the six second-order activities. Two humps are

TABLE II
Characteristic Values of EEG Component Waves in Fig. 14 ($M=13$) (50)

		f_N	f_D	f_R	ζ	$\dfrac{T_1}{T_2}$ (sec)	C	P_1, P_2	Q	λ (deg)	I (bit)	K	G
δ	δ_0	—	—	—	—	0.260	—	168.35 39.04%	—	—	1.40	0.143	7.013
	δ_1	3.47	3.15	2.79	0.419	0.109	0.34	73.27 16.99%	−33.43	−24.52	1.73	0.091	10.941
θ		7.76	7.60	7.44	0.201	0.102	0.77	54.59 12.66%	22.17	22.10	0.72	0.369	2.711
	α_2	10.05	10.03	10.01	0.066	0.240	2.41	104.85 24.31%	−36.90	−19.39	0.98	0.257	3.883
β	β_1	15.32	15.47	15.63	0.178	0.092	1.42	11.22 2.60%	0.85	4.33	0.49	0.506	1.976
	β_2	19.13	19.41	19.70	0.306	0.088	1.71	11.34 2.63%	−0.67	−0.339	0.96	0.265	3.766
	β_3	22.10	23.99	—	0.938	0.058	1.39	7.64 1.77%	5.80	37.20	1.73	0.091	10.941

For abbreviations, see Table I.

observed, respectively, on the right and left slopes of the peak at 0 and 10 Hz " resonance frequency " in the whole power spectral density in Fig. 14P, each of which corresponds, respectively, to the peaks of delta (δ_1) and theta (θ) component waves in the second-order activities. The " damping frequencies " of their autocovariance curves are 3.15 and 7.60 Hz, respectively, as shown in Fig. 14 and Table II. In the frequency range of beta waves higher than 14 Hz, three low peaks are suggested in the whole power spectral density. They correspond to the component peaks of the second-order activities called β_1, β_2, and β_3 with " damping frequencies " of 15.47, 19.41, and 23.99 Hz, respectively, in the autocovariance curves. Their " damping coefficients " are 0.18, 0.31, and 0.94 (>0.707), respectively, so that the duration of the damped oscillation of β_1 wave is the longest, whereas that of β_3 wave is the shortest. As each of the second-order activities of the component wave is shown between the respective power spectral densities and autocovariance curves, the first coefficient φ_{1w} of β_1, β_2, and β_3 component waves has negative values, -0.59, -1.22, and -1.41, respectively. Therefore, all of them are " high frequency active (high pass) types " of the second-order processes. As the " damping coefficient " of the β_3 wave excesses the " critical damping coefficient " (0.707), the peak in its power spectral density is located in the vicinity of the highest frequency $f_{max}=25(=1/2\times0.02)$ Hz, since this process is a " high frequency active type."

2. Frequency of EEG Component Waves

It is well known that EEG component waves are classified by their frequencies, *i.e.*, the frequency ranges of delta, theta, alpha, and beta waves, respectively, are less than 3.5, 4–7, 8–13, and 14–30 Hz (*51*). Bipolar EEGs led from the Fz-Pz regions of ninety normal adult males in a relaxed state in a supine position on a comfortable armchair with eyes closed and during hyperventilation for 3 min were MT recorded by FM modulation. A sample of the 1,025 (N) discrete digital time series of 20.48 sec (1,024 × 20 msec) duration at every 20-msec (τ) interval was obtained from each EEG by AD modulation for AR and component analyses (*40, 50, 55, 56*). During hyperventilation, each time series was taken from about 160 to 180 sec after initiation of hyperventilation. As some of the EEGs were contaminated with artifacts, data were sampled from a neighboring trace without contamination at about 2 min after hyperventilation. Several modes of distribution were observed in the frequency distribution of " natural," " damping " and " resonance " frequencies of the EEG second-

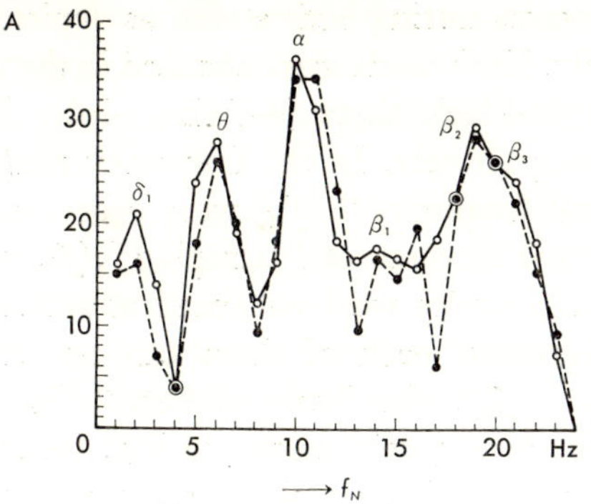

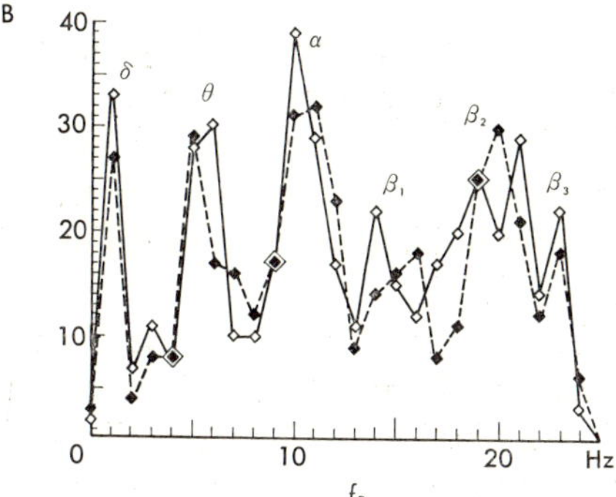

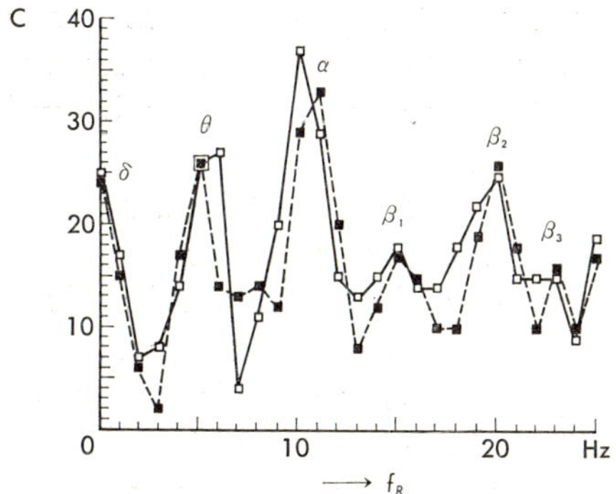

Fig. 15. Frequency polygons of natural (A : f_N), damping (B : f_D), and resonance frequencies (C : f_R) of second-order component waves in 90 EEGs (F_Z-P_Z bipolar lead) in the resting relaxed state and during hyperventilation. Empty points and full line, resting relaxed state ; filled points and broken line, during hyperventilation (*50*).

order oscillatory component waves during both states, as depicted in Fig. 15A, B, and C (*50*), respectively. Each mode was enhanced in the frequency ranges of well known delta (0–4 Hz), theta (4–7 Hz), alpha (8–13 Hz), and beta waves (13–25 Hz), respectively. In the frequency range of beta waves, two modes of " natural frequency " (f_N) were observed at 14–16 and 19 Hz, respectively, whereas in cases of " damping " (f_D) and " resonance " frequencies (f_R) three modes were enhanced at around 15, 20, and 23 Hz, respectively. Therefore, each of them can be termed, for convenience, the " first beta " (β_1), the " second beta " (β_2), and the " third beta " (β_3) waves, respectively, as observed in Fig. 14. In the distribution of the " natural frequency," the alpha wave mode was at 10–11 Hz and the two components termed α_1 (" slow alpha ") and α_2 (" typical alpha ") in Table I seem to be suggested. Though the β_1 wave mode was not so marked in the relaxed state, it appeared distinctly during hyperventilation. However, no significant changes caused by hyperventilation were observed in the modes of theta and α_2 waves, but the delta wave mode at 2 Hz was reduced slightly in its height and that of the alpha wave at 10 Hz became broader to 10–11 Hz.

The delta wave mode in the distribution of the " damping frequency " was enhanced at 1 Hz, which was lower in both states than that of the " natural frequency " at 2 Hz. During hyperventilation, the theta wave frequency mode became lower by 1 Hz, whereas that of the alpha, β_1, and β_2 waves became higher by 1, 2, and 1 Hz, respectively. This evidence suggests that the " decrementing process " which causes decay in the amplitude of oscillation would be augmented by hyperventilation in the second-order process of the theta wave, whereas it would be depressed in that of α_1 and α_2 waves.

The delta wave frequency mode in the distribution of the " resonance frequency " was at 0 Hz, which suggests that the damping coefficients of many delta components would exceed the " critical damping coefficient " (0.7071). Though not so marked, a mode was also enhanced at 25 Hz, which seems to verify components of " high frequency active " type. The modes of theta and alpha waves at 5 and 10 Hz in the relaxed state, respectively, shifted leftward to 4 Hz and rightward to 11 Hz in hyperventilation. These changes are the same as observed in the distribution of " damping frequency ." The cause of each change would also be the same.

As shown by Eq. 4. 25, the autocovariance curve of the second-order component process is composed of an " oscillatory " and a " decrement-

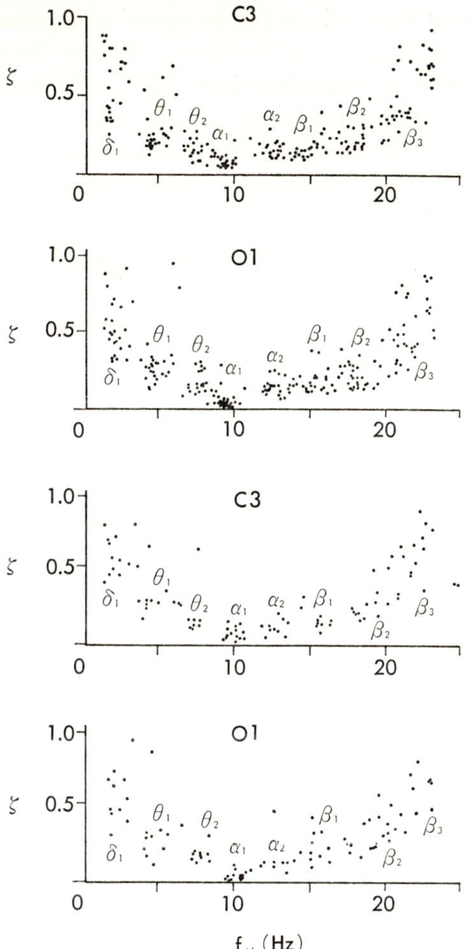

Fig. 16. Damping coefficients with respect to "natural frequency" of the second-order components of EEGs of adults and children (57). Upper two (infant) and lower two (adult), respectively, are the damping coefficients of autocorrelation (autocovariance) curves of EEG components of 30 children (5–8 years old) and 14 adults (20–23 years old). Each EEG was led monopolarly from the bilateral central (C_3-A_1 and C_4-A_2) and occipital (O_1-A_1 and O_2-A_2) regions of resting relaxed subjects with their eyes closed in a supine position. (A_1 and A_2; left and right ear lobes, respectively.) Component waves were obtained by AR- and its component analyses of the discrete digital time series of each EEG sampled at 20 msec intervals for the length of 20 sec (2,000 data=N). Abscissa, natural frequency f_N; ordinate, damping coefficient ζ.

ing " processes to display a damped oscillation as already noted. The former process is characterized by the natural frequency and the latter one by the damping coefficient or the damping time. The lesser the damping coefficient and the longer the damping time are, the longer the oscillation of the the second-order process lasts and *vice versa*. Figure 16 depicts the damping coefficients of the second-order component waves of EEGs led monopolarly from the left central (C_3) and occipital (O_1) regions of children and adults with respect to the natural frequency. The least coefficients appeared in the frequency ranges of 8.5–10 Hz in children and of 9–10.5 Hz in adults, respectively. These frequency ranges have counterparts in the ranges of the damped frequency (9.5–11.4 Hz) of the α_2 wave. The higher or lower the natural frequency of the component waves shifted from these frequency ranges of the α_2 wave, the more the coefficients tended to increase. Therefore, it was verified that the " typical alpha " (α_2) wave has the longest duration in the second-order component waves.

Figure 17 displays the damping frequency, damping coefficient and percent power of EEG component waves led monopolarly from the left occiput (O_1) of 41 adults and 78 children. The alpha component waves at 8–12 Hz in adults and 8–11 Hz in children had the least damping coefficient and the highest power. In addition, the same relationship as observed in Fig. 16 also appeared between the damping frequency and the damping coefficient. It would be recognized, therefore, that the most powerful and durable component wave is the α_2 wave, so that this wave can be termed the " typical alpha " one, as already noted above. In the 4–7 Hz theta waves frequency in adults, the component waves were separated into two groups, each of which was at about 6 and 4 Hz, respectively. The damping coefficient of the former was less than that of the latter. In children, the component waves of alpha waves in the frequency range of 8–11 Hz flocked together around 8 and 9.5 Hz, separately forming two groups α_1 and α_2, respectively. The damping coefficient of the former group was larger than that of the latter. Delta waves of less than 3 Hz in children flocked together around three different damping coefficients of 0.4, 0.5, and 0.6, respectively, to form three component groups, each of which has longer, moderate and shorter durations, respectively.

3. *Pattern Discrimination of EEG Component Waves*

In Table III, the second-order component waves of EEGs led bipolarly from the Fz-Pz regions of 359 normal adult males were classified by their " damping frequencies " f_D into eight component waves, as al-

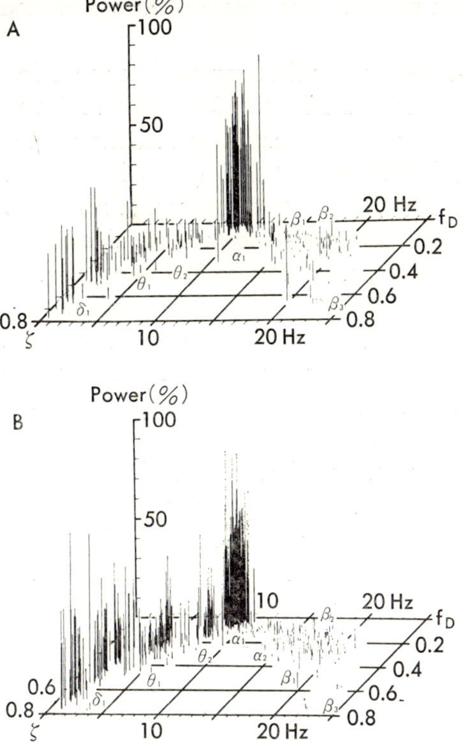

Fig. 17. Damping frequency, damping coefficient, and percent power of the second-order oscillatory component EEG waves of adults and children (57). Abscissa, damping frequency f_D; front to back axis, damping coefficient ζ; ordinate, percent power. A: EEG of 41 resting, relaxed adults with closed eyes led from the left occiput monopolarly (O_1-A_1). B: of 78 children. Component waves were obtained from the discrete time series of 2,000 data (N) of each EEG sampled at intervals of 20 msec for the length of 40 sec.

ready noted in the former section, such as " delta " (δ) waves in the frequency range of 0.5–3.4 Hz, " theta " (θ) waves of 3.5–7.4 Hz, " slow alpha " (α_1) waves of 7.5–9.4 Hz, " typical alpha " (α_2) waves of 9.5–11.4 Hz, " fast alpha " (α_3) waves of 11.5–13.4 Hz, the " first beta " (β_1) waves of 13.5–17.4 Hz, the " second beta " (β_2) waves of 17.5–21.4 Hz, and the " third beta " (β_3) waves of 21.5–30 Hz, respectively. However, the " damping coefficients " of theta waves and the " time constant " of delta waves flocked into two groups as depicted in Figs. 16 and 17, each of which may be, therefore, divided into two component waves, respectively, so that it would be worthwhile to take this evidence into consideration in the next step.

All of the second-order component waves illustrated in Table III amount to 1,641 in number. If all EEGs of 359 adult subjects were composed of the eight second-order component waves, the total sum of the component waves should amount to 2,872 (359×8) in number. It is obvious, therefore, that many EEGs are composed of fewer second-order component waves than eight. The well-known basic alpha rhythm, which is a counterpart of the α_2 wave, amounted to 325 in number and 90.53% in its appearance percentage ($8N_j/2,875$), respectively, which are the highest in the eight component waves. Their powers are usually higher than those of other second-order component waves, though not always necessarily (see Fig. 12D, Tables I and II). The third beta (β_3) wave amounted to nearly the highest in number (323) and appearance percentage (89.97%), respectively. Its power, however, was far less than that of the typical alpha wave, as seen in Figs. 12D and 17. It is suggested, therefore, that a powerful typical alpha and a weak third beta wave will be contained in most EEGs led from the Fz-Pz regions. Next to them, the theta wave amounted to 282 in number and 78.55% in appearance percentage, suggesting that not a few of the EEGs contain this wave. The first beta and delta and the second beta waves were contained in about half of all EEGs.

As explained previously in paragraph IV. 3, each of the above eight EEG component waves is capable of being described by a second-order

TABLE III

Second-order Component Waves of Normal Adult Male EEGs (Fz-Pz lead) from 359 Cases *(40)*

Component No. j	Name	Damping frequency f_D (Hz)	Number of cases		
			N_j	%	APP %
1	δ	Less than 3.4	176	10.72	49.02
2	θ	3.5–7.4	282	17.18	78.55
3	α_1 (Slow α)	7.5–9.4	98	5.97	27.30
4	α_2 (Typical α)	9.5–11.4	325	19.80	90.53
5	α_3 (Fast α)	11.5–13.4	94	5.73	26.18
6	β_1	13.5–17.4	182	11.09	50.70
7	β_2	17.5–21.4	161	9.81	44.85
8	β_3	21.5–25	323	19.68	89.97

Sum 1,641

$\% = N_j/1,641$; APP $\% = 8N_j/2,875$ (appearance percentage); $2,875 = 359 \times 8$.

AR-process such that

$$y_{jkt} = a_{1jk}y_{jk,t-1} + a_{2jk}y_{jk,t-2} + n_{jkt},$$
$$(j = 1, 2, \cdots, 8; \ k = 1, 2, \cdots, N_j), \qquad (4.23\text{-}1)$$

which is already given by Eq. 4.23. It is obvious, therefore, that the essential characteristics of the j-th component EEG wave are determined by the first and second AR-coefficients, a_{1jk} and a_{2jk}. These coefficients obtained from the 1,641 EEG component waves of 359 normal adults

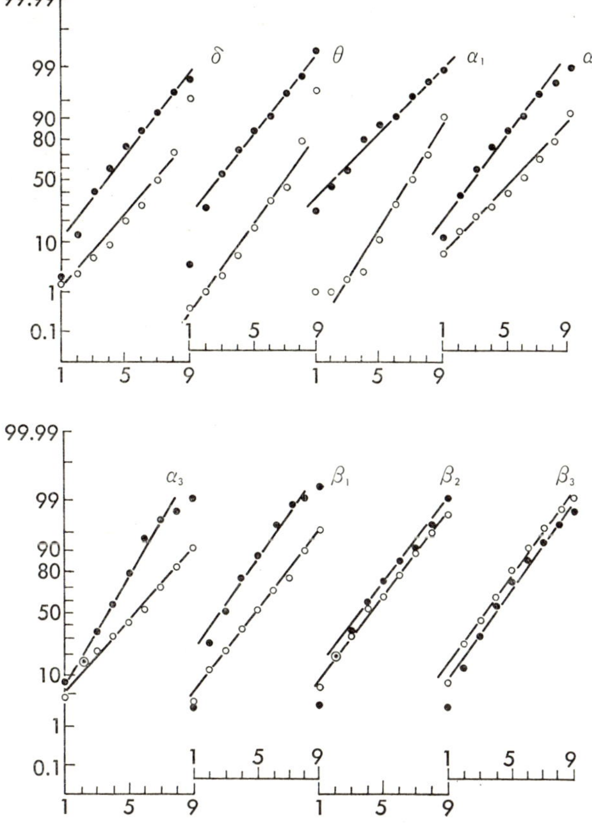

Fig. 18. Percentage cumulative frequency distributions of the first and second AR-coefficients of the component waves (δ, θ, α_1, α_2, α_3, β_1, β_2, β_3) of 359 EEGs in Table III plotted on normal probability paper. Empty white ($\bigcirc$) and filled black ($\bullet$) circles; the first and second AR-coefficients a_1 and a_2, resepectively. Unit of abscissa is taken as one tenth of the difference between the maximal and minimal values in each AR-coefficient.

shown in Table III are distributed displaying oblique straight lines, at least for the first approximation, on normal probability paper, as depicted in Fig. 18. Consequently, the vector of both the coefficients of the j-th component wave, say $\boldsymbol{A}_{jk}$ such that

$$\boldsymbol{A}_{jk} = (a_{1jk}, a_{2jk})', \qquad (j = 1, 2, \cdots 8; \; k = 1, 2, \cdots, N_j) \qquad (5.1)$$

can be considered, at least for the first step, as a sample vector obtained from a bivariate normal distribution population given by the case of $K=2$ in Eq. 3.5.

There is a relationship between both the coefficients a_{1jk} and a_{2jk} such that (23)

$$a_{2jk} = -(\operatorname{cosec}^2 2\pi f_{Dj}/4)a_{1jk}{}^2, \qquad (a_{2jk} < 0). \qquad (5.2)$$

Hence, points (a_{1jk}, a_{2jk}) of both the coefficients of a component wave with a fixed " damping frequency " scatter displaying the locus of a parabola on the a_1-a_2 plane, as demonstrated in Fig. 19. The lower the " damp-

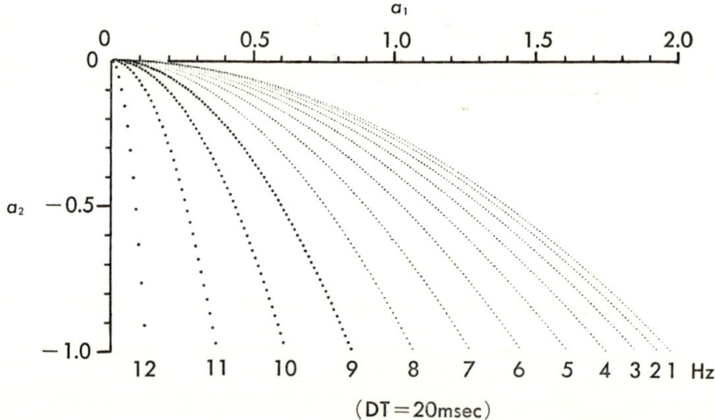

Fig. 19.   Parabolas showing the relationship between the first (a_1) and second (a_2) AR-coefficients of a second-order bio-phenomenon. Abscissa, the first AR-coefficient a_1; ordinate, the second AR-coefficient a_2 (<0). The points (a_1, a_2) of a second-order damped oscillatory bio-phenomenon draw a locus of parabola on the a_1-a_2 plane, provided the frequency of the phenomenon is fixed. Though the parabolas were drawn only in the case of positive a_1 on the right side of the ordinate, those in the case of negative a_1 display parabolas on the left side of the ordinate as mirror images of those on the right side with respect to the ordinate. Interval of discrete digital time series of each second-order bio-phenomenon is taken as 20 msec. DT, sampling interval of the EEG time series.

ing frequency" of the component wave is, the more rightward the parabola locates and *vice versa*. The points of component waves with damping frequency higher than 13 Hz will be scattered displaying parabolas on the left side of the ordinate (a_2-axis) as mirror images of those on the right side with respect to the ordinate.

Figure 20 demonstrates the scatter diagram of these points of the above 1,641 component waves. Points of delta waves are distributed displaying right-end parabolas of about 1 and 2 Hz frequencies, those of theta waves are crowded on parabolas of about 5 and 6 Hz, those of alpha waves swarm on parabolas of about 10 Hz scattering leftward on parabolas of about 12 and 13 Hz, and those of beta waves are flocked together around parabolas of from about 15 to 24 Hz, respectively.

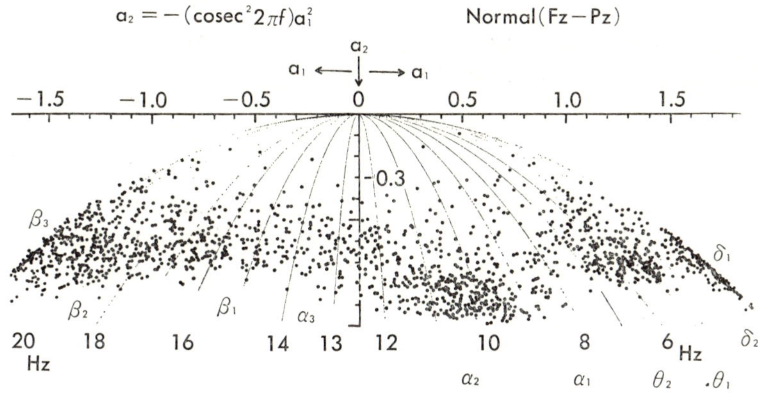

Fig. 20. Bivariate scatter diagram of the first and second AR-coefficients of the second-order component waves of EEGs led from the Fz-Pz regions of 359 normal adult males (*40*). Abscissa (a_1) and ordinate (a_2): the first and second AR-coefficients, respectively. Numerals beneath the parabolas displaying Eq. 5. 2 indicate the "damping frequency" in Hz of the component waves.

Let here the average vector and unbiased variance-covariance coefficient matrix of the j-th component waves be $\bar{A}_j$ and S_j, respectively, then

$$\bar{A}_j = (\bar{a}_{1j}, \bar{a}_{2j}), \quad j = 1, 2, \cdots, 8, \tag{5.3}$$

where

$$\left. \begin{aligned} \bar{a}_{1j} &= (1/N_j) \sum_{k=1}^{N_j} a_{1jk} \\ \bar{a}_{2j} &= (1/N_j) \sum_{k=1}^{N_j} a_{2kj} \end{aligned} \right\} \tag{5.4}$$

and

$$S_j = \begin{pmatrix} S_{11j} & S_{12j} \\ S_{21j} & S_{22j} \end{pmatrix} \tag{5.5}$$

respectively, where

$$\left. \begin{aligned} S_{iij} &= (1/(N_j-1)) \sum_{k=1}^{N_j} (a_{ijk} - \bar{a}_{ij})^2, \quad i = 1, 2, \\ S_{12j} &= S_{21j} = (1/(N_j-1)) \sum_{k=1}^{N_j} (a_{1jk} - \bar{a}_{1j})(a_{2jk} - \bar{a}_{2j}). \end{aligned} \right\} \tag{5.6}$$

Let further the first and second AR-coefficients of an arbitrary second-order component wave be a_1 and a_2, respectively, and its coefficient vector be

$$A = (a_1, a_2), \tag{5.7}$$

then the "Mahalanobis' generalized distance" of this vector from the j-th average vector, say $D^2(A, \bar{A}_j)$, is given as

$$D^2(A, \bar{A}_j) = (A - \bar{A}_j)' \cdot S_j^{-1} \cdot (A - \bar{A}_j). \tag{3.7-1}$$

Provided here such a "null hypothesis" is accepted that

H_0: *An arbitrary second-order component wave characterized by the above coefficient vector A does not differ from the group of the j-th component waves, but belongs to this group,* (5.8)

then the following F_A:

$$F_A = \{N_j - 2)N_j/2(N_j+1)\} D^2(A, \bar{A}_j) \tag{5.9}$$

will follow the well known F-distribution with degrees of freedom

$$n_1 = 2 \quad \text{and} \quad n_2 = N_j - 2. \tag{5.9-1}$$

If two "levels of significance" or "risk" are taken here to have sufficiently small α_1 and α_2 $(<\alpha_1)$, such as 0.05 and 0.01, respectively, then the above "null hypothesis" H_0 cannot be rejected, but can be accepted, when

$$F_A \leqq F(n_1, n_2, \alpha_1), \tag{5.10}$$

so that the component wave to be tested can be considered as one of the j-th component waves, whereas the "null hypothesis" H_0 can be rejected and such an "alternative hypothesis" H_1 that

H_1: *The second-order component wave to be tested differs from the*

> *j-th component wave and it does not belong to the group of the
> j-th component waves,* $\qquad$ (5. 8-1)

can be accepted, when

$$F_A > F(n_1, n_2, \alpha_2). \qquad (5.\,10\text{-}1)$$

However, not only the " null hypothesis " H_0 cannot be rejected, but the " alternative hypothesis " H_1 also cannot be accepted, when

$$F(n_1, n_2, \alpha_1) < F_A \leqq F(n_1, n_2, \alpha_2). \qquad (5.\,10\text{-}2)$$

Consequently,

$$(A - \bar{A}_j)' \cdot S_j^{-1} \cdot (A - \bar{A}_j) = c, \qquad (5.\,11)$$

where

$$c = 2F(2, N_j - 2, \alpha)(N_j + 1)/N_j(N_j - 2): \quad \text{const.} \qquad (5.\,11\text{-}1)$$

gives the *critical ellipse* or *confidence ellipse* at the significant level of α (*58*) of the coefficients a_1 and a_2 of the second-order j-th component wave.

Consider here an arbitrary sample group of the j-th component waves amounting to N_j in number and let the sample averages of the first and second coefficients be $\bar{a}_1$ and $\bar{a}_2$, respectively. Let further the sample average coefficient vector and variance-covariance matrix of the average coefficients be $\bar{A}$ and S_{j*}, respectively, then

$$\bar{A} = (\bar{a}_1, \bar{a}_2) \qquad (5.\,12)$$

and S_{j*} can be considered as

$$S_{j*} = S_j/N_j. \qquad (5.\,13)$$

Hence, the " critical ellipse " at the level of α of the average coefficients $\bar{a}_1$ and $\bar{a}_2$ can be given (*59*) as

$$(\bar{A} - \bar{A}_j)' \cdot S_{j*}^{-1} \cdot (\bar{A} - \bar{A}_j) = c_*, \qquad (5.\,14)$$

where

$$c_* = 4F(2, 2N_j - 3, \alpha)N_j(N_j - 1)/N_j^2(2N_j - 3): \quad \text{const.} \qquad (5.\,14\text{-}1)$$

In Fig. 21 are demonstrated the " critical ellipses " at the level of 0.01 of the EEG component waves classified into eight kinds in Table III, wherein those of the first and second AR-coefficients and their averages are displayed by dotted and black ellipses, respectively (*60*). Though the dotted

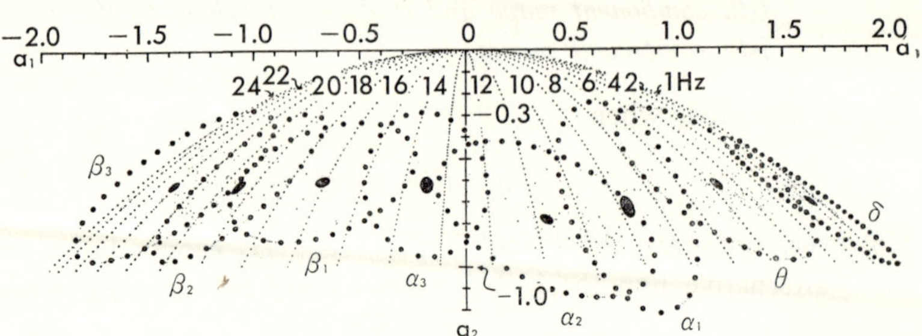

Fig. 21. Critical ellipses of the first and second AR-coefficients (dotted curves) and those of the average coefficients (black ellipses) of the second-order component waves of EEGs led from the Fz-Pz regions of 359 normal adult males (*60*). For abbreviations, see Fig. 20.

ellipses are intersected by other with their adjacent ones on both sides, black ellipses are markedly separated from each other. Therefore, the classification of EEG component waves in eight kinds with respect to their damping frequency ranges can be recognized at least for the first step.

An example of pattern discrimination of the second-order EEG component waves is shown in Figs. 22 and 23. Markedly low alpha wave peaks are seen around 10 Hz in the group of power spectral densities of " congenital color blindness " (" rod-monochromatism ") in Fig. 22, whereas those of a normal control subject are high and sharp (*61*). As demonstrated in the scatter diagram of the first and second average co-efficients (Fig. 23 bottom), fast alpha (α_3) and the third beta (β_3) component waves of the patient did not differ from those of the normal control, but slow alpha (α_1), typical alpha (α_2), and the first and second beta (β_1 and β_2) component waves of the patient were different at the level of 0.01 and the delta component wave at the level of 0.05. In the scatter diagram of the first and second coefficients of component waves (Fig. 23 top), this evidence is suggested in their different locations of both the point (a_1, a_2) groups, but not so distinct as verified in the diagram of the average coefficients. It would be recognized, therefore, that the evidence observed in the whole AR-power spectral density of EEG and the scatter diagrams of the coefficients and their average coefficients of the component waves can be affirmed in more detail by an appropriate pattern discrimination of the component waves.

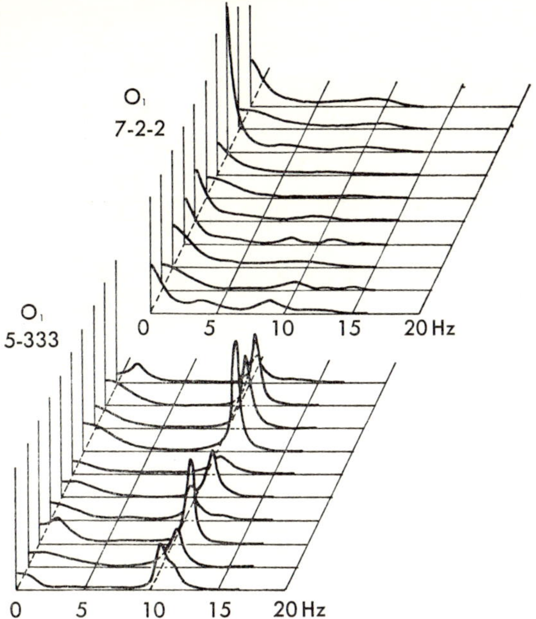

Fig. 22. AR-power spectral densities of left occipital EEGs of a congenital color blindness (rod-monochromatism) (top) and a normal adult control (bottom) (*61*). By visual inspection of ink-written EEGs of the patient and the control subject, 10 sections without contamination due to artifacts were selected from each of the EEGs, wherein intervals between two sections were 5–15 sec. Discrete digital time series of 1,000 data for AR- and component analyses (*55, 56*) were sampled in each section at every 20 msec interval for 10.24 sec length of MT recorded EEGs. Similar spectral patterns were also obtained from the right occipital EEGs of the patient and control subject. Further explanation in the text.

VI. MULTIVARIATE HIGHER ORDER ACTIVITIES OF BIO-SYSTEMS

1. Multivariate AR-process in Bio-phenomena

There are many human and animal bio-systems that exhibit two or more bio-phenomena, each state of which fluctuates more or less at any time around its average, suggesting its level of "homeostasis." The numbers of various simultaneously measured time series not only of these bio-phenomena but also the natural phenomena can be denoted by J and S, respectively. And let the amounts of the sways of the j-th and k-th $(j \neq k)$ bio-phenomena around their respective averages at an arbitrary time $t-p\tau$ be $y_{j,t-p}$ and $y_{k,t-p}$, respectively, where j and $k=1, 2, \cdots, J, (j \neq k), p=0,$

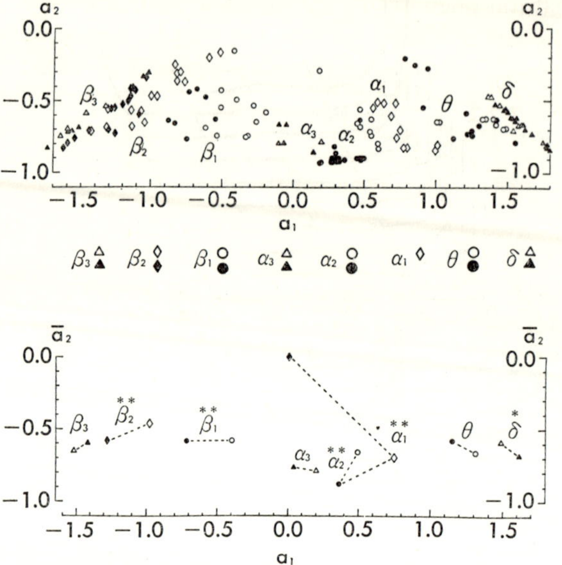

Fig. 23.   Scatter diagram of the first and second AR-coefficients (top) and
that of the average coefficients (bottom) of the second-order component waves
of EEGs of a congenital color blindness and a normal control subject (*61*).
Abscissa, the first AR-coefficient a_1 and its average coefficient $\bar{a}_1$; ordinate,
the seoncd AR-coefficient a_2 and its average coefficient $\bar{a}_2$. Empty ($\triangle$, $\diamond$, $\bigcirc$),
the patient; black ($\blacktriangle$, $\blacklozenge$, $\bullet$), the normal control subject. $*$ significant 0.05;
$**$ significant 0.01.

1, 2, $\cdots$ and τ is the sampling time interval. Let further the amount of the
sway of the s-th environmental natural phenomenon around its average
be $x_{s,t-p}$ at time $t-p\tau$, where $s=1, 2, \cdots, S$. Then the j-th bio-phenomenon can be described, for the first step, by a multivariate AR-process
given as an extension of the monovariate AR-process of Eq. 2. 2, *i.e.*, by
the weighted sums of previous values of bio- and natural phenomena and
a value indicating an ensemble state of all other unmeasured bio- and
natural phenomena, enormous in their numbers (*10*), such that

$$y_{jt} = \sum_{m=1}^{M}(a_{jjm}y_{j,t-m} + \sum_{k=1}^{J} a_{jkm}y_{k,t-m} + \sum_{s=1}^{S} a_{jsm}x_{s,t-m}) + n_{jt},$$
$$(j, k = 1, 2, \cdots, J; \; j \neq k) \tag{6.1}$$

where a_{jjm}, a_{jkm}, and a_{jsm} are weights termed AR-coefficients dependent
on the j-th, k-th bio-, s-th environmental natural phenomena, respectively.
And n_{jt} is the ensemble sums of the weighted values of all other unmeas-

ured bio- and natural phenomena, enormous in their numbers. Therefore, considering " the law of large numbers," n_{jt} can be assumed, for the first step, to be a purely random process with average zero, quite the same as given by Eq. 2. 3, such that

$$E[n_{jt}] = 0,$$
$$E[n_{j,t-p} \cdot n_{jt}] = 0, \quad \text{when} \quad p \neq 0,$$
$$= \sigma_{nj}^2 \neq 0, \quad \text{when} \quad p = 0. \tag{6.2}$$

The value of the order M in Eq. 6. 1 can be estimated as indicating the " $(J+S)$-variate minimal final prediction error " FPE of Akaike (20–22). Substituting the " backward shift operator " B given by Eq. 2. 4 into the above Eq. 6. 1, we obtain

$$y_{jt} = \sum_{k=1}^{J} y_{kt}(g_{jk}(B) \cdot g_{jj}(B)) + \sum_{s=1}^{S} x_{st}(h_{js}(B) \cdot g_{jj}(B)) + n_{jt} g_{jj}(B),$$
$$(j = 1, 2, \cdots, J), \tag{6.1-1}$$

where

$$g_{jj}(B) = 1/a_{jj}(B) = 1/(1 - \sum_{m=1}^{M} a_{jjm} B^m) \tag{6.3}$$

$$g_{jk}(B) = \sum_{m=1}^{M} a_{jkm} B^m \tag{6.3-1}$$

$$h_{js}(B) = \sum_{m=1}^{M} a_{jsm} B^m. \tag{6.3-2}$$

The block diagram in Fig. 24 illustrates Eq. 6. 1-1 graphically, wherein the rectangular block drawn by the broken line indicates the j-th subsystem of the bio-system having $(J+S)$-variate M-th order activity to exhibit the j-th phenomenon amounting to y_{jt} at an arbitrary time t. Consequently, this bio-subsystem can be termed " $(J+S)$-variate M-th order one." And $(g_{jk}(B)g_{jj}(B))$, $(h_{js}(B)g_{jj}(B))$, and $g_{jj}(B)$ in Eq. 6. 1-1, respectively, are the " M-th order (transforming) activities " (19) of the j-th bio-subsystem with respect to the k-th, s-th bio-phenomena and the random natural stimulation, $i.e.$, y_{kt}, y_{st}, and n_{jt}, respectively.

It can be recognized from Eq. 6. 1-1 and the block diagram in Fig. 24 that the amount of the j-th bio-phenomenon y_{jt} will be given as the total sum of the responses of the bio-subsystem caused by other bio-phenomena y_{kt} and natural phenomena x_{st}, each of which is $(J-1)$ and S in their number, respectively, and random stimulation n_{jt} (10, 19). Let each of them be $y_{j/k,t}$, $y_{j/s,t}$ and $y_{j/n,t}$, respectively, then

$$y_{j/k,t} = y_{kt}(g_{jk}(B) \cdot g_{jj}(B)), \quad (k = 1, 2, \cdots, J; \, j \neq k) \tag{6.4}$$
$$y_{j/s,t} = y_{st}(h_{js}(B) \cdot g_{jj}(B)), \quad (s = 1, 2, \cdots, S) \tag{6.4-1}$$

and

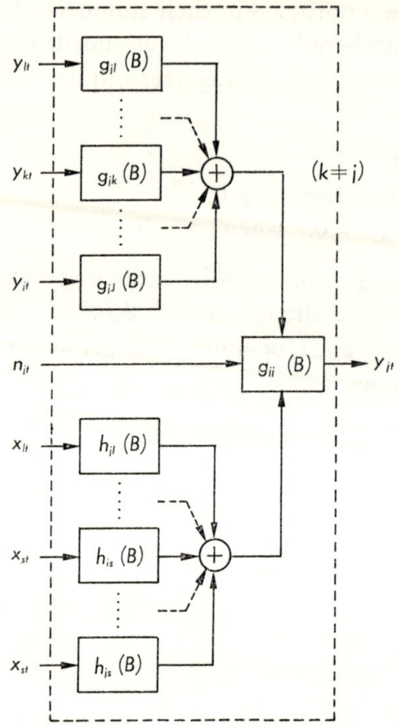

Fig. 24. Block diagram of bio-subsystem exhibiting a multivariate bio-phenomenon y_{jt} as its response caused by various other bio-phenomena y_{kt}, $k=1, 2, \cdots, J$, $k \neq j$, natural phenomena x_{st}, $s=1, 2, \cdots, S$ and a random phenomenon n_{jt} at an arbitary time t.

$$y_{j/n,t} = n_{jt} \cdot g_{jj}(B). \qquad (6.4\text{-}2)$$

Then the j-th bio-phenomenon y_{jt} will be given as

$$y_{jt} = \sum_{k=1}^{J} y_{j/k,t} + \sum_{s=1}^{S} y_{j/s,t} + y_{j/n,t}. \qquad (6.1\text{-}2)$$

If the second and consecutive terms on the right side of Eq. 6.1 were collected into a single term and let it be n_{jt}' such that

$$n_{jt}' = \sum_{m=1}^{M} \left(\sum_{k=1}^{J} a_{jkm} y_{k,t-m} + \sum_{s=1}^{S} a_{jsm} x_{s,t-n} \right) + n_{jt}, \qquad (6.5)$$

then this can be considered as being a purely random process, for the first step, since the random process n_{jt} in Eq. 6.5 is the weighted sum of bio- and natural phenomena too enormous to measure, so that the properties of the former two terms enclosed in the parentheses will be buried in the randomness of n_{jt}. Thence, Eq. 6.1 becomes a monovariate AR-process of order M already given by Eq. 2.2 in Sect. II such that

$$y_{jt} = \sum_{m=1}^{M} a_{jjm} y_{j,t-m} + n'_{jt}. \tag{6.6}$$

It can be recognized, therefore, that simultaneously measuring the bio- and environmental natural phenomena of $(J-1)$ and S in their numbers, respectively, is nothing but sorting out these phenomena from the random process n_{jt} in Eq. 2.2 (*17, 19*).

2. *Impulse and Frequency Responses of a Multivariate Higher Order Bio-system*

As already noted in a previous paragraph, each of the various bio-phenomena can be considered as a response of a multivariate bio-system to other bio- or environmental natural phenomena. As shown by Eq. 6.4 and the block diagram in Fig. 24, the response $y_{j/k,t}$ of the j-th system due to the k-th bio-phenomenon other than the j-th one $(k \neq j)$ is given graphically by the following block diagram:

$$y_{kt} \longrightarrow \boxed{g_{jk}(B)} \longrightarrow \boxed{g_{jj}(B)} \longrightarrow y_{j/k,t}. \tag{6.7}$$

Substituting Eqs. 6.3 and 6.3-1 into Eq. 6.4, the response $y_{j/k,t}$ is rewritten as

$$y_{j/k,t} = \sum_{m=1}^{M} (a_{jjm} y_{j/k,t-m} + a_{jkm} y_{k,t-m}). \tag{6.8}$$

Let here the k-th phenomenon $y_{k,t-m}$ be

$$\begin{aligned} y_{k,t-m} &= 1, \quad \text{when} \quad t = m, \\ &= 0, \quad \text{when} \quad t \neq m, \quad (t = 0, \pm 1, \pm 2, \cdots; \ m = 1, 2, \cdots, M) \end{aligned} \tag{6.9}$$

then $y_{k,t-m}$ in this case indicates a *unit impulse stimulus* delivered to the k-th bio-system of the j-th bio-system at time $t=m$. Substituting this "unit impulse stimulus" into Eq. 6.8, we obtain the *unit impulse response* of the j-th (multivariate) higher order bio-system caused by the k-th bio-phenomenon. Let it be $u_{j/k,t}$, then

$$y_{j/k,t} \quad \text{and} \quad n_{j/k,t} = 0, \quad \text{when} \quad t = 0 \tag{6.10}$$

provided the "unit impulse stimulus" is considered to be delivered at $t=0$. Hence, the "unit impulse response" at times $t=1, 2, \cdots$ will be given as

$$\left. \begin{aligned} u_{j/k,1} &= a_{jk1}, \\ u_{j/k,2} &= a_{jj1} u_{j/k,t} + a_{jk2}, \end{aligned} \right\} \quad (j \neq k) \tag{6.11}$$

etc., and generally *(19)*

$$
u_{j/k,t} = \sum_{m=1}^{t-1} a_{jjm} u_{j/kj-m} + a_{jkt},
$$
$$
(m = 1, 2, \cdots, M; \ t = 1, 2, \cdots; \ j \neq k),
$$
$$
a_{jjm}, a_{jkm} = 0, \quad \text{when} \quad m > M.
$$
$$\tag{6.12}$$

Practically, $u_{j/k,t}$ gives the amount of the " unit impulse response " at time $t\tau$, where τ is the sampling time interval of the measured time series of bio- and environmental natural phenomena.

The above " unit impulse response " displays a time-pattern of higher order activity of the j-th bio-system concerned with the k-th bio-phenomenon given by $(g_{jk}(B)g_{jj}(B))$ in Eq. 6.4. The frequency pattern of this activity is termed the *frequency response* of the j-th bio-subsystem due to the k-th bio-phenomenon, which will be given as the Fourier transform of the " unit impulse response." Let this transform be $u_{j/k}(f)$, where f indicates circular frequency, then

$$
u_{j/k}(f) = \sum_{t=-\infty}^{\infty} u_{j/k,t} \exp(-i2\pi ft\tau). \tag{6.13}
$$

Generally, the above " unit impulse response " given by Eq. 6.12 will converge to zero with increasing t, so that $u_{j/k,T}=0$ provided T is taken as an appropriately large positive integer. Therefore, considering Eq. 6.10, $u_{j/k}(f)$ can be rewritten as

$$
u_{j/k}(f) = \sum_{t=1}^{T} u_{j/k,t} \exp(-i2\pi ft\tau)
$$
$$
= \sum_{t=1}^{T} u_{j/k,t} (\cos 2\pi ft\tau - i \sin 2\pi ft\tau). \tag{6.13-1}
$$

Let further the complex conjugate function of this be $u_{j/k}{}^{*}(f)$

$$
u_{j/k}{}^{*}(f) = \sum_{t=1}^{T} u_{j/k,t} \exp(i2\pi ft\tau)
$$
$$
= \sum_{t=1}^{T} u_{j/k,t} (\cos 2\pi ft\tau + i \sin 2\pi ft\tau). \tag{6.13-2}
$$

Consequently, the " frequency response " of the j-th higher order bio-system concerning the k-th bio-phenomenon *(19)*, say $U_{j/k}(f)$, will be given as

$$
U_{j/k}(f) = u_{j/k}(f) \cdot u_{j/k}{}^{*}(f) = |u_{j/k}(f)|^{2}. \tag{6.14}
$$

In quite the same procedures as above, the " unit impulse response " of the j-th higher order bio-system due to the s-th environmental natural phenomenon, say $u_{j/s,t}$, which displays a time-pattern of the higher order activity $(h_{js}(B) \cdot g_{jj}(B))$ given in Eq. 6.4-1 will be given as

$$u_{j/s,t} = \sum_{m-1}^{t=1} a_{jjm} u_{j/s,t-m} + a_{jsm}, \quad \text{when} \quad t > 0,$$
$$= 0, \quad \text{when} \quad t \leqq 0, \tag{6.15}$$

where

$$a_{jjm} \quad \text{and} \quad a_{jsm} = 0, \quad \text{when} \quad m > M. \tag{6.15-1}$$

And the " frequency response " of this system, say $U_{j/s}(f)$, will be given as

$$U_{j/s}(f) = u_{j/s}(f) \cdot u_{j/s}{}^{*}(f) = |u_{j/s}(f)|^2, \tag{6.16}$$

where

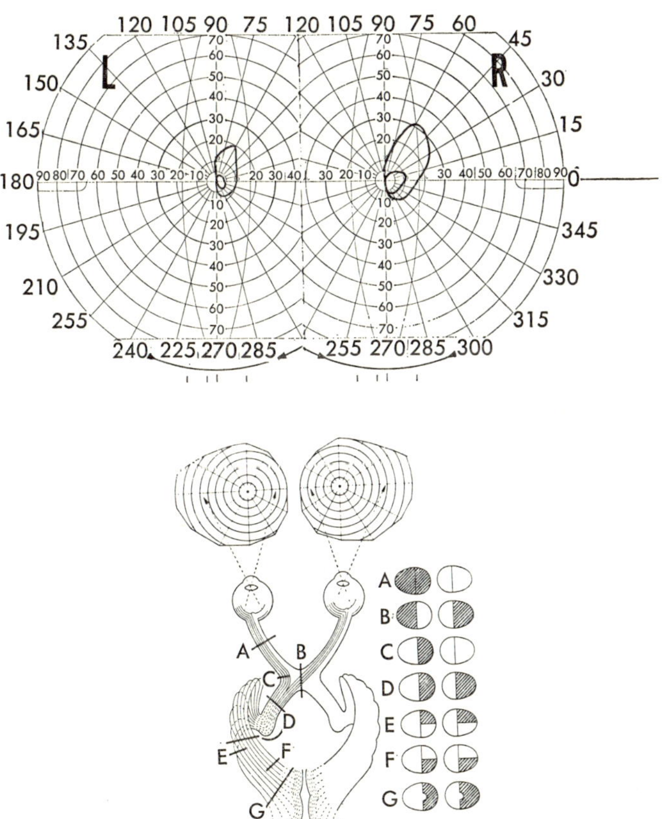

Fig. 25. Visual field showing left homonymous hemianopsia of a female patient (54-year-old) with left hemiplegia (top) (*62, 63*) and visual pathway indicating sites of lesions as related to various anopsia (bottom).

$$u_{j/s}(f) = u_{j/s,t} \exp\left(-i2\pi f t\tau\right) \tag{6.17}$$

and

$$u_{j/s}{}^{*}(f) = u_{j/s,t} \exp\left(i2\pi f t\tau\right). \tag{6.17-1}$$

It is well known that the " unit impulse response " will also be given as the inverse Fourier transform of the " frequency response."

3. *Human Corticocortical Impulse Responses*

An example of a visual field indicating a marked bilateral strangulation and hemianopsia in a patient (*62, 63*) with left hemiplegia is illustrated by the upper in Fig. 25. Such a defect in the visual field suggests some

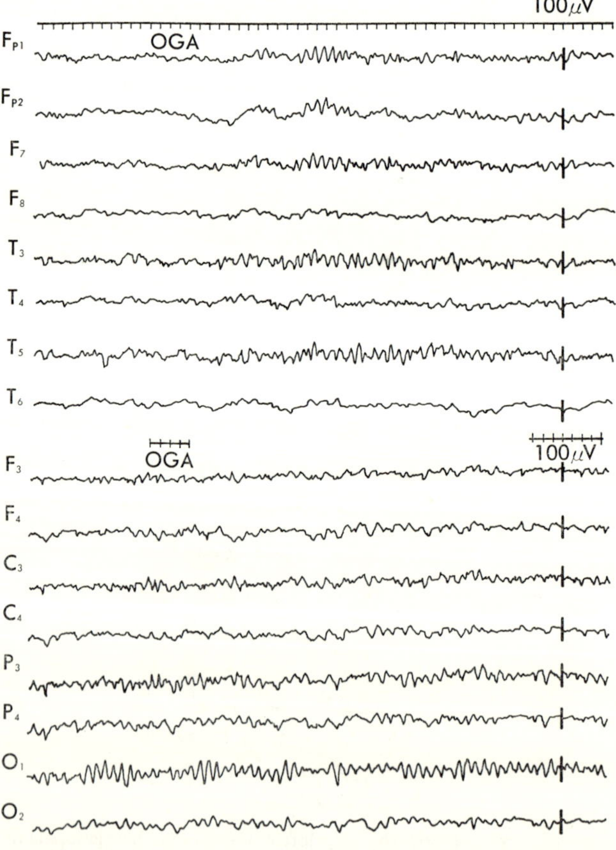

Fig. 26.   EEG of the patient with left homonymous hemianopsia and hemi-plegia in Fig. 25. Time mark on the top, 0.1 sec (*62, 63*).

infractions around the right lateral geniculate body, as can be recognized in " D " in the lower. EEGs of this patient led monopolarly from various regions (Fig. 26) depict marked depression of alpha waves in the right temporal T_6 and T_4 and frontal F_8 regions decided by the " Ten-Twenty International Method." In other regions, such as the right occipital O_2, parietal P_4 regions, *etc.*, this wave is also lower in its amplitude than that of respective contralateral regions. AR-power spectral densities of these EEGs in Fig. 27 elucidate more distinct evidence. The alpha wave peak at around about 10 Hz in the above temporal and frontal regions is impossible to recognize, but seems to be shifted to about 8 Hz, while the peak is observed in the respective contralateral regions T_5, T_3, and F_7, respectively. In addition, beta wave peaks at higher frequencies than about 15 Hz in all of the left regions are more powerful than those in the respective right regions.

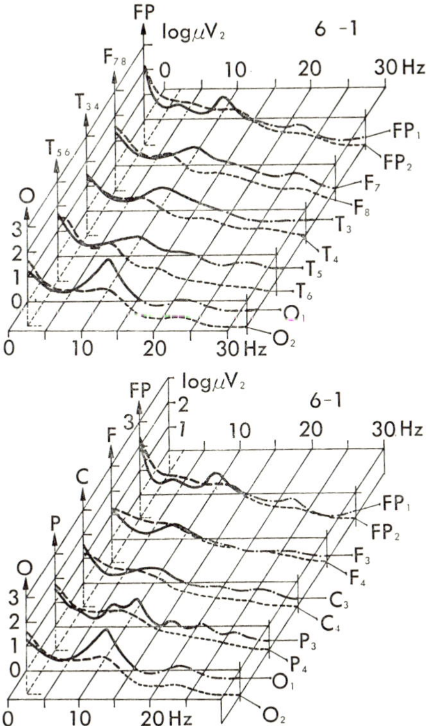

Fig. 27. AR-power spectral densities of EEGs of the patient shown in Figs. 25 and 26 (*62, 63*).

" Corticocortical unit impulse responses " (*10, 63*) in bilateral occiputs O_1 and O_2 due to a " unit impulse " originating from other anothers region, illustrated by solid line and broken line curves in Fig. 28, respectively, were obtained from eight EEGs traced simultaneously as the case of $J=8$ and $S=0$ in Eqs. 6. 1 and 6. 12. Those responses at the left occiput O_1 displayed relatively regular second-order damped oscillatory time patterns of about 11 Hz, suggesting an equivalent to the most powerful peak in the spectral density, whereas those at the contralateral occiput O_2 did not, but displayed some transient time configurations, suggesting highly deteriorated second-order or some first-order responses caused by the above noted cerebral infraction. The responses in other regions, *e.g.*, in the parietal P_3 and P_4, and frontal regions, F_4 and F_4, displayed irregular and less durable patterns even in the left regions. However, slightly regular responses were suggested in the left parietal regions P_3. Though the impulse response in the left occiput ($F_3 \rightarrow O_1$) due to a " unit impulse " initiated at a time origin in the left frontal region depicted the highest amplitude and the longest duration of about 1 sec among

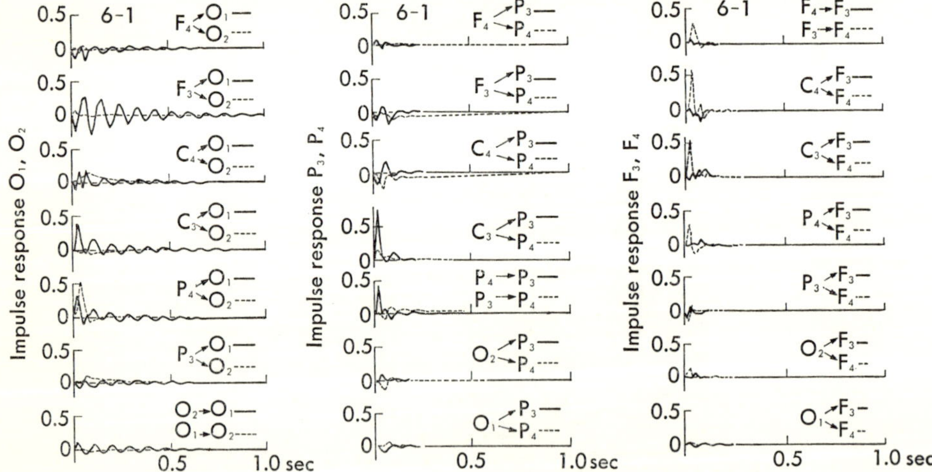

Fig. 28. Impulse responses of the occipital (left), parietal (middle), and frontal regions (right) obtained from the EEGs (Fig. 26) of a patient with left hemianopsia (Fig. 25) and left hemiplegia (*10, 63*). Full and broken line curves show impulse responses of left and right regions, respectively, due to a unit impulse initiated in one of other regions, *e.g.*, $F_4 \rightarrow O_1$ on the left top is the impulse response of the left occiput O_1 elicited by a unit impulse initiated by a unit impulse in the right frontal region F_4.

all responses, the reverse impulse response ($O_1 \rightarrow F_3$) in the left frontal region displayed far less amplitude and duration in its oscillatory pattern. There were observed similar differences not only in the " corticocortical impulse responses " between the left occipital and parietal regions ($P_3 \rightarrow O_1$ and $O_1 \rightarrow P_3$), but also in those responses between the left occipital and right parietal regions ($P_4 \rightarrow O_1$ and $O_1 \rightarrow P_4$). It would be worthwhile to consider similar evidence obtained between the visual and association areas in cats, as shown in the next paragraph.

4. *Geniculocortical, Corticogeniculate, and Corticocortical Impulse Responses in Cats*

EEG potentials led simultaneously from the lateral geniculate body (GL), middle lateral (L3, visual area) and middle suprasylvian giri (SS3, association area) in cats are demonstrated in the upper of Fig. 29. And in the lower are shown the " impulse responses " of the L3, GL, and SS3 elicited by a " unit impulse " initiated at time origin in one of the other structures *(64, 65)*. At the top left in the lower in Fig. 29 is displayed a " corticopetal impulse response " (GL $\rightarrow$ L3) of L3 due to an afferent unit impulse initiated in GL. This response is the largest in size and displayed a damped oscillatory configuration of about 3 Hz frequency lasting for about 2 sec. Though this oscillation started from a surface positive deflection, a small but relatively sharp negative deflection was superimposed on the positive deflection. On the contrary, the " corticofugal impulse response " in GL (L3 $\rightarrow$ GL) due to a unit impuse initiated at time origin in L3 was also observed, as depicted at the top right in the lower of Fig. 29. This response, however, displayed repetitive deflections with lower amplitude, greater irregularity, higher frequency of 10–12 per sec, and shorter duration of about 1 sec in comparison with the above " corticofugal impulse response " of L3.

The existence of a corticogeniculate projection pathway has been verified by anatomical studies *(66–68)*. Physiological studies not only have revealed presynaptic inhibitory and/or occlusive corticofugal influences of the lateral gyrus upon the lateral geniculate body *(69–74)*, but also have demonstrated facilitory corticogeniculate effects *(69, 73–77)*. Therefore, manifold neuronal network systems would be organized by geniculocortical and corticogeniculate connections. The above verified corticopetal and corticofugal impulse responses of L3 (GL $\rightarrow$ L3) and GL (L3 $\rightarrow$ GL) probably manifest, respectively, some of the dynamic higher order activities of these systems.

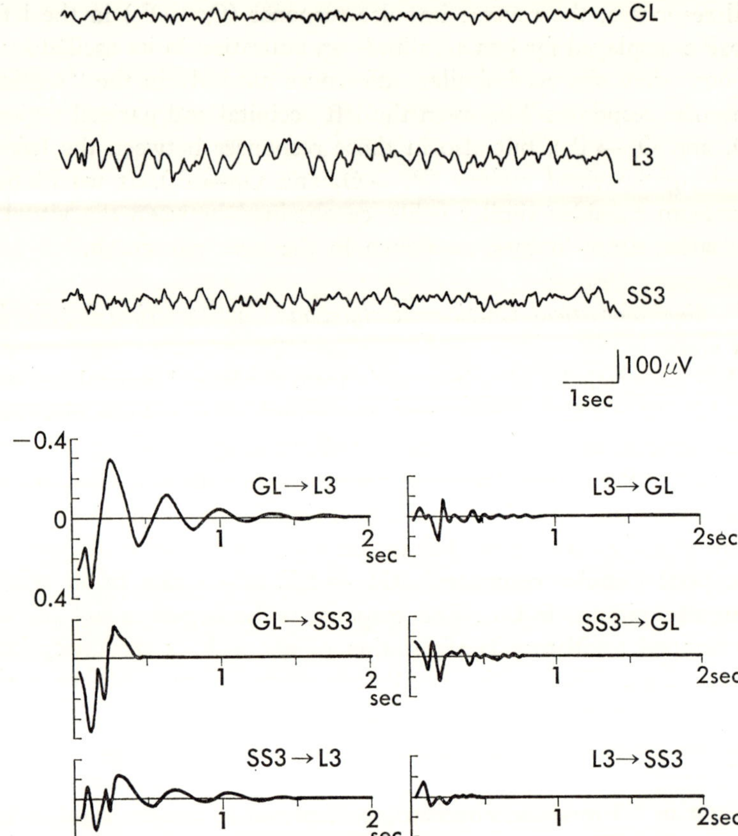

Fig. 29.   EEGs (upper) and impulse responses (lower) in the lateral geniculate body (GL), the middle lateral (L3), and middle suprasylvian giri (SS3) of a cat immobilized by Flaxedil. GL → L3 and GL → SS3, geniculocortical impulse responses of the L3 and SS3, respectively, estimated to be elicited by a corticopetal unit impulse initiated in the GL. L3 → GL and SS3 → GL, corticogeniculate impulse responses of the GL estimated to be elicited, respectively, by a corticofugal unit impulse initiated, respectively, in the L3 and SS3. SS3 → L3 and L3 → SS3, corticocortical impulse responses of the L3 and SS3, respectively, estimated to be elicited by a unit impulse initiated in the SS3 and L3, respectively.

The " corticopetal impulse response " of SS3 (GL → SS3) and the " corticofugal response " of GL (SS3 → GL) elicited by a unit impulse initiated at a time origin in GL and SS3, respectively, are also manifested in the left and right tracings, respectively, in the lower of Fig. 29. The former response is composed of a surface positive deflection followed by a negative one, both of which are relatively large in size lasting for about a half sec, and in the decreasing phase of the former smaller and sharper negative-positive biphasic deflections are intercalated. The relatively large positive-negative deflections seem to be a damped oscillatory response of about 2.5 Hz with decay too intense to follow subsequent oscillatory deflections. The intercalated biphasic deflections also seem to be followed by a negative deflection, suggesting a damped oscillatory component response with a frequency higher than 2.5 Hz.

Evoked responses in the suprasylvian gyrus have also been elicited by a photic stimulus to the eyes and/or an electric shock to the optic nerve, which suggests direct or indirect geniculo-cortical projections to the suprasylvian gyrus (73, 78, 79). In addition, the evidence indicating the corticofugal facilitory and occlusive effects of the suprasylvian gyrus upon the GL have also been demonstrated (70). Between the GL and SS3 would also be organized manifold neuronal network systems that exhibit both of the above " corticopetal " and " corticofugal impulse responses."

The " corticocortical impulse responses " of the L3 (SS3 → L3) and SS3 (L3 → SS3) elicited by a unit impulse initiated at time origins in SS3 and L3, respectively, are also verified by the bottom tracings on the left and right in the lower in Fig. 29. Corticocortical connections through association fibers and/or *via* subcortical structures, *e.g.*, the thalamic nuclei and reticular formation, would also be organized as manifold neuronal network systems to exhibit these " impulse responses."

The " corticocortical impulse response " of visual area L3 (SS3 → L3) due to a unit impulse initiated at the association area SS3 is higher in amplitude in comparison with the response of SS3 (L3 → SS3). This evidence may coincide with the the higher and longer impulse response in amplitude and duration, respectively, of the human left occiput O_1 due to a unit impulse initiated in each of the left frontal (F_3) and parietal regions (P_3) in comparison with each impulse response of these regions due to a unit impulse initiated in O_1, as demonstrated in Fig. 28.

The EEG potentials in the upper of Fig. 29 were estimated to be a trivariate seventh-order AR-process as the case of $J=3$, $S=0$, and $M=7$ in Eq. 6. 1. Thus each of the impulse responses in the lower is also the seventh-order one given as the case of $M=7$, j, $s=1$, 2, 3, and $j \neq s$ in Eq. 6. 12. The seventh-order " corticopetal impulse response " of the L3 (GL → L3) elicited by a " centripetal unit impulse " initiated in the GL was decomposed of three second- and one first-order " component im-

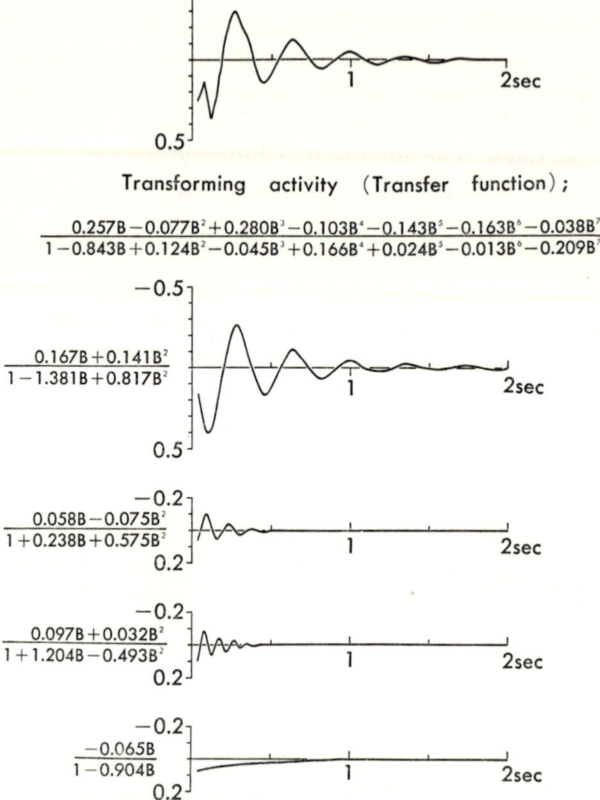

$$\frac{0.257B-0.077B^2+0.280B^3-0.103B^4-0.143B^5-0.163B^6-0.038B^7}{1-0.843B+0.124B^2-0.045B^3+0.166B^4+0.024B^5-0.013B^6-0.209B^7}$$

$$\frac{0.167B+0.141B^2}{1-1.381B+0.817B^2}$$

$$\frac{0.058B-0.075B^2}{1+0.238B+0.575B^2}$$

$$\frac{0.097B+0.032B^2}{1+1.204B-0.493B^2}$$

$$\frac{-0.065B}{1-0.904B}$$

Fig. 30.   Impulse response GL → L3 and its component impulse responses
Top: seventh-order geniculocortical impulse responses GL → L3 of the L3
characterized by the seventh-order activity ("transforming activity" or
"transfer function") given underneath the impulse response, in which B is
the "backward shift operator." From the second to fourth, the second-
order component activities and impulse responses. Bottom: the first-order
component activity and impulse response.

pulse responses" (*65*), as shown in Fig. 30 and Table IV. Each of the
second-order component responses oscillated with a damped frequency
of about 2.8, 6.9, and 10 Hz lasting for about 2, 0.6, and 0.5 sec, respectively.
    The second-order response with a damping frequency of about 2.8 Hz
is the most predominant, forming the main pattern of the seventh-order

TABLE IV

Characteristics of Component Impulse Responses of the Lateral Gyrus GL → L3 due to the Lateral Geniculate Body in a Cat (*19, 65*).

M	f_N	f_D	ζ	T_1, T_2	C	G	I
2	2.82	2.79	0.14	0.41	1.12	7.117	1.416
2	6.77	6.88	0.13	0.18	1.24	1.529	0.306
2	9.93	10.35	0.55	0.03	0.30	3.774	0.956
1	—	—	—	0.40	—	5.482	1.227

M, order; f_N and f_D, natural and damping frequencies in Hz; ζ, damping coefficient; T_1, time constant of the damped exponential first-order component response in sec; T_2, damping time (time constant of the envelope of damped oscillatory second-order component response) in sec; C $(=f_D T_2)$, continuity of the second-order component response (number of swells per damping time); G, value of activity; I, bio-informing activity amount in bit.

impulse responses GL → L3 and is initiated by a surface positive deflection, which was the highest, longest, and largest in its amplitude, damping time (0.41 sec) and bio-informing activity (1.4 bit), respectively. However, the remaining two second-order responses with damping frequencies of 6.9 and 10.3 Hz, respectively, are far smaller in their sizes and durations with damping times of 0.18 and 0.03 sec, respectively, so that they are both buried in the predominant response except for an initial negative sharp deflection in the seventh-order response. The first-order component response was a surface negative small damped exponential lasting for about 1 sec with its time constant of 0.40 sec, which was, however, invisible in the seventh-order response, though its bio-informing activity was relatively large, amounting to 1.23 bit.

As shown on the top left in Fig. 31, the highest and sharpest peak is pierced upward at a resonance frequency (f_R) of about 3 Hz in the geniculocortical " frequency response " of GL → L3, which coincides with the predominant damped oscillatory impulse response. In the next, a relatively high peak of the first-order response is enhanced at 0 Hz in accordance with the high amount of its bio-informing activity of 1.23 bit. The remaining two very low peaks appeared at resonance frequencies of about 8 and 13 Hz and coincide with the two impulse responses having small amounts of bio-informing activities of 0.31 and 0.96 bit, respectively.

Though each of other impulse responses, L3 → GL, SS3 → GL, SS3 → L3, GL → SS3, and L3 → SS3, displays a far more complex pattern and is far smaller in its size than that of GL → L3, the " frequency response " of L3 → GL displays a very low monomodal peak at around

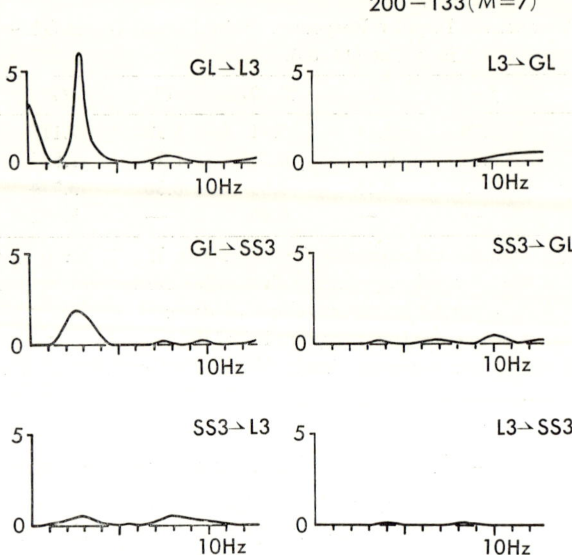

Fig. 31.   Geniculocortical, corticogeniculate, and corticocortical "frequency responses" of a cat (*19, 65*). Abscissa, resonance frequency (f_R) in Hz; ordinate, power. For abbreviations GL → L3, L3 → GL, *etc.*, see Figs. 29 and 30.

a resonance frequency of about 12 Hz, while those of SS3 → L3 and L3 → SS3 display very low bimodal peaks at about 3 and 8 Hz in the former and about 4 and 8 Hz in the latter in comparison with the highest peak at about 3 Hz in the "frequency response" of GL → L3. In addition, a moderately high and broad peak is observed at the "resonance frequency" of about 3 Hz in the "frequency response" of GL → SS3 coinciding with the relatively high amplitude but short duration of the "impulse response."

5.   *Feedback Control Activity of Bio-systems*

The simultaneous tracings of two sways of human body gravity and that of ankle joint flexion during erect standing posture shown in Fig. 7 are an example of the bivariate AR-process given as the case of $J=2$ and $S=0$ in Eq. 6.1. Let both the sways from each average state, respectively, be y_t and x_t at an arbitrary time t, then each of them coincides with y_{1t} and y_{2t} in Eq. 6.1, respectively. Thence, they can be described, respectively, as

$$y_t = \sum_{m=1}^{M}(a_{yym}y_{t-m}+a_{xym}x_{t-m})+n_{yt},$$

and

$$x_t = \sum_{m=1}^{M}(a_{xxm}x_{t-m}+a_{yxm}y_{t-m})+n_{xt},$$

(6. 18)

where a_{yym}, a_{xym}, a_{xxm}, and a_{yxm} are AR-coefficients. And n_{yt} and n_{xt} are assumed to be purely random processes, for the first step, termed " natural stimulations," each of which is composed of enumerable values caused by natural and other bio-phenomena in the external and internal environments of the bio-system. It is obvious here that y_t and x_t can describe not only postural sways, but also those of various other bio-phenomena in general.

Substituting the " backward shift operator " B given by Eq. 2. 6 into Eq. 6. 18 and letting

$$a_{yy}(B) = 1-\sum_{m=1}^{M}a_{yym}B^m,$$
$$g_{yy}(B) = 1/a_{yy}(B),$$
$$g_{xy}(B) = \sum_{m=1}^{M}a_{xym}B^m,$$
$$a_{xx}(B) = 1-\sum_{m=1}^{M}a_{xxm}B^m,$$
$$g_{xx}(B) = 1/a_{xx}(B)$$

and

$$g_{yx}(B) = \sum_{m=1}^{M}a_{yxm}B^m.$$

(6. 19)

The above Eq. 6. 18 will be rewritten as

$$y_t = x_t \cdot (g_{xy}(B) \cdot g_{yy}(B))+n_{yt} \cdot g_{yy}(B),$$
$$x_t = y_t \cdot (g_{yx}(B) \cdot g_{xx}(B))+n_{xt} \cdot g_{xx}(B),$$

(6. 18-1)

which will coincide with the case of $J=2$ and $S=0$ in Eq. 6. 1-2 and indicates that a bio-phenomenon y_t is composed of two responses of the bio-system, one is caused by the other one x_t and the other by the natural stimulation n_{yt}. Bio-phenomenon x_t is quite the same. Let each of the two responses in the sway y_t due to x_t and n_{yt} be y_{xt} and y_{nt}, respectively, then

$$n_{yt} \cdot g_{yy}(B) = y_{nt},$$

(6. 20)

$$x_t \cdot (g_{xy}(B) \cdot g_{yy}(B)) = y_{xt},$$

(6. 20-1)

thence

$$y_t = y_{nt}+y_{xt}.$$

(6. 20-2)

In quite the same, we obtain

$$n_{xt} \cdot g_{xx}(B) = x_{yt} \tag{6.21}$$
$$y_t \cdot (g_{yx}(B) \cdot g_{xx}(B)) = x_{nt} \tag{6.21-1}$$

and

$$x_t = x_{nt} + x_{yt}. \tag{6.21-2}$$

Consequently, the above obtained Eq. 6.18-1 gives a *feedback control mechanism* of a bio-system, which exhibits two bio-phenomena, as illustrated in Fig. 32. Therefore, the *feedback (control) activities* of the bio-

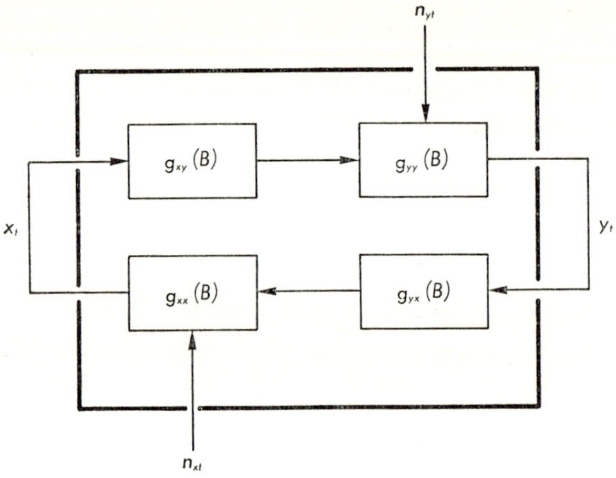

Fig. 32. Block diagram of feedback control activities of a bio-system exhibiting two bio-phenomena x_t and y_t at time t *(46)*, $g_{xx}(B)$ and $g_{yy}(B)$, higher order activities of the bio-system to exhibit its responses x_t and y_t, respectively, due to the respective natural stimulations n_{xt} and n_{yt}. $g_{yx}(B) \cdot g_{xx}(B)$ and $g_{xy}(B) \cdot g_{yy}(B)$, feedback control activities from x_t to y_t and from y_t to x_t, respectively. B, backward shift operator.

system exhibiting y_{xt} and that exhibiting x_{yt}, respectively, can be given by the following $g_{y/x}(B)$ and $g_{x/y}(B)$ such that

$$g_{y/x}(B) = g_{xy}(B) \cdot g_{yy}(B), \tag{6.22}$$
$$g_{x/y}(B) = g_{yx}(B) \cdot g_{xx}(B). \tag{6.22-1}$$

The above Eqs. 6.20 and 6.21 coincide with the case of $J=2$ in Eq. 6.4, Eqs. 6.20-1 and 6.21-1 coincide with the same case in Eq. 6.4-2. Therefore, the " unit impulse response " at time t of the bio-system exhibiting y_t due to x_t, say $u_{y/x,t}$, will be given as the case of replacing j and k in Eq. 6.12 by y and x such that

$$u_{y/x,t} = \sum_{m=1}^{t-1} a_{yym} u_{y/x,t-m} + a_{xyt},$$
$$u_{y/x,0} = 0; \quad m = 1, 2, \cdots, M; \quad t = 1, 2, \cdots$$
$$a_{yym}, a_{xym} = 0, \quad \text{when} \quad m > M. \tag{6.23}$$

In quite the same way, the " unit impulse response " of the bio-system exhibiting x_t due to y_t will be given as

$$u_{x/y,t} = \sum_{m=1}^{t-1} a_{xxm} u_{x/y,t-m} + a_{yxt}$$
$$u_{x/y,0} = 0; \quad m = 1, 2, \cdots, M; \quad t = 1, 2, \cdots$$
$$a_{xxm}, a_{yxm} = 0, \quad \text{when} \quad m > M. \tag{6.23-1}$$

The *frequency response activities* of the bio-system exhibiting y_t due to x_t and exhibiting x_t due to y_t, respectively, will be given as the power spectral densities of the " unit impulse responses " given by Eqs. 6.23 and 6.23-1, respectively. Each of them will also be given as power spectral densities of the respective " feedback control activities " $g_{x/y}(B)$ and $g_{y/x}(B)$, respectively. Let each of them be $U_{x/y}(f)$ and $U_{y/x}(f)$, respectively, then

$$U_{x/y}(f) = |\sum_{t=1}^{T} u_{x/y,t} \exp(-i2\pi ft)|^2 \tag{6.24}$$
$$= g_{x/y}(\exp(-i2\pi f)) \cdot g_{x/y}(\exp(i2\pi f))$$
$$= |g_{x/y}(\exp(-i2\pi f))|^2 \tag{6.24-1}$$

and

$$U_{y/x}(f) = |\sum_{t=1}^{T} u_{y/x,t} \exp(-i2\pi ft)|^2 \tag{6.25}$$
$$= |g_{y/x}(\exp(-i2\pi f))|^2, \tag{6.25-1}$$

where T is taken as a positive integer large enough to satisfy $U_{x/y,T} \doteq 0$ and $U_{y/x,T} \doteq 0$.

The " unit impulse responses " of the bio-system due to the natural stimulations n_{yt} and n_{xt} can also be obtained from Eqs. 6.20 and 6.21, respectively. Let each of them be $u_{y/n,t}$ and $u_{x/n,t}$, respectively, then from Eq. 6.21 we obtain

$$u_{x/n,t} = \sum_{m=1}^{t-1} a_{xxm} u_{x/n,t-m} + a_{xxt},$$
$$u_{x/n,0} = 0; \quad m = 1, 2, \cdots, M; \quad t = 1, 2, \cdots$$
$$a_{xxm} = 0, \quad \text{when} \quad m > M, \tag{6.26}$$

and from Eq. 6.20 we obtain

$$u_{y/n,t} = \sum_{m=1}^{t-1} a_{yym} u_{y/n,t-m} + a_{yyt}$$
$$u_{y/n,0} = 0; \quad m = 1, 2, \cdots, M; \quad t = 1, 2, \cdots$$
$$a_{yym} = 0, \quad \text{when} \quad m > M. \tag{6.26-1}$$

Therefore, "frequency responses" of the bio-system due to the natural stimulations will also be given by the power spectral densities of the above $u_{x/n,t}$ or $g_{xx}(B)$ and $u_{y/n,t}$ or $g_{yy}(B)$, respectively. Let, therefore, each of them be $U_{x/n}(f)$ and $U_{y/n}(f)$, respectively, then

$$U_{x/n}(f) = |\sum_{t=1}^{T} u_{x/n,t} \exp(-i2\pi ft)|^2 \qquad (6.27)$$
$$= |g_{x/n}(\exp(-i2\pi f))|^2 \qquad (6.27\text{-}1)$$

and

$$U_{y/n}(f) = |\sum_{t=1}^{T} u_{y/n,t} \exp(-i2\pi ft)|^2 \qquad (6.28)$$
$$= |g_{y/n}(\exp(-i2\pi f))|^2, \qquad (6.28\text{-}1)$$

respectively, where T is also taken as a positive integer large enough to satisfy $u_{x/n,T} \doteqdot 0$ and $u_{y/n,T} \doteqdot 0$.

The (*power*) *contribution ratios* in the frequency domain of "feedback control activities" can be defined by the above "frequency responses." Let this "contribution ratio" of y_t in the response x_t be $R_{x/y}(f)$, then

$$R_{x/y}(f) = U_{x/y}(f)/(U_{x/y}(f)+U_{x/n}(f)). \qquad (6.29)$$

In quite the same way, the ratio of x_t in the response y_t, say $R_{y/x}(f)$, is given as

$$R_{y/x}(f) = U_{y/x}(f)/(U_{y/x}(f)+U_{y/n}(f)). \qquad (6.30)$$

6. *Impulse and Frequency Responses of Human Posture Holding System*
"Unit impulse" and "frequency responses" of dorsi-plantar flexion of the human ankle joint due to anterior sway of the body gravity and those of antero-posterior sway of body gravity due to dorsal flexion of the ankle joint during erect posture are demonstrated in Fig. 33. Each of the responses were obtained, respectively, by inverse Fourier and Fourier transforms of the respective "feedback control activities" $g_{x/y}(B)$ and $g_{y/x}(B)$ given by Eq. 6.22-1 and 6.22, respectively. The "unit impulse response" of the ankle joint was initiated in the same direction as the unit impulse of bodily sway and suggested a second-order damped oscillatory response with about 10 Hz damping frequency lasting for about 2.2 sec, whereas that of bodily sway was initiated in the reverse direction as the unit impulse of the ankle joint flexion displaying slower and similar responses with about 7 and 9 Hz damping frequencies, respectively, and lasted for about 1.5 sec.

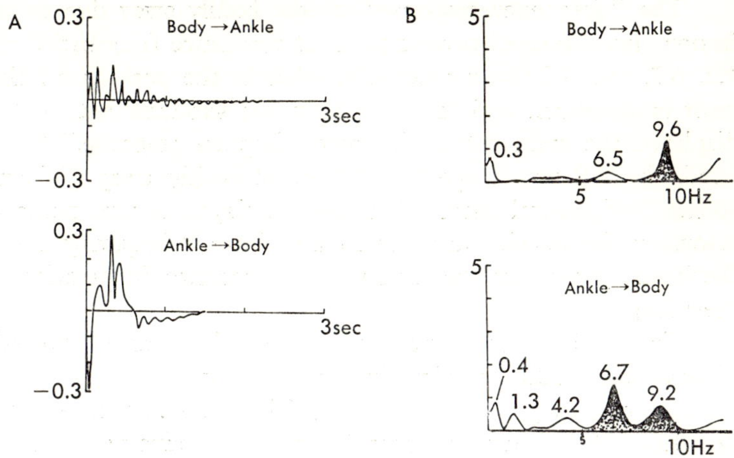

Fig. 33. Unit impulse and frequency responses of ankle joint flexion and bodily sway. A: unit impulse response of ankle joint flexion (top) due to a unit impulse of the bodily sway and that of bodily sway due to a unit impulse of the ankle joint flexion (bottom). B: frequency responses of ankle joint (top) and bodily sway (bottom), respectively, due to bodily sway and ankle joint flexion. Both responses were obtained from the simultaneous recording of ankle joint flexion and bodily sway in Fig. 7 (*10, 46*). $M=11$. (H. Noguchi).

In the " frequency response " of ankle joint flexion due to bodily sway several peaks of the second-order component responses at each resonance frequency of 0.3, 2.5, 4.2, 6.5, 9.6, and 15 Hz, respectively, were observed among which the peak at 9.6 Hz was the highest in height, which indicates a damped oscillatory " impulse response " with about 10 Hz damping frequency as noted above, that at 0.3 Hz was relatively lower, that at 6.5 Hz was markedly lower than the highest one and those at 2.5 and 4.2 Hz were the lowest. The AR-power spectral density of the ankle joint flexion shown in Fig. 8 also displayed peaks at similar resonance frequencies forming a row in such a nearly reverse pattern to the " frequency response " that the peak at 9.5 Hz was the lowest and that at less than about 2 Hz was the highest. The AR-power spectral densities of the sways of ankle joint flexion and bodily gravity, respectively, are nothing but the " frequency response " of the " posture holding system " exhibiting these two sways x_{nt} in Eq. 6.21 and y_{nt} in Eq. 6.20 as its response due to the respective natural stimulations n_{xt} and n_{yt}, respectively, given in Eq. 6.18, since both the sways were described as bivariate AR-processes.

The "frequency response" of the bodily sway due to ankle joint flexion also displayed several peaks at resonance frequencies of 0.4, 1.3, 4.2, 6.7, and 9.2 Hz, respectively, wherein the peak at 6.7 Hz was the most predominant and that at 9.2 Hz was subdominant, both of which displayed the main pattern in their "impulse response." On the other hand, the AR-power spectral density of bodily sway ("frequency response" of natural stimulation) also displayed several peaks at similar resonance frequencies to those in the above "frequency response" of the bodily sway, but the higher their resonance frequencies, the lower their height.

Figure 34 demonstrates another example of antero-posterior sways of head and bodily gravity during erect posture of a normal adult with the recording system, while Fig. 35 depicts their AR-power spectral densities ("frequency responses" due to respective natural stimulations) and "frequency responses" of the head and body, in which several peaks were also observed at similar and/or relatively similar resonance frequencies displaying the same or similar patterns as those observed in the former example (see Figs. 8 and 33).

Though the peak at 9.2 Hz was subdominant and the one at 6.7 Hz was predominant in the "frequency response" of the body due to ankle joint flexion (Fig. 33), the peak at 8.1 Hz, which seemed to correspond to the former peak at 9.2 Hz, was predominant and no peak corresponding to the latter one at 6.7 Hz appeared.

7. *Impulse Responses of Human Pulmonary Circulation System*

Venous blood pumped out from the left ventricle into the pulmonary artery flows into the lungs, wherein gas exchange between the blood and alveolar air takes place, and arterial blood flows out from the lungs into the pulmonary veins, so that the pulmonary arterial pressure and venous *wedge* pressure may be regarded as the input (cause, stimulation) and output (effect, response), respectively, of the *pulmonary circulation system*.

As demonstrated by the schematic representation in the upper in

→ Fig. 34. Recording system and tracings of head and bodily sways. Top left: Goniometer (*48*) to the occipital protuberance and neck on the spine of VIIth cervical vertebra. Top right: diagram verifying linearity of the Goniometer. Middle: block diagram of experimental system for recording antero-posterior sways of head and bodily gravity during erect posture through the static sensograph (see Fig. 7). Bottom: simultaneous tracings of sways of head and bodily gravity. Scales of A. 10 and P. 10 on the left of the tracings indicate, respectively, anterior and posterior sways amounting to 10 mm (*80*).

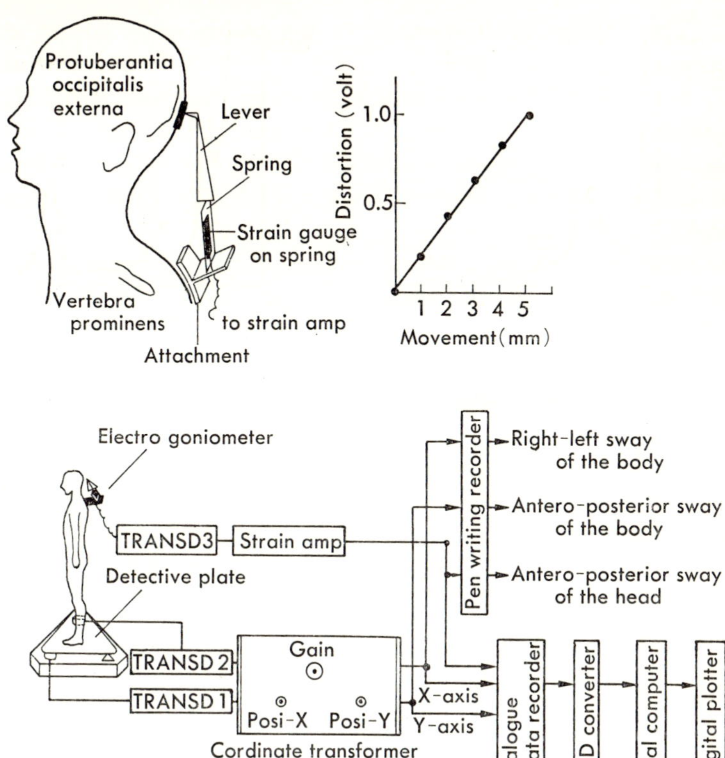

Antero-posterior sway of head

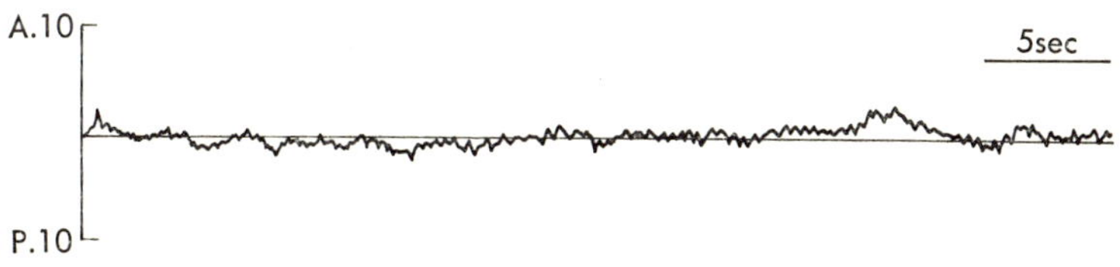

Antero-posterior sway of body

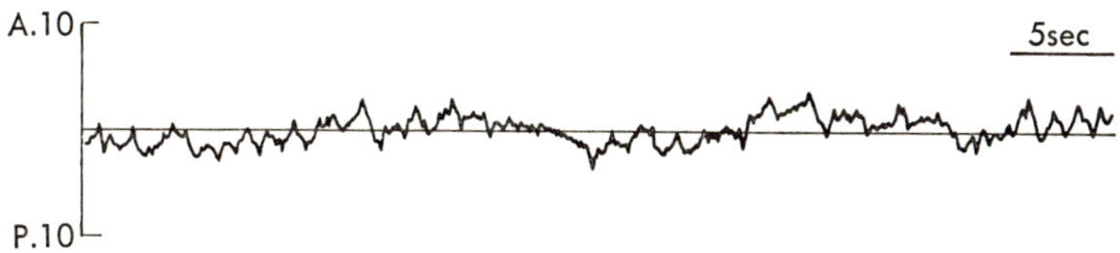

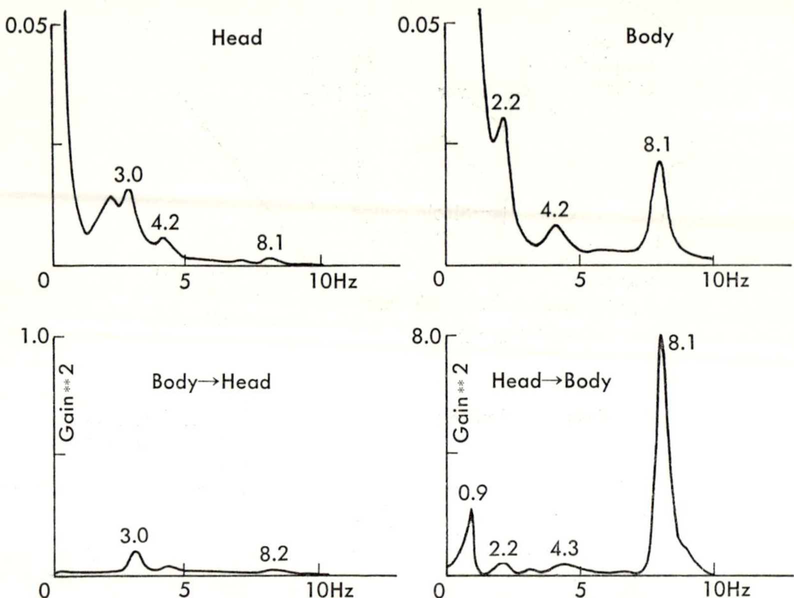

Fig. 35. AR-power spectral densities (frequency responses due to natural stimulations) (top) and frequency responses (bottom) of antero-posterior sways of head (left) and bodily gravity (right) (*80*). Body → head and head → body, frequency responses of the sway of head due to bodily sway and that of bodily sway due to sway of the head.

Fig. 36, one of two catheters (USI 8F, U.S.A.) inserted from the right femoral vein of a human subject was introduced into the pulmonary artery *via* the right ventricle, while the other was *wedged* into a pulmonary

→ Fig. 36. Schematic representation of blood pressure measurement (top), pulmonary arterial pressure and pulmonary venous wedge pressure curves (A), frequency distributions and cumulative frequency distributions of these time series (B and C). BA, bronchial artery; IVC, inferior vena cava; LPA, left pulmonary artery; LPV, left pulmonary vein; PAP, pulmonary arterial pressure; PVWP, pulmonary venous wedge pressure; RA, right atrium; RV, right ventricle; recorder, Mingograph (EM34, ELMA-Schönander, Sweden); transducer, strain guage (Statham P-23Db, U.S.A). A catheter was wedged in a pulmonary vein and the other was placed in the pulmonary artery. AV, average; S.D., standard deviation. A: time series at every 0.02 sec in the length of ten cardiac cycles for autoregressive analysis. B and C: frequency distribution (●) and cumulative frequency distribution (○) of pulmonary arterial pressure (left B, PAP) and pulmonary venous wedge pressure curves (right, C, PVWP).

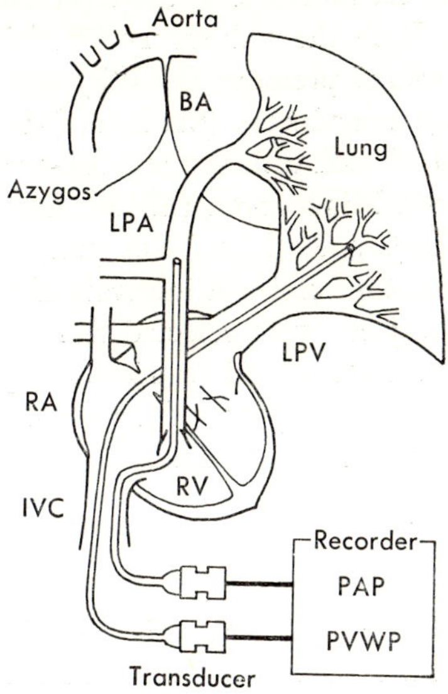
Aorta
BA
Lung
Azygos
LPA
LPV
RA
RV
IVC
Recorder
PAP
PVWP
Transducer

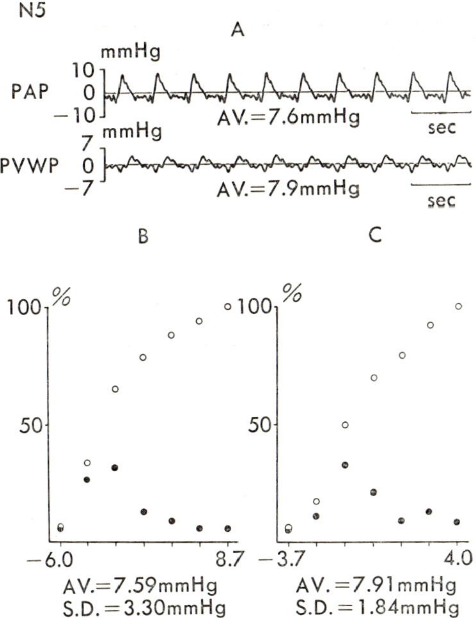
N5
A
mmHg
PAP   10
        0
      −10
AV.=7.6mmHg
sec
mmHg
PVWP   7
        0
       −7
AV.=7.9mmHg
sec
B
%
100
50
−6.0   8.7
AV.=7.59mmHg
S.D.=3.30mmHg
C
%
100
50
−3.7   4.0
AV.=7.91mmHg
S.D.=1.84mmHg

vein to exclude direct influence of the left cardiac beat on the pulmonary venous pressure. Both the pulmonary arterial pressure (PAP) and pulmonary venous wedge pressure (PVWP) were measured simultaneously using strain-gauge transducers (Statham P-23Db, U.S.A.) connected to the proximal end of each catheter, and ink-written records were obtained simultaneously by a Mingograph (EM34, Elma-Schönander, Sweden) at a speed of 500 mm/sec. Both analogue pressure curves were transformed into digital discrete time series at intervals of every 0.02 sec for ten heart beats (*80, 81*). Subjects were selected from patients scheduled for routine cardiac catheterization, and restricted to patients with atrial defects and control subjects with patent foramen ovale but whose pulmonary circulation was verified to be normal by catheterization, as shown in Table V.

TABLE V

Nonpulsatile Static Parameters on the Pulmonary Circulation of Control Subjects and Patients with Atrial Septal Defects (*82*)

Case	Age, sex	Surface area (m²)	Diagnosis	Rate (per sec)	Mean pressure (mmHg)		Mean flow (cm³/sec)	PVR
					PA	LA		
N1	15M	1.66	Normal	1.21	11.7	5.2	103	84
N2	20M	1.61	Normal	1.43	11.7	4.4	147	66
N3	20M	1.54	Normal	1.30	12.3	4.9	180	55
N4	23M	1.73	Normal	0.79	10.5	7.3	102	42
N5	39M	1.45	Normal	1.62	7.6	4.6	108	37
A1	18M	1.68	ASD	1.53	14.2	8.8	350	21
A2	18F	1.43	ASD	1.49	18.2	9.1	313	39
A3	25F	1.47	ASD	1.91	36.0	8.7	390	93
A4	39M	1.47	ASD	0.91	16.3	6.1	180	76
A5	41M	1.44	ASD	1.47	22.7	6.4	363	60
A6	47F	1.34	ASD	1.17	17.0	4.9	270	60

N1–N5 and A1–A6, case numbers of the control subjects and patients, respectively; M, male; F, female; PA, pulmonary artery; LA, left atrium; PVR, pulmonary vascular resistance in dyn/sec/cm^{-5} ((PA-LA)/mean flow); ASD, atrial septal defect.

The time series of the sways in the PAP and PVWP of most subjects were found, in at least the first step, to follow a normal distribution, since the percent cumulative frequency distribution plotted on normal probability paper formed a straight line, as depicted in Fig. 37. When this distribution did not form a straight line, FFT (fast Fourier transform) and processing for " low " and " high frequency cutting " digital filters

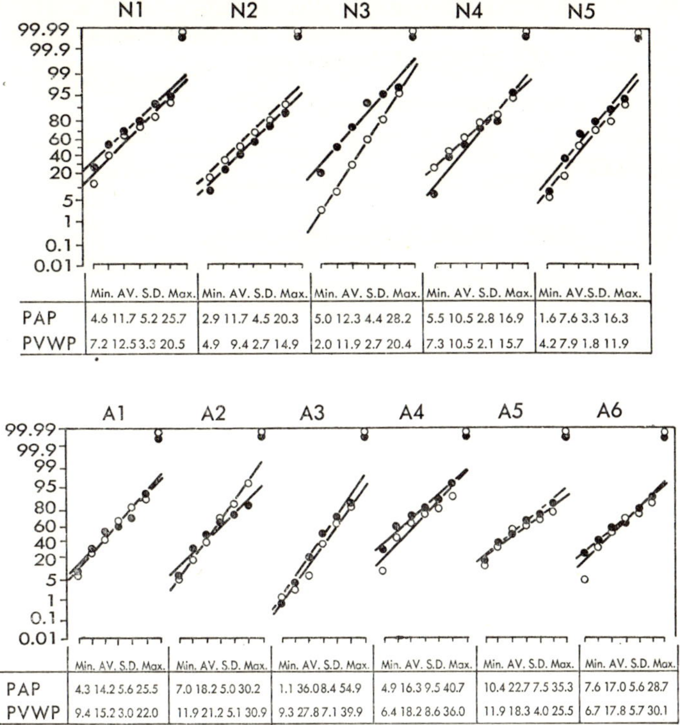

Fig. 37. Percent cumulative frequency distributions of pulmonary arterial pressures and pulmonary venous wedge pressures of normal control subjects (top) and patients with atrial septal defects (bottom) plotted on normal probability paper. N1-N5 and A1 A6, case numbers of the control subjects and patients, respectively; closed black circles (●), pulmonary arterial pressure; opened white circles (○), pulmonary venous wedge pressure; PAP and PVWP, pulmonary arterial and pulmonary venous wedge pressures in mmHg, respectively. Max., maximum; Min., minimum. AV., average; S.D., standard deviation.

were applied to both the time series. The sway of the time series of high frequency obtained by the processing of " low cut digital filter " formed a straight line distribution on the probability paper, whereas that of low frequency obtained by applying the " high cut digital filter " displayed a rhythmic curve synchronized with the electrocardiogram (ECG), as seen in Fig. 38 (81). Consequently, only some essential information on the dynamic activities of the " lung circulation system " would be hidden in the time series of low frequency, but most of them would be in those

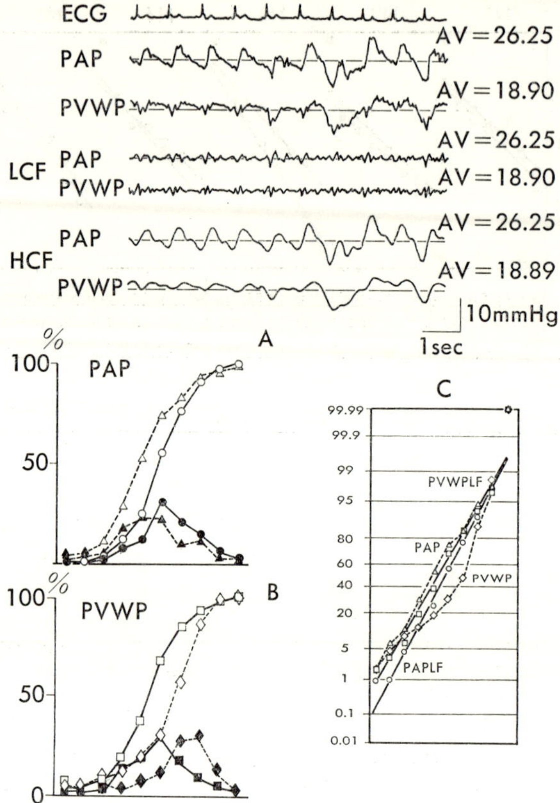

Fig. 38.   Pulmonary arterial pressure (PAP) and pulmonary venous wedge
pressure curves (PVWP) of a patient with bronchiectasis (top) and distributions of sways in these pressure curves (bottom). ECG, electrocardiogram,
LCF and HCF, pressure curves of PAP and PVWP of high frequency obtained by applying the low frequency cutting digital filter and those of low
frequency obtained by applying the high frequency cutting digital filter,
respectively, A and B (bottom left), percentage frequency distributions and
cumulative frequency distributions of PAP (▲, △), PALP (●, ○, PAP of
high frequency), PVWP (◆, ◇), and PVWPLF (■, □, PVWP of high frequency; C (bottom right), percentage cumulative frequency distributions of
these four pressure curves plotted on normal probability paper. Each PAP
and PVWP did not form a straight line, but each PAPLF and PVWPLF
did.

of high frequency. It is obvious here that the PVWP at an arbitrary time
is related more or less in stochastic fashion not only to the earlier pressure

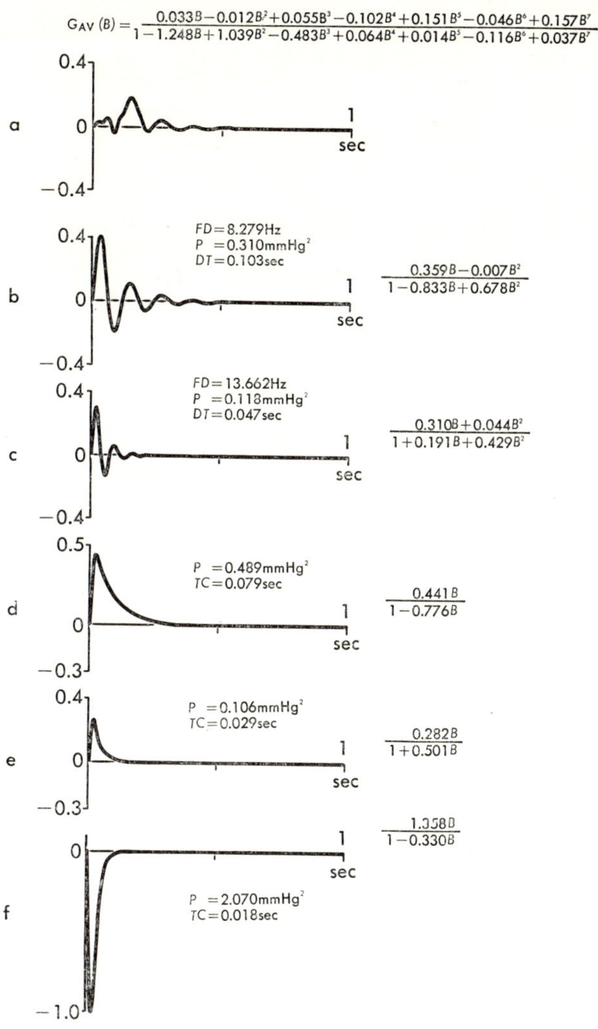

Fig. 39. Pulmonary blood pressure regulatory activity, impulse response and its component responses of the control subject N5 (*81*, *83*). G_{AV}(B), seventh-order pulmonary blood pressure regulartory activity. a: seventh-order impulse response of pulmonary venous wedge pressure due to an impulse-like abrupt rise in pulmonary arterial pressure. b and c: second-order component impulse responses of the above regulatory activity. d, e, and f: first-order component impulse responses. *B*, backward shift operator; *FD*, damping frequency in Hz; *P*, power in mmHg²; *DT*, damping time in sec; *TC*, time constant in sec.

TABLE VI

Characteristics of the First- and Second-order Component Impulse Activities of All Subjects (*80, 82*)

Case	Order (*M*)	First-order activity			
N1	6	P (mmHg^2)	2.1130	1.3878	0.0366
		TC (sec)	0.0109	0.0387	0.0372
N2	8	P (mmHg^2)	1.8089	1.3551	
		TC (sec)	0.0424	0.0076	
N3	7	P (mmHg^2)	0.3503		
		TC (sec)	0.0369		
N4	8	P (mmHg^2)	1.2749	0.2158	
		TC (sec)	0.0486	0.0137	
N5	7	P (mmHg^2)	2.0699	0.4888	0.1058
		TC (sec)	0.0181	0.0790	0.0289
A1	8	P (mmHg^2)	0.1104	0.0002	
		TC (sec)	0.1135	0.1051	
A2	7	P (mmHg^2)	0.1482		
		TC (sec)	0.0044		
A3	7	P (mmHg^2)	0.2503		
		TC (sec)	0.0068		
A4	9	P (mmHg^2)	0.0250		
		TC (sec)	0.0990		
A5	6	P (mmHg^2)			
		TC (sec)			
A6	7	P (mmHg^2)	0.1838		
		TC (sec)	0.0514		

N1–N5, case numbers of normal control; A1–A6, case numbers of pressure activity; *FD*, damping frequency in Hz; *P*, power in mmHg^2; components were classified by their damping frequencies into three

but also to the earlier PAP, the earlier states of elasticity, radius, branching, *etc.* of pulmonary vessel networks, and the earlier conditions of the sur-

Responses Obtained from the Pulmonary Blood Pressure Regulatory

	Second-order activity					
	P_1		P_2		P_3	
FD (Hz)			9.1083			
P (mmHg^2)			0.1685			
DT (sec)			0.0738			
FD (Hz)		4.8038	10.3783			19.3984
P (mmHg^2)		0.4053	0.0069			0.1395
DT (sec)		0.0757	0.2640			0.0499
FD (Hz)	1.5406		9.6429		15.5517	
P (mmHg^2)	0.2927		0.0142		0.0376	
DT (sec)	0.0518		0.2941		0.1002	
FD (Hz)	3.5886			11.5644		19.0611
P (mmHg^2)	0.5771			0.0442		0.0228
DT (sec)	0.0940			0.0661		0.0852
FD (Hz)			8.2793	13.6616		
P (mmHg^2)			0.3100	0.1180		
DT (sec)			0.1027	0.0473		
FD (Hz)		4.4545	9.3892		15.7086	
P (mmHg^2)		0.1442	0.0323		0.0006	
DT (sec)		0.1534	0.1217		0.1144	
FD (Hz)	2.3091			13.3376		19.2288
P (mmHg^2)	0.1920			0.0283		0.0405
DT (sec)	0.1751			0.1139		0.0541
FD (Hz)	3.2357			11.2993		20.8998
P (mmHg^2)	0.0369			0.0628		0.0191
DT (sec)	0.0863			0.0464		0.0822
FD (Hz)		5.1095	10.7163		16.2003	21.5664
P (mmHg^2)		0.2830	0.0106		0.0118	0.0100
DT (sec)		0.3402	0.1169		0.1315	0.1560
FD (Hz)	3.0126			12.3075		20.9503
P (mmHg^2)	0.1433			0.0163		0.1071
DT (sec)	0.0392			0.2179		0.0402
FD (Hz)	3.0075			13.0254		21.0598
P (mmHg^2)	0.1513			0.1245		0.1166
DT (sec)	0.1286			0.0369		0.0282

of patients with atrial septal defects; M, order of the pulmonary blood
DT, damping time in sec; TC, time constant in sec. Second-order
groups, namely P_1 (0–7.5 Hz), P_2 (7.5–15 Hz), and P_3 (15–22.5 Hz).

rounding tissues and autonomic nervous activities, *etc*. Moreover, this
would be similar to the PAP, so that both the PVWP and PAP can be

described by a bivariate AR-process, as given in paragraph VI. 5.

In the case of a normal control subject (N5), for instance, a seventh-order "pressure regulatory activity" of the pulmonary circulation system was obtained such as is shown on the top in Fig. 39, from which can be lead the "unit impulse response" (a) of the PVWP due to the PAP. Due to an impulse-like abrupt rise in unity in the PAP at the time origin, the PVWP manifested its maximal rise amounting to only about one-fifth of unity with delay of about 0.16 sec, which decayed to a tracing of a damped oscillatory curve with about 8 Hz damping frequency pre-ceeded by two minute positive swells, a negative deflection and a hump in the initial rise of the damped oscillatory curve. The maximal rise in PVWP in other normal subjects and patients also had similar amounts.

Expanding the above seventh-order "activity" in partial fractions, we obtained two second-order (b and c) and three first-order component activities and the respective impulse responses (d, e, and f). Except for the one first-order component response (f), all others were initiated at the prompt rise in PVWP in accordance with the abrupt rise in PAP, whereas the first orders one was initiated at the reverse direction suggesting dilation of the lung capillary vessels by the thrust of blood flow.

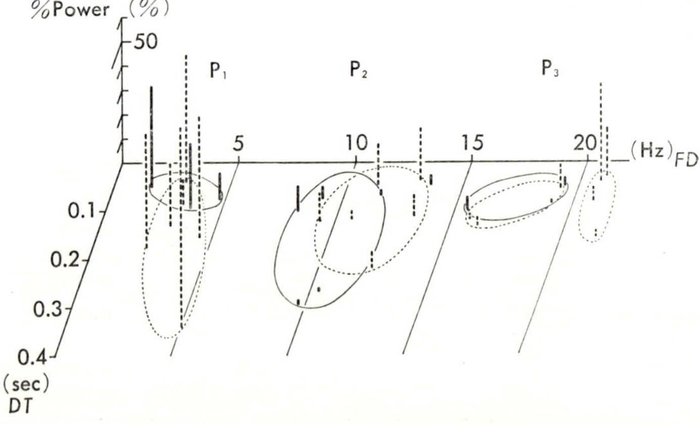

Fig. 40. Percent powers of the second-order component impulse responses in relation of the damping frequency (*FD*) and damping time (*DT*). %Power, the percentage of the power of each component response to the total power; solid and broken perpendicular lines, control subjects and patients with atrial septal defects, respectively. Component responses were classified into three groups, namely, P_1, P_2, and P_3 in the frequency ranges of less than 7.5 Hz, 7.5–15.0 Hz, and 15–22.5 Hz, respectively.

The second-order component responses were classified with respect to the damping frequency into three groups, *i.e.*, P_1 (0.5–7.5 Hz), P_2 (7.5–15 Hz), and P_3 (15–22.5 Hz), respectively (see Table VI and Fig. 40). Although the P_2 response was observed in all subjects, no remarkable

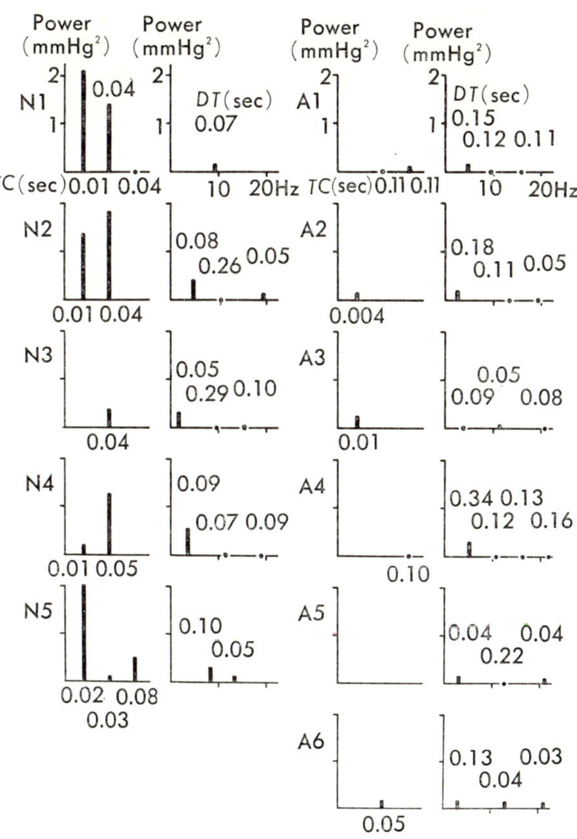

Fig. 41. Powers of the first- and second-order component impulse responses, as related to the time constant in sec in the former, and to the damping frequency in Hz. Numerals on the perpendiculars of power of the second-order component responses indicate the damping time (DT) in sec. In all of the control subjects (N1–N5), the total powers of the first-order components were not only far greater than those of the second order components, respectively, but also far greater than those of the first-order in the patients' group (A1–A6), whereas reverse relations were manifested in the patients with atrial septal defects except for A3.

difference was verified between the control and patients' groups, whereas P_3 responses with damping frequencies higher than 20 Hz were observed only in the patients' group and not in the control group. No difference was observed between the two groups in the frequency of the P_1 response, but a difference was manifested in the damping time. In the control group it was less than 0.1 sec, while that of the patients' group was longer than 0.1 sec except in two subjects.

The input impedance curve with respect to the frequency has been found to be shifted to the right in dogs with pulmonary hypertension induced by an intravenous injection of serotonin (83), and a similar change has been observed between patients with and without pulmonary hypertension (84). These findings would be consistent with the evidence that the component response P_3 with a damping frequency higher than 20 Hz was observed only in the patients' group.

In the control group, the total power of the first-order component responses was far greater than that of the second component responses P_1, P_2, and P_3, whereas the reverse relation was manifested in the patients' group except for only one patient A3 (see Fig. 41). The time constant of the first-order component responses was usually less in the control group in many cases than in the patients' group. The average in the former was 0.034 sec and the latter was 0.064 sec. In addition, the power of the first-order component responses in the control group was far greater than that in the patients' group. A significant difference was noted in the average power at the significance level of 0.05. The prominent, dominant total power of the first-order responses in the control group would suggest that the pulmonary vessels are capable of expanding and thereby able to absorb the abrupt rise in the pulmonary arterial pressure due to cardiac pulsation, whereas this ability in the patients' group would be reduced markedly due to stiffening of the vessels. The prolonged damping time of the second-order response P_1 with a damping frequency higher than 20 Hz in the patients' group would also suggest stiffened pulmonary vessels.

8. *Impulse Responses of Glucoregulatory System in Dogs*

Constancy in the blood glucose level in humans and animals is well known as typical evidence manifesting the *homeostasis* of their *internal environment* (9). However, sways are also elicited to a greater or lesser degree on this level by many causes, such as various types of stress, the physiological consumption of glucose by the muscular and nervous systems, the intes-

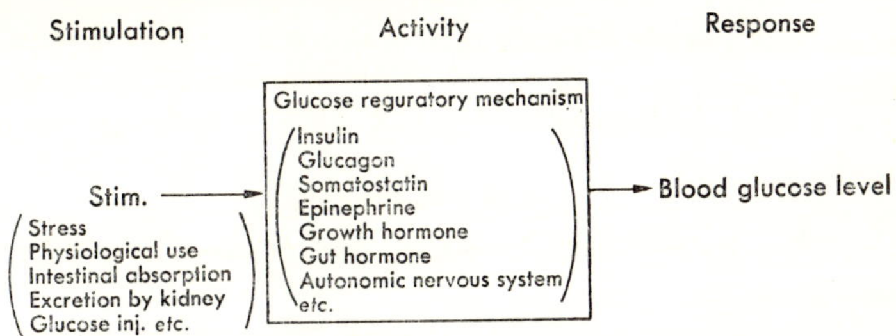

Fig. 42. Glucoregulatory system having bio-mechanism (activity) to maintain "homeostasis" of blood glucose level, *i.e.*, to minimize sways (response) in the level due to various causes (stimulation).

tinal absorption of glucose by feeding, excretion of glucose through the kidneys, *etc.* These sways are regulated to maintain homeostasis by a variety of hormones, *e.g.*, insulin, glucagon, somatostatin, epinephrine, growth hormone, *etc.*, and by the autonomic nervous system, *etc.* The aforementioned causes can be regarded as stimulations delivered to the *glucoregulatory system* shown in Fig. 42 that elicit a sway in the blood glucose level as its response, whereby this system acts to minimize the amount of the sway. Consequently, it is obvious that the sway in the blood glucose level is also a bio-phenomenon of humans and animals such that at an arbitrary time it is related in stochastic fashion not only to the blood glucose and serum insulin in each preceding time interval, but also to external glucose administration and an enormous number of other bio- and natural phenomena. Consequently, the time series of the blood glucose level measured at a time interval of appropriate length can be described, for the first step, as an AR-process, in which will be hidden some dynamic activities of the "glucoregulatory system."

Not only the time series of the blood glucose, but also that of serum insulin of dogs (17–20 kg body weight) anesthetized with Nembutal under normal conditions and in an Alloxan diabetic state were obtained simultaneously from each sample of arterial blood amounting to 1.5 ml, which was taken out 120 times from the femoral artery at intervals of 5 min for 10 hr (*85, 86*). Moreover, impulse-like rapid injections of glucose in small amounts (0.04 g/kg) to avoid unphysiological accumulation were administered into the femoral vein at random time intervals determined by the normal random numbers in a computer library, by which the sway of the blood glucose level was amplified like those of human standing posture and EEG potential. Thus, as shown in Fig. 43,

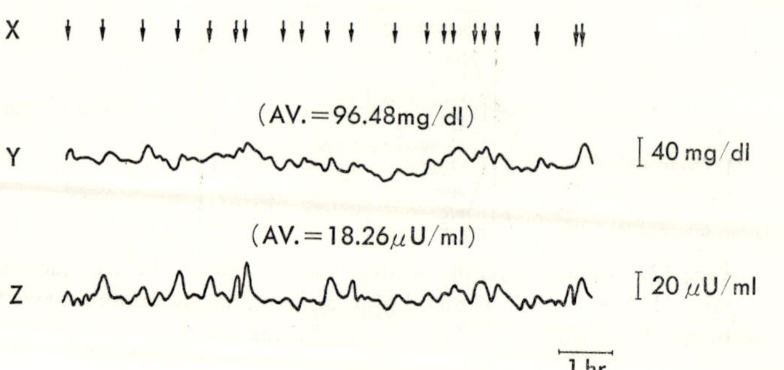

Fig. 43.   Sways in the levels of blood glucose (Y) and serum insulin (Z) in the femoral artery of a dog during random impulse-like rapid injections of glucose (X) into the femoral vein in the length of 10 hr (*86*). C3, case number of dog; AV., average level, which suggests the level of "homeostasis." Arrows in X indicate time point of random impulse-like injections, each of which was carried out within 1 sec.

trivariate time series of blood glucose (y), serum insulin (z), and impulse-like random injection of glucose (x) of dogs under normal and Alloxan diabetic states were obtained, each of which can be described by a trivariate AR-process given by the case of $J=2$ and $S=1$ in Eq. 6. 1.

Let the time series of random impulse-like glucose injection, sways in blood glucose and serum insulin levels at an arbitrary time t be x_t, y_t, and z_t, respectively. Then Eq. 6. 1 is rewritten as

$$
\begin{aligned}
y_t &= \sum_{m=1}^{M}(a_{yym}y_{t-m}+a_{zym}z_{t-m}+a_{xym}x_{t-m})+n_{yt} \\
&= x_t(g_{xy}(B)\cdot g_{yy}(B))+z_t(g_{zy}(B)\cdot g_{yy}(B))+n_{yt}\cdot g_{yy}(B) \\
&= y_{xt}+y_{zt}+y_{nt}
\end{aligned}
\tag{6.31}
$$

and

$$
\begin{aligned}
z_t &= \sum_{m=1}^{M}(a_{zzm}z_{t-m}+a_{yzm}y_{t-m}+a_{xzm}x_{t-m})+n_{zt} \\
&= x_t(g_{xz}(B)\cdot g_{zz}(B))+y_t(g_{yz}(B)\cdot g_{zz}(B))+n_{zt}\cdot g_{zz}(B) \\
&= z_{xt}+z_{yt}+z_{nt},
\end{aligned}
\tag{6.31-1}
$$

where

$$
\left.
\begin{aligned}
g_{yy}(B) &= 1/a_{yy}(B) = 1/(1-\sum_{m=1}^{M}a_{yym}B^m) \\
g_{xy}(B) &= \sum_{m=1}^{M}a_{xym}B^m \\
g_{zy}(B) &= \sum_{m=1}^{M}a_{zym}B^m
\end{aligned}
\right\}
\tag{6.32}
$$

$$g_{zz}(B) = 1/a_{zz}(B) = 1/(1 - \sum_{m=1}^{M} a_{zzm}B^m)$$

$$g_{xz}(B) = \sum_{m=1}^{M} a_{xzm}B^m$$

$$g_{yz}(B) = \sum_{m=1}^{M} a_{yzm}B^m,$$

$$(6.32\text{-}1)$$

y_{xt}, y_{zt}, and y_{nt} are the sways in the blood glucose level caused by the random glucose injection, serum insulin and natural stimulation n_{yt}, respectively, while z_{xt}, z_{yt}, and z_{nt} are the sways in the serum insulin level due to the random injection, blood glucose and the respective natural stimulation n_{zt}, respectively. Thus, the characteristics of the "glucoregulatory system" will be given by the block diagram in Fig. 44.

In quite the same procedure as leading Eq. 6.12, the "unit impulse response" of the sways in the blood glucose and serum insulin, respectively, exhibited by the "glucoregulatory system" will be obtained, as demonstrated in Fig. 45. By a rapid impulse-like injection of glucose into the femoral vein at the time origin, the blood glucose level in the femoral artery was enhanced within the first 5 min to a maximum value and then decayed with irregular sways to converge to the base line amount-

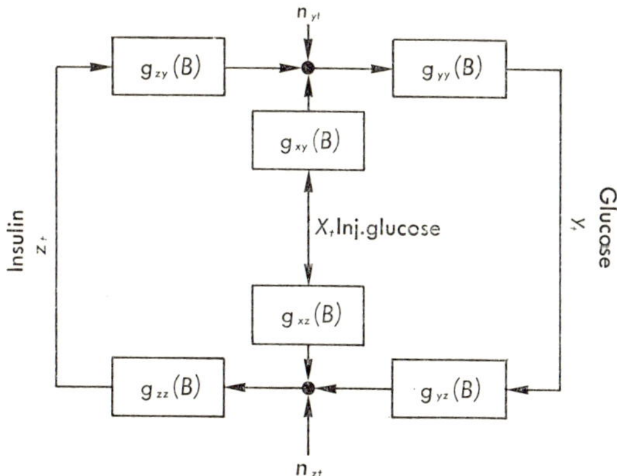

Fig. 44. Block diagram of glucoregulatory system. x_t, y_t, and z_t, random impulse-like glucose injection into the femoral vein, blood glucose and serum insulin in the femoral artery at time t, respectively; n_{yt} and n_{zt}, random natural stimulations concerning the blood glucose and serum insulin, respectively; $g_{yy}(B)$ and $g_{zz}(B)$, glucose and insulin activities, respectively; $g_{xy}(B) \cdot g_{yy}(B)$, and $g_{xz}(B) \cdot g_{zz}(B)$, glucose and insulin activities due to random glucose injection. $g_{zy}(B) \cdot g_{yy}(B)$, and $g_{yz}(B) \cdot g_{zz}(B)$, feedback control activity of blood glucose due to serum insulin and that of serum insulin due to blood glucose, respectively.

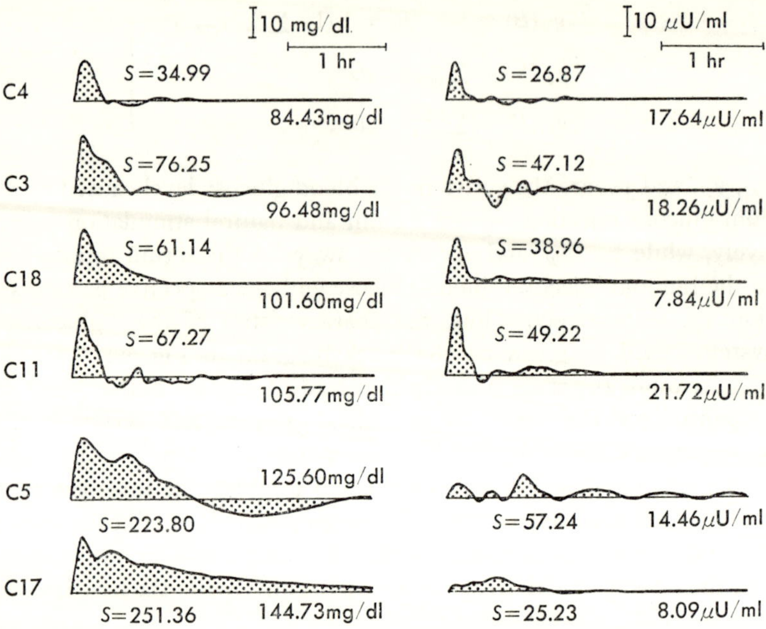

Fig. 45. Impulse responses of blood glucose (left) and insulin (right) due to impulse-like glucose injection (*86*). S, control area (area of deviation from the base line). Value underneath the base line indicates the amount of the base line (average level of sway), which suggests the level of "homeostasis."

ing to about 100 mg/dl (84.43–105.77 mg/dl) within about 1 to 2 hr and the *control area* given as the area of deviation from the base line was less than 80.0 (34.99–76.25) in most dogs (Fig. 45, left, C4, C3, C18 and C11). Regarding the *settling time*, the length of time for recovery to the base line, it seems that the shorter this time, the better the *glucoregulatory activity* is, as shown in Fig. 42. In some dogs, as in the cases of C5 and C17 in Fig. 45, the decay was far slower to converge to the base line, which was higher than about 100 mg/dl (125.60 and 144.73 mg/dl), spent a longer "settling time" than 3 hr and showed a far greater "control area" than 80.0 (223.80 and 251.36). This evidence suggests poorer "glucoregulatory activity" of the system than normal.

Though the "impulse response" of serum insulin due to the rapid glucose injection also verified a similar time pattern in most dogs, the initial rise and decay were more rapid, and the subsequent sway occurred at higher frequencies than those of blood glucose. However, no initial

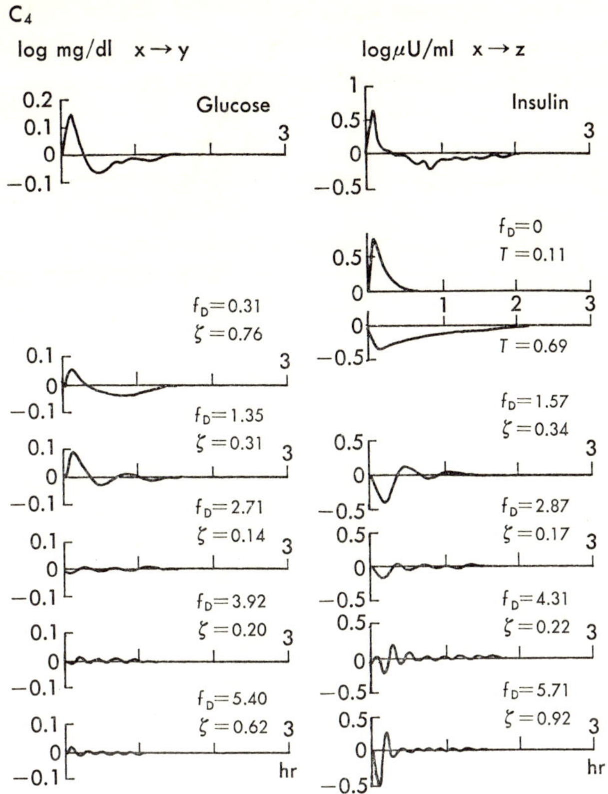

Fig. 46. An example of the impulse responses of blood glucose ($x \rightarrow y$) and serum insulin ($x \rightarrow z$) of case C4 (top) and their component responses (the second to bottom). x, sudden impulse-like glucose injection at time origin; y, blood glucose; z, serum insulin. Abscissa, hours; ordinate, left ($x \rightarrow y$) and right ($x \rightarrow z$) columns are mg/dl and μU/ml in logarithmic scale, respectively. T, time constant in hour of the first-order component damped exponential response; f_D and ζ, damping frequency (cyc/hr) and damping coefficient of the second-order damped oscillatory response, respectively.

rise in the serum insulin response occurred and its sway was lower in frequency and amplitude in the dogs which suggested poorer "glucoregulatory activity."

As demonstrated in the lower of Fig. 46, the "impulse response" of blood glucose due to the impulse-like injection ($x \rightarrow y$) consisted of several second-order damped oscillatory component responses, whereas that of serum insulin ($x \rightarrow z$) not only consisted of several second-order damped oscillatory responses but also included one or two first-order damped exponential responses. Oohtkins *et al.* (*88*) demonstrated oscillations in the glucose and insulin concentrations in the arterial blood of conscious starved dogs, verifying a dominant and a subdominant peak at 1 cyc/200 min (0.30 cyc/hr) and 1 cyc/50 min (1.2 cyc/hr), respectively. The former peak in the glucose spectrum will correspond to the component response of 0.31 cyc/hr damping frequency, whereas the component responses of glucose and serum insulin with damping frequencies of 1.35 and 1.57 cyc/hr, respectively, seem to coincide to the latter peak. Other oscillations with 1 cyc/9 min (6.7 cyc/hr) on average were also demonstrated by Goodner *et al.* (*89*) in the plasma concentration of glucose, insulin, and glucagon in overnight-starved rhesus monkeys, which would correspond to the component responses of glucose and insulin with damping frequencies of 5.4 and 5.71 cyc/hr, respectively.

There were observed not only the second-order damped oscillatory component responses of glucose with an initial rise, which would be elicited by the glucose injection, but also those with an initial falling phase. On the other hand, most of the second-order serum insulin responses were initiated in their falling phases, suggesting negative feedback regulation against the initial rapid rise in the insulin level followed by a marked enhancement of the blood glucose level. The rapid rise in the first-order damped exponential component response of serum insulin would indicate that there is an initial release of insulin from β cells in the pancreas following the initial enhancement of the blood glucose level elicited by glucose injection, while the transient relatively rapid fall of serum insulin displaying another first-order component response would suggest some depression of insulin release due to some hormonal activity, *e.g.*, insulin release itself, somatostatin release from D cells in the pancreas, *etc.*, or other humoral and neuronal factors.

It has been demonstrated by Niijima (*90, 91*) in rabbits that intra-arterial injection into the carotid artery or intravenous injection of glucose caused a decrease in the efferent discharge rate of the adrenal nerve, whereas it caused an increase in that of the pancreatic branch of the vagus nerve. In addition, hypoglycaemia due to insulin administration increased the former nervous activity, but decreased the latter. Glucose administration also depressed the discharge rate of afferent nerve fibers from the glucoreceptors in the liver of guinea pigs (*91*). In rat and cat hypothalamus, several workers (*e.g. 92–94, etc.*) have verified glucosensitive neurons and Oomura *et al.* (*95*) demonstrated that such neurons in the ventromedial

(VMH) and lateral hypothalamic nuclei (LH), which are known as centers regulating feeding behavior, respectively increase and decrease their discharge rates upon administration of glucose, whereas reverse results were observed upon insulin application. The neurons in both the hypothalamic nuclei regulate each other reciprocally (96), and it was suggested that there seems to be a pathway from the hypothalamic area to the adrenal nerve cells in the spinal cord, but not to the renal neurons (96). This evidence would suggest a variety of neural, humoral, and neuro-humoral negative " control activities " in the " glucoregulatory sys-

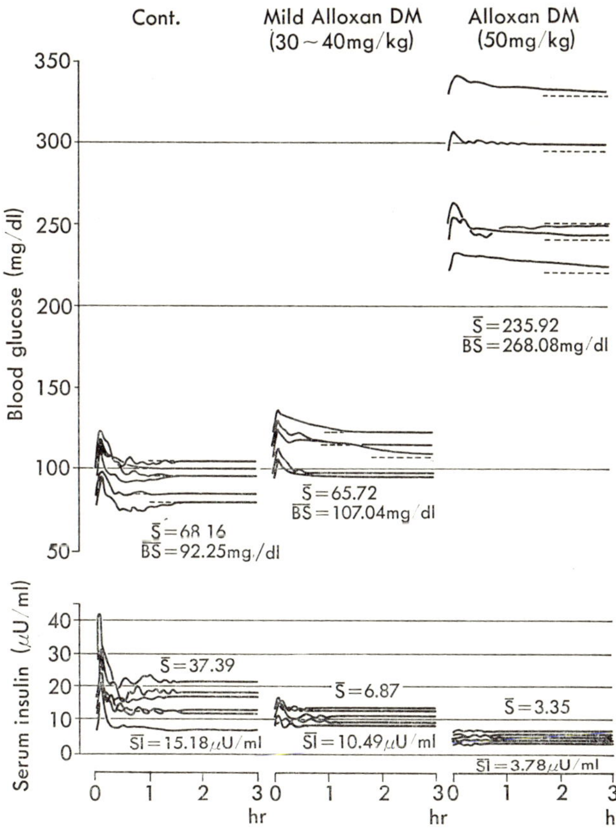

Fig. 47. Impulse responses of blood glucose (top) and serum insulin (bottom) of Alloxan administered dogs. left, control dogs; middle, mild diabetic state by 30–40 mg/kg Alloxan; right, diabetic state by 50 mg/kg Alloxan; $\bar{S}$, control area; $\overline{BS}$ and $\overline{SI}$, levels of blood glucose and serum insulin, respectively (87).

tem " to exhibit several component impulse responses of glucose ($x \rightarrow y$) and insulin ($x \rightarrow z$) due to the impulse-like glucose injection, respectively.

Two weeks after administration of 30–40 mg/kg Alloxan some dogs exhibited a slightly higher blood glucose level, a longer " settling time " and a wider " control area " in the glucose " impulse response " than those of control dogs, as depicted in the middle of the upper in Fig. 47, but the responses of all these dogs manifested a marked depression in their initial rise and in the subsequent damped oscillatory sways, which converged to a slightly higher average blood glucose level (107.04 mg/dl) than that of control dogs (92.25 mg/dl) depicted on the upper left in the uppper, but no significant difference was observed in their " control areas." However, two weeks after administration of 50 mg/kg Alloxan the glucose level of all dogs was enhanced prominently (Fig. 47, upper right). In addition, not only were the initial rise and the subsequent damped oscillatory sways in the " impulse responses " depressed most prominently but also spent far longer " settling times " than 3 hr and the average " control areas " (235.92) were much wider than those of controls (68.16) and of dogs administered less Alloxan (65.72).

The insulin " impulse responses " (Fig. 47, bottom) manifested a lower (10.49 μU/ml) and the lowest levels on average, respectively, in animals administered 30–40, and 50 mg/kg Alloxan than in controls (15.18 μU/ml). The initial rapid rise was depressed markedly in the former group, whereas it did not appear in the latter group. The subsequent oscillatory sways were also depressed to a greater and the greatest degrees in these groups, respectively. The " control areas " on average were also markedly less (6.87) and the least (3.35) in both the groups, respectively, in comparison with that of the controls (37.39). Liver functions of all dogs in both the groups administered Alloxan were verified to be in the normal range. The urine was negative in the former group, whereas it was afirmed to be two plus in the latter group. A slight decrease in β cells in the pancreas of the former group was observed, but not only a prominent decrease in these cells in number but also similar changes occurred in β granules of the β cells in the latter group. Consequently, the mild Alloxan diabetic and Alloxan diabetic states were affirmed in both groups, respectively (87). Various evidence observed in both the " impulse responses " of both groups is therefore recognized as accounting for the above histological and functional changes caused by Alloxan administration. It was also suggested, as demonstrated in Fig. 48, that the damped oscillatory second-order component responses in both the " impulse re-

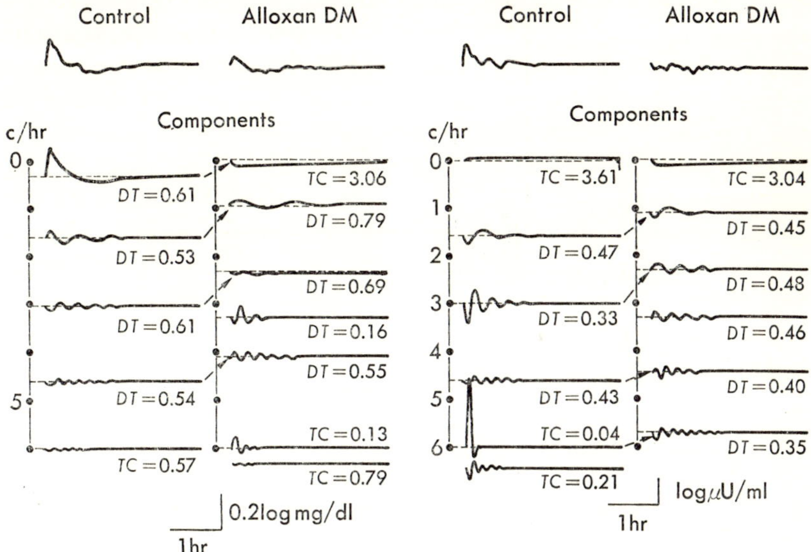

Fig. 48. An example of impulse responses of blood glucose (left) and serum insulin (right) and their component responses (components) in the control (control) and Alloxan diabetic (Alloxan DM) states (case C11). Location of each component response is at its damped frequency (cyc/hr) shown by the ordinate. Abscissa, time in hours; DT and TC, damping time and time constant of the scond- and first-order component responses, respectively.

sponses " seemed to be shifted more or less lower in their damping frequencies in the Alloxan diabetic state.

The initial rapid rise, relatively rapid fall and damped oscillatory sways with initial falling and/or rising phaeses in the insulin " impulse responses " play important roles in the " homeostasis " of blood glucose, since mild Alloxan diabetic and Alloxan diabetic states induced lower and the lowest insulin levels, respectively, as well as the depression in the rapid rise and fall of insulin concentration and subsequent oscillatory sways in the insulin " impulse responses," which accounted for a marked decrease in the number of β cells and β granules in the pancreas of Alloxan diabetic dogs. The dapmed oscillatory component " impulse responses " of the blood glucose are the same.

The ratios (S_I: S_G) of the " control area " of the insulin " impulse responses " (S_I) to that of the blood glucose (S_G) were displayed as related to both the levels of the blood glucose and serum insulin, respectively, as shown in Fig. 49. The ratios were the highest in the control

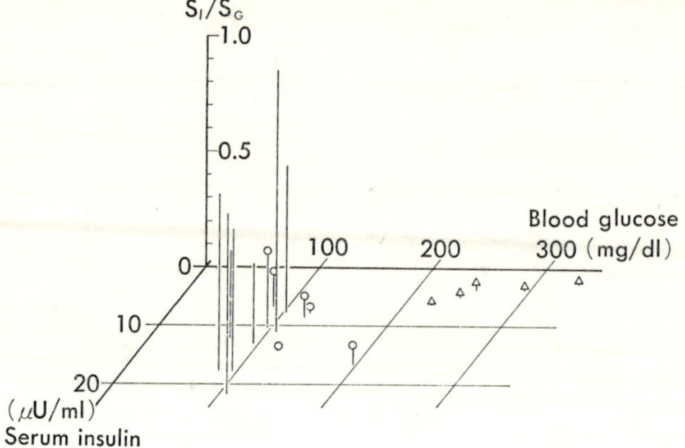

Fig. 49.   Ratio of the control area in the impulse response of the serum insulin (S$_I$) to that of the blood glucose (S$_G$) as related to the blood glucose and serum insulin in dogs. Solid vertical, normal control state; empty c'rcle (◯), mild Alloxan diabetic state; empty triangle (△), Alloxan diabetic state.

group and the lowest in the Alloxan diabetic group. The former clustered around about 100 mg/dl glucose and about 10–20 μU/ml levels, whereas the latter were scattered in an area of higher than 200 mg/dl glucose and lower than 10 μU/ml insulin levels, respectively. The ratios in the mild Alloxan diabetic group were higher than those in the Alloxan diabetic group but lower on average than those in the control group, though some of them were not different from lower ones in the control group. They clustered in an area between those of the other two groups, but near to that of the control group. The ratios in the Alloxan diabetic group were less than 0.1. Such a very low ratio is derived from a far narrower " control area " of the serum insulin level in comparison with that of the blood glucose level, as manifested in Fig. 48, so that the worst and lowest " glucoregulatory activity " will also be suggested in the lowest ratio.

SUMMARY

Humans and animals consist of a variety of bio-systems exhibiting various bio-phenomena over the course of time, from the past to the present and into future, up to just before their death. Each state of a bio-phenomenon at any time is related in stochastic fashion not only to its past history, but those of many other bio- and natural phenomena, enormous in number, in their internal and external environments.

Most states of these bio-phenomena sway more or less around respective averages, which suggest their levels of *homeostasis*, essentially important for maintaining life. In the above past history of sway was hidden an essential characteristic, *i.e.*, *dynamic higher-order actitity*, of the bio-system, whereas the bio- and natural phenomena in the environments act to drive, *i.e.*, *stimulate*, as an ensemble, the bio-system to exhibit the bio-phenomena as its *responses*. From this new point of view, mono- and multivariate dynamic *stimulation-system (activity)-response* relations in stochastic fashion can be seen as an extension leading from of one of the most fundamental *static laws of excitability*, that is the *threshold stimulus-excitability-unit response* relation in physiology.

The dynamic mono- and multivariate *higher-order activities*, each of which consisted of some *first-* and *second-order component activities*, can be described in the frequency and time-patterns as the *power spectral densities* or *frequency responses* and (*unit*) *impulse responses*, respectively. Some of these "dynamic activities" were manifested in the brain system of humans and cats, the human "posture holding system," "the pressure regulatory system" in the human pulmonary circulation and the "glucoregulatory system" of dogs, respectively.

Acknowledgments

The author wishes to express his sincere thanks to his collaborators in Nagasaki University, School of Medicine, and to Prof. Y. Katsuki for giving him the opportunity to write the present review.

REFERENCES

1 C. Bernard, "An Introduction to the Study of Experimental Medicine," trans. by H. C. Green, Dover Publ., New York, p. 229 (1949).
2 K. Sato, *Kybernetik*, **4**, 195 (1968).
3 L. Lapicque, "L'excitabilité en Funtion du Temps," Paris (1926).
4 K. Sato, *Japan. J. Physiol.*, **9**, 327 (1959).
5 K. Sato, *Kybernetik*, **9**, 45 (1971).
6 K. Sato, *Kybernetik*, **6**, 146 (1969).
7 K. Sato, K. Mimura, T. Ozaki, Y. Yamamoto, S. Masuya, and N. Honda, *Japan. J. Physiol.*, **7**, 181 (1957).
8 S. S. Stevens, "The Psychophsics of Sensory Function," ed. by W. A. Rosenblith, John Wiley & Sons, New York, p. 1 (1961).
9 W. B. Cannon, "The Wisdom of the Body," W. W. Norton, New York (1932).
10 K. Sato, K. Ono, and K. Fukuda, *Int. Symp. Med. Inform. System, MEDIS'* 78, 95 (1978).

11 K. Sato, *Japan. J. EEG EMG*, **7**, 169 (1979).

12 K. Sato, *Biokybernetik*, **5**, 40 (1975).

13 K. Sato, " Informationsverbreitung im Zentralnervensystem, Limbisches System-Motivation-Datenverarbeitungs Probleme," ed. by W. Haschke, VEB Gustag Fischer Verlag, Jena, p. 213 (1976).

14 K. Sato, *Bull. Neuroinform. Lab. Nagasaki Univ.*, **2**, 1 (1975).

15 K. Sato, *Japan. J. Neurosci. Res. Assoc.*, **1**, 140 (1975).

16 K. Sato, *Math. Sci.*, **153**, 32 (1975).

17 K. Sato, *Bull. Neuroinform. Lab. Nagasaki Univ.*, **3**, 111 (1976).

18 K. Sato, G. Chiba, K. Fukata, K. Ono, and C. Morisada, *Int. J. Neurosci.*, **7**, 115 (1977).

19 K. Sato, *Bull. Neuroinform. Lab. Nagasaki Univ.*, **5**, 1 (1978); "Higher Order Activities of Biological Systems and Bio-informations," Memorial Assoc. for Retire of Prof. K. Sato, Second Dept. Physiol., Nagasaki Univ. Sch. Med. Nagasaki, p. 249 (1979).

20 H. Akaike, *Ann. Inst. Stat. Math.*, **21**, 243 (1969).

21 H. Akaike and T. Nakagawa, " Statistical Analysis and Control of Dynamic System," Science Co., Tokyo, p. 189 (1972).

22 H. Akaike, *Math. Sci.*, **153**, 5 (1976).

23 G. E. P. Box and G. M. Jenkins, " Time series Analysis Forcasting and Control," Holden-Day, San Francisco, p. 553 (1970).

24 C. E. Shannon and W. Weaver, " The Mathematical Theory of Communication," Univ. of Illinois Press, Urbana, p. 117 (1949).

25 K. Sato, *Bull. Neuroinform. Lab. Nagasaki Univ.*, **2**, 14 (1975).

26 M. G. Saunders, *Electroenceph. Clin. Neurophysiol.*, **15**, 761 (1963).

27 R. Ellul, *Science*, **164**, 328 (1969).

28 K. Sato, K. Mimura, H. Sata, N. Ochi, and T. Ishino, *Japan. J. Physiol.*, **21**, 167 (1970).

29 W. R. Adey, *Int. J. Neurosci.*, **3**, 271 (1972).

30 L. H. Zetterberg, *Math. Biosci.*, **5**, 227 (1969).

31 A. Wennberg and L. H. Zetterberg, *Electroceph. Clin. Neurophysiol.*, **31**, 457 (1971).

32 L. H. Zetterberg and K. Ahlin, *Med. Biol. Eng.*, **13**, 272 (1975).

33 R. H. Jones, D. H. Crowell and L. E. Kapuniani, *Electroenceph. Clin. Neurophysiol.*, **27**, 436 (1969).

34 W. Gersch, *Math. Biosci.*, **7**, 205 (1970).

35 K. Ono, *Bull. Neuroinform. Lab. Nagasaki Univ.*, **2**, 5 (1975).

36 K. Sato, G. Chiba, and K. Fukata, *Bull. Neuroinform. Lab. Nagasaki Univ.*, **2**, 17 (1975).

37 T. Tasaki and N. Yoshitani, *Japan. J. EEG EMG*, **3**, 403 (1975).

38 K. Sato, C. Morisada, K. Fukata, G. Chiba, and K. Ono, *Japan. J. EEG EMG*, **3**, 250 (1975).

39 K. Sato, K. Ono, G. Chiba, and K. Fukata, *Int. J. Neurosci.*, **7**, 201 (1977).

40 K. Sato, M. Oshima, M. Matsuo, T. Umezawa, K. Yajima, M. Sasaki, K. Ono, G. Chiba, K. Kita, and M. Akagi, " High Speed Exact Analysis of EEG Group (III)," Research Foundation on Traffic Medicine, Tokyo, p. 73 (1978).

41 R. B. Blackman and J. W. Tukey, "The Measurement of Power Spectra," Dover Publ., New York, p. 190 (1959).

42 K. Sato, H. Akaike, G. Chiba, K. Fukata, and K. Ono, *Bull. Neuroinform. Lab. Nagasaki Univ.*, **2**, 38 (1975).

43 K. Sato and K. Mimura, *Japan. J. Physiol.*, **6**, 206 (1956).

44 R. W. Southworth, "Mathematical Methods for Digital Computers," ed by A. Ralston and H. S. Wilf, John Wiley & Sons, New York, p. 213 (1966).

45 C. Morisada, K. Sato, R. Suzuki, G. Chiba, K. Fukata, K. Chiwata, and K. Ono, *Bull. Neuroinform. Lab. Nagasaki Univ.*, **2**, 108 (1975).

46 C. Morisada, K. Sato, R. Suzuki, G. Chiba, K. Chiwata, K. Fukata, and K. Ono, *Bull. Neuroinform. Lab. Nagasaki Univ.*, **2**, 114 (1975).

47 C. Morisada, K. Sato, R. Suzuki, G. Chiba, K. Fukata, K. Chiwata, and K. Ono, *Bull. Neuroinform. Lab. Nagasaki Univ.*, **2**, 123 (1975).

48 C. Morisada, *Bull. Neuroinform. Lab. Nagasaki Univ.*, **2**, 130 (1975).

49 S. Mori, "Neurophysiology Studied in Man," ed. by G. G. Somjen, Excerpta Medica, Amsterdam, p. 401 (1972).

50 K. Sato, K. Ono, G. Chiba, and K. Fukata, *Int. J. Neurosci.*, **7**, 239 (1977).

51 W. G. Walter, "The Living Brain," Gerald Duckwerth, London, p. 216 (1953).

52 R. Cohn, *J. Neurophysiol.*, **11**, 31 (1948).

53 K. Sato and K. Nakane, *Folia Psych. Neurol. Japonica*, **3**, 44 (1948).

54 K. Imahori and K. Suhara, *Folia Psych. Neurol. Japonica*, **3**, 137 (1949).

55 K. Ono, K. Sato, G. Chiba, and K. Fukata, *MBE, Electrocomm. Assoc.*, **76** (19), 41 (1976).

56 K. Ono, *Bull. Neuroinform. Lab. Nagasaki Univ.*, **3**, 19 (1976).

57 G. Chiba, *Clin. EEG*, **19**, 258 (1977).

58 Res. Assoc. Statist. Sci., "Tables of Statistical Values," Seisan Gijutsu Center, Tokyo, p. 214 (1976).

59 B. W. Bolch and C. J. Huang, "Multivariate Statistical Methods for Business and Economics," Prentice-Hall, Englewood Cliffs, New Jersey (1974), transl. by K. Nakamura, Japanese Ed., Morita Shuppan, Tokyo, p. 286 (1976).

60 K. Sato, *Japan. J. EEG EMG*, **7**, 169 (1979).

61 K. Nakatsuka, *Bull. Neuroinform. Nagasaki Univ.*, **2**, 133 (1975).

62 K. Sato, S. Masuya, K. Ono, G. Chiba, and K. Fukata, *Bull. Neuroinform. Lab. Nagasaki Univ.*, **2**, 54 (1975).

63 K. Sato, *Adv. Neurol. Sci.*, **23**, 19, 1061 (1975).

64 K. Sato, K. Ono, and K. Fukata, *Neurosci. Lett.*, Suppl. **2**, S 19 (1979).

65 K. Ono, K. Sato, and K. Fukata, *Proc. Japan Acad.*, **54**, Ser. B, 375 (1978).

66 M. Berkley, E. Wolf, and M. Glickstein, *Exp. Neurol.*, **19**, 188 (1967).

67 R. W. Guillery, *J. Comp. Neurol.*, **130**, 197 (1967).

68 K. Niimi, S. Kawamura, and S. Ishimaru, *J. Comp. Neurol.*, **143**, 279 (1977).

69 C. Ajmon-Marsan and A. Morillo, *Electroenceph. Clin. Neurophysiol.*, **13**, 553 (1961).

70 L. Widen and C. Ajmon-Marsan, *Exp. Neurol.*, **2**, 468 (1960).

71 K. Iwama and T. Kasamatsu, *Japan. J. Physiol.*, **15**, 310 (1965).

72 H. Suzuki and E. Kato, *Tohoku J. Exp. Med.*, **86**, 277 (1965).

73 R. E. Kalil and R. Chase, *J. Neurophysiol.*, **33**, 459 (1970).

74 D. Richard, Y. Gioanni, A. Kotsikis, and P. Buser, *Exp. Brain Res*, **22**, 235 (1975).
75 A. Angel, F. Magni, and P. Strata, *Arch. Ital. Biol.*, **105**, 104 (1967).
76 E. Hull, *Vision Res.*, **8**, 1285 (1968).
77 E. F. Vastola, *Vision Res.*, **7**, 599 (1967).
78 R. W. Doty, *J. Neurophysiol.*, **21**, 437 (1958).
79 E. F. Vastola, *J. Neurophysiol.*, **24**, 469 (1961).
80 A. Iwanaga, G. Chiba, T. Fujiwara, and K. Sato, *Bull. Neuroinform. Lab. Nagasaki Univ.*, **3**, 53 (1976).
81 G. Chiba, A. Iwanaga, K. Sato, and T. Fujiwara, *Bull. Neuroinform. Lab. Nagasaki Univ.*, **3**, 65 (1976).
82 A. Iwanaga, *Nagasaki Med., J.* **53**, 149 (1973).
83 D. H. Bergel and W. R. Milnor, *Circ. Res.*, **16**, 401 (1965).
84 W. R. Milnor, C. R. Conti, K. B. Lewis, and M. F. O'Rourke, *Circ. Res.*, **25**, 637 (1969).
85 K. Takebe, *Bull. Neuroinform. Lab. Nagasaki Univ.*, **3**, 37 (1976).
86 G. Chiba, T. Takebe, K. Sato, R. Tsuchiya, and S. Eto, *Bull. Neuroinform. Lab. Nagasaki Univ.*, **3**, 29 (1976).
87 K. Sato, G. Chiba, R. Tsuchiya, K. Takebe, and S. Eto, *Proc. Japan Acad.*, **54**, Ser. B, 69 (1978).
88 M. Oohtkins, D. J. Marsh, S. W. Smith, R. N. Bergman, and F. E. Yates, *Am. J. Physiol.*, **226**, 910 (1974).
89 C. J. Goodner, B. C. Walike, D. J. Koeker, J. W. Ensinck, A. C. Brown, E. W. Chideckel, J. Palmer, and L. Kalnasy, *Science*, **195**, 177 (1977).
90 A. Niijima, *Brain Res.*, **87**, 195 (1975).
91 A. Niijima, *Pharm. Biochem. Behav.*, **3**, Suppl. 1, 139 (1975).
92 B. K. Anand, G. S. C. Chhina, K. N. Sharm, S. Dua, and B. Singh, *Am. J. Physiol.*, **207**, 1146 (1964).
93 Y. Oomura, K. Kimura, H. Ooyama, T. Maeo, M. Iki, and N. Kuniyoshi, *Science*, **143**, 484 (1964).
94 Y. Oomura, H. Ooyama, T. Yamamoto, T. Ono, and N. Kobayashi, *Ann. N.Y. Acad. Sci.*, **157**, 642 (1969).
95 Y. Oomura, M. Sugimori, T. Nakamura, and Y. Yamada, "The Limbic System in Physiological Regulations and Behavior—J. A. F. Stevenson Memorial Vol.," ed. by G. J. Mogenson and F. R. Calaresu, Toronto Univ. Press, Toronto, p. 377 (1975).
96 Y. Oomura, *Adv. Biophys.*, **5**, 65 (1973).
97 A. Niijima, *J. Physiol.*, **251**, 231 (1975).

Received for publication December 27, 1979.

Adv. Biophys., Vol. 14, pp. 139–204 (1981)

TWO-DIMENSIONAL NMR SPECTROSCOPY: AN APPLICATION TO THE STUDY OF FLEXIBILITY OF PROTEIN MOLECULES

KUNIAKI NAGAYAMA

Department of Physics, Faculty of Science, University of Tokyo, Tokyo, Japan

Nuclear magnetic resonance (NMR) spectroscopy has been a unique tool in the field of organic chemistry owing to the advantages it offers in the study of the structures of organic molecules in solution. Now the target of NMR study has largely been shifted to biological materials. When we recall the history of the development of NMR applications to biological systems, we observe two strong motivations for technical advances. One motivation was to get good resolution in the NMR spectra. Another was to obtain high sensitivity in NMR measurements. The introduction of a super conducting magnet (SCM) with a very high field was an epoch-making break-through that solved these two problems simultaneously. Besides efforts to improve the performance of NMR devices themselves, an enormous amount of effort has also been involved developing new measurement principles. The pulse Fourier transformation (FT) technique is a good example of such development of measurement principles. This technique has actually borne fruit in biological applications of NMR together with the use of the SCM system. Recently pulse FT NMR has taken one more step toward higher dimensions in the form of two-dimensional (2D) NMR spectroscopy. This article deals with aspects of 2D NMR spectroscopy from its principles to its biological applications.

This paper is divided into two main parts. From Sec. I to Sec. III, the principles, the experimental realization, and the practical aspects of data handling of 2D NMR are explained. From Sec. IV to Sec. VI, an example of this method as applied to a biological system is described, so that one can see how the characteristic advantages of this new technique are utilized in the study of conformations of proteins in solution. In the first part many pages are devoted to elaborate the physical background of this technique. Readers who are more interested in the application range and the practicle use of this method may begin with Sec. IV.

I. GENERAL ASPECTS OF NMR

1. *Measurements of NMR*

If a nucleus with a magnetic moment μ is placed in an external magnetic field H_0, a Zeeman interaction is brought about. Its energy is given by μH_0. The interaction energy is quantum mechanically expressed as $\gamma H_0 m$ using a magnetic quantum number m $(m = -I \cdots I)$ for a nucleus of spin I. NMR spectroscopy primarily measures the energy differences of split energy levels by irradiating a radio frequency (r.f.) field with resonance frequencies and observing its response. At the birth of the development of NMR spectroscopy, two ways were proposed for observing or detecting the resonance phenomenon. A method proposed by Purcell *et al.* (*1*) utilized an electromagnetic device Q meter called a " Pound box " with an understanding of the NMR phenomenon analogous to the absorption of optical light by matter. Another method proposed by Bloch *et al.* (*2*) detected, on the other hand, a current induced in a resonant coil by the resulting response to the input r.f. field. In the latter method, the resonance phenomenon was explained as the movement of a macroscopic magnetization in an external magentic field, that is the rotation of the magnetization vector about the fied H_0 relaxing to the thermal equilibrium of the initial state prior to r.f. excitation. This so-called nulcear induction method has now been modified and extended to the pulse FT method (*3*), where the excitation of the magnetic transitions and the detection of the resonance are completely separated in the experimental time course.

The exepriments described so far make a big contrast to the trigger type experiments routinely carried out for the detection of NMR in the field of nuclear physics (*4, 5*), where the behavior of one spin is the point of interest. On the other hand the low sensitivity of NMR detection by electromagnetic means automatically makes it necessary to use bulk sample materials, since it is only possible to observe the behavior of the statistically averaged macroscopic magnetization (classical spin) in this type of NMR experiments. At thermal equilibrium, the magnetization in the direction of the H_0 field (z-direction) is simply given by assuming a Boltzman distribution of the populations of energy levels and taking the high temperature approximation.

$$M_z = \frac{N\gamma^2\hbar^2 I(I+1)}{3kT} H_0 = \chi_0 H_0 \tag{1}$$

This very small magnetization and low magnetic susceptibility (10^{-3} smaller than the normal paramagnetic susceptibilities) as detected by static measurement, however, are not the subjects to be observed in NMR measurements. In NMR we observe the magnetic components M_x or M_y perpendicular to the z-direction under inequilibrium conditions of spin systems. This transverse magnetization rotates in the external magnetic field with a Larmor precession frequency γH_0, which is equal to the resonance absorption frequency, and creates an oscillating magnetic field to be picked up by a probe coil as an electric signal $i(t)$.

$$i(t) \propto M_x(t) \quad \text{or} \quad M_y(t)$$
$$M_x(t) = M_x(0)f(t) \cos \omega_0 t, \qquad M_y(t) = M_y(0)f(t) \cos \omega_0 t$$
$$\omega_0 = \gamma H_0 \tag{2}$$

Here $f(t)$ is a time function which represents magnetic relaxation of the spin system. Depending on the orientation of the pick up coil, we may detect a magnetic component of $M_x(t)$ or $M_y(t)$. With the above classical picture for nuclear spins, the NMR measurement process can be schematically drawn as shown in Fig. 1.

2. Classical and Quantum Mechanical View for the NMR Experiment

As mentioned before, what is observed in pulse FT NMR is the movement of a spin system returning to the equilibrium state from an excited initial inequilibrium state. The relaxation process leads to a damping of the oscillating signal that is picked up by the electromagnetic devices (free induction decay). It is important to notice that in various NMR techniques excitation and detection can be separated in time. In Fig. 1c this situation is illustrated by a $\theta°$ pulse representing the excitation, and by a free induction decay immediately followed after the excitation as a damping oscillation. Here we concentrate our arguments on the second

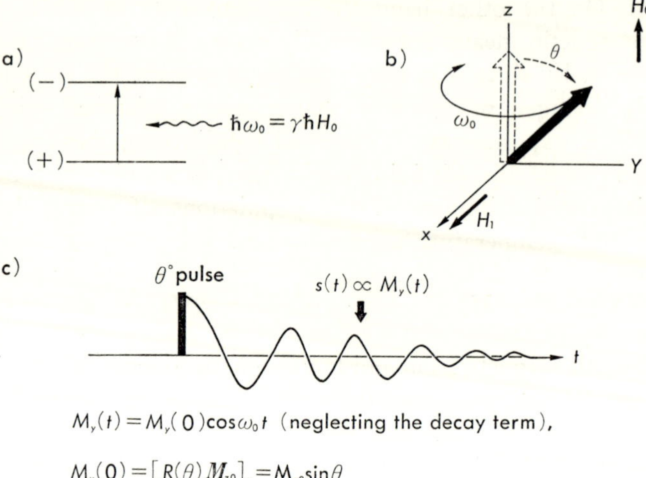

$$M_y(t) = M_y(0)\cos\omega_0 t \quad (\text{neglecting the decay term}),$$

$$M_y(0) = [R(\theta) M_{z0}]_y = M_{z0}\sin\theta$$

Fig. 1. Schematics of the principle of NMR measurement for a classical spin. a) Energy level scheme of nuclear magnetic resonance ; b) movement of magnetization in an external magnetic field; c) detected induced current (free induction decay) after irradiation with a radio frequency pulse.

part of the NMR experiment and investigate the physical picture of the motion of the nuclear magnetization from a quantum mechanical approach.

Motions of classical spins (magnetization of ensembles of nuclei) in external magnetic fields are usually described by the Bloch equation in an analogy to the motion of macroscopic magnetic moments placed in magnetic fileds (6). While this classical description is simple minded and is legitimated by statistical approaches (7), it has a limitation in that it only covers the description of an ensemble of spins which are interacting with surroundings at random. Spin systems where mutual nuclear interactions lead to well defined energy states cannot be covered by the Bloch equation. Especially, nonlinear properties of the spin system like the connectivity of various transitions in interacted spin systems, which is the central subject to be examined in 2D spectroscopy, are not fully grasped by a phenomenological view of this sort.

In the previous section, the macroscopic magnetization in the equilibrium state is shown to be expressed by the Boltzman population distribution of the spin at various energy levels. For an ensemble of N identical nuclei with nuclear spin $I=1/2$, this is explicitly written as

$$M_{z0} = \sum_{i=1}^{N} \langle \mu_{zi} \rangle = \gamma \hbar N \langle \bar{I}_z \rangle = \frac{\gamma \hbar N}{2}(p_+ - p_-), \qquad (3)$$

where p_+ and p_- are the relative populations of eigen states $|+1/2\rangle$ and $|-1/2\rangle$ that correspond to the upper energy level and the lower energy level in Fig. 1, respectively. These two relative populations are related to the probability amplitudes a_+ and a_- of two eigen states

$$p_+ = \overline{|a_+|^2}, \qquad p_- = \overline{|a_-|^2}. \qquad (4)$$

Here statistical averages are taken over all the spins in the system by assuming the Boltzman distribution. With this quantum mechanical description, the transverse magnetization components are expressed as follows.

$$M_x = \gamma \hbar N \langle I_x \rangle = \frac{1}{2} \gamma \hbar N (\overline{a_+{}^* a_-} + \overline{a_+ a_-{}^*}),$$

$$M_y = \gamma \hbar N \langle I_y \rangle = \frac{1}{2} \gamma \hbar N (\overline{a_+{}^* a_-} - \overline{a_+ a_-{}^*}). \qquad (5)$$

At equilibrium states, of course, the statistical averages $\overline{a_+{}^* a_-}$ and $\overline{a_+ a_-{}^*}$ should become zero, otherwise the system would experience a spontaneous magnetization perpendicular to the static magnetic filed H_0. This implies that even if the cross-term of a_+ and a_- has finite values for each of the spin constituents in the system, the summation of the contributions from all the spins over the system vanishes. This averaging-out phenomenon is usually called perfect dephasing of transverse magnetizations of many spins. The cross-term $a_+ a_-{}^*$, called the coherence of states a_+ and a_-, does not contribute to the macroscopically observable components, here M_y or M_x. In the pulse NMR experiment, however, the coherence $\overline{a_+ a_-{}^*}$ is not perfectly dephased after the excitation but rather partially aligned in a fixed direction, e.g., in the y-direction. The value of $\overline{a_+ a_-{}^*}$ is proportional to the population difference $(p_+ - p_-)$ at the equilibrium state. Then the coherence evolves in the experimental time course as governed by the spin Hamiltonian of the system and finally tends to zero by the relaxation mechanisms. Therefore we need to know the behavior of the coherences $\overline{a_+{}^* a_-}$ or $\overline{a_+ a_-{}^*}$ through the time course but not the populations $\overline{|a_+|^2}$ or $\overline{|a_-|^2}$ to predict the movement of the observed transverse magnetizations M_x or M_y. This is an essential clue to understanding the principle of 2D NMR spectroscopy because 2D NMR is thought to give a way to visualize the evolution of such coherences in interacting spin systems by mapping the frequencies present in evolution on a two-dimensional frequency plane.

In order to trace the movement of nuclear magnetizations, then, coherences (and often also populations) of multi-states should be followed in the experimental time course. The density matrix (density operator) treatment has been successfully introduced for that purpose (8).

$$\sigma = |m\rangle\langle m'|, \qquad \sigma_{mm'} = a_m a_{m'}{}^*, \tag{6}$$

where m and m' represent quantum states of the system, and a_m and $a_{m'}$ indicate the probability amplitudes of states m and m'. The density operator σ includes all the necessary information to describe the various properties of the system. To know the time course of the system, it is enough to know the evolution of the density operator.

$$\frac{d\sigma}{dt} = -i[\mathcal{H}, \sigma] \tag{7}$$

Here $\mathcal{H}$ is the spin Hamiltonian of the system measured in the frequency unit. The macroscopic transverse magnetization is given by

$$M_y = \mathrm{Tr}\,[F_y \sigma], \tag{8}$$

where F_y is the total spin for the system in the y-direction

$$F_y = \sum_{i=1}^{N} I_{yi}, \qquad I_{yi}: y \text{ component of the } i \text{ spin.}$$

The formal solution of Eq. (7) is easily given for the case of a time independent Hamiltonian

$$\begin{aligned}
\sigma(t) &= \exp\,(-i\mathcal{H}t)\sigma(0)\exp\,(i\mathcal{H}t) \\
&= \exp\,(-i\hat{\hat{\mathcal{H}}}t)\sigma(0).
\end{aligned} \tag{9}$$

Here a Liouville operator $\exp\,(i\hat{\hat{\mathcal{H}}}t)$ is introduced to make the operatorization process transparent and to emphasize the origin of the driving force on the system. $\hat{\hat{\mathcal{H}}}$ is an operator imposed on the density operator $\sigma(0)$. It is called a superoperator (9) and has a matrix representation of rank 4, $(\hat{\hat{\mathcal{H}}})_{ij,lm}$. In the representation where the Hamiltonian $\hat{\hat{\mathcal{H}}}$ becomes a diagonal matrix, $\hat{\hat{\mathcal{H}}}$ has nonzero values only for components (transition frequencies) $(\hat{\hat{\mathcal{H}}})_{ii,ll}=(\omega_{ii}-\omega_{ll})$, where ω_{ii} and ω_{ll} correspond to the frequencies (energies) at the eigen states i and l of Hamiltonian $\mathcal{H}$. A Larmor precession of the nuclear magnetization and its frequency corresponds to one of these transition frequencies which allow it to be detected by NMR through nuclear induction.

Due to the nature of the observable F_y, one type of transitions and therefore one type of transition frequencies can be detected among various possible transitions which are all implied in the density operator $\sigma(t)$ during its time evolution. That is, the frequencies corresponding to the single quantum transitions in which the magnetic quantum number is changed by 1 or $|m-m'|=1$. With use of Eqs. (8) and (9) this is easily verified and is schematically shown in Fig. 2.

We have not yet given any explicit form to the density operator $\sigma(0)$ at $t=0$. The content of $\sigma(0)$ depends on how the initial inequilibrium state is prepared, in other words, how the system of ensembles of spins is

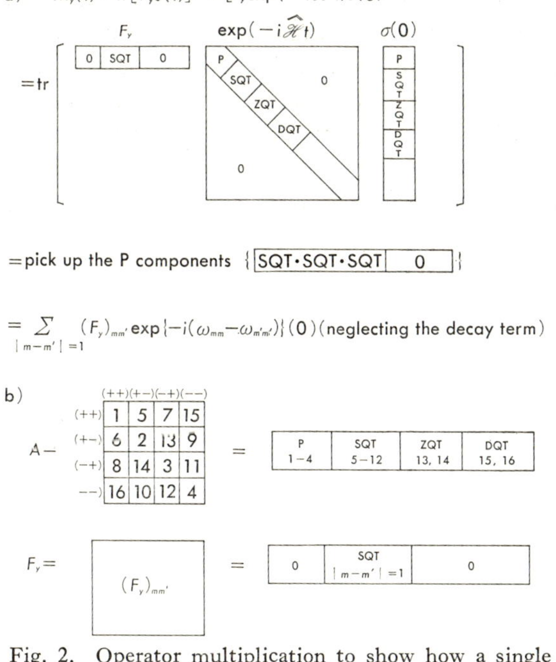

Fig. 2. Operator multiplication to show how a single quantum transition (SQT) is picked up by an observable F_y from the evolution of the density matrix $\sigma(t)$. a) Calculation to obtain the motion of magnetization component $M_y(t)$. Here m and m' represent the magnetic quantum numbers; b) reshuffling of the components of the represented operator into the vectorial array. An operator A is represented by the eigen functions for the vector sum of two half spins. P, population in each state; ZQT, zero quantum transition; DQT, double quantum transition.

excited by external r.f. fields. With the same formula as given above, the initial density operator $\sigma(0)$ is explicitly written for the case of preparing the transverse magnetization by a strong and short r.f. field $H_1 \cos \omega_0 t$ shown in Fig. 1c.

$$\sigma(0) = \exp\left(-i\theta F_x\right)\sigma_0 \exp\left(i\theta F_x\right) = \hat{\hat{R}}\bar{F}_z$$
$$= \cos \bar{F}_z + \sin \bar{F}_y$$

$$R = \exp\left(-i\theta F_x\right), \qquad \sigma_0 = \bar{F}_z = \langle \sum_{i=1}^{N} I_{zi} \rangle,$$

$$\bar{F}_y = \exp\left(-i\frac{\pi}{2}F_x\right)\bar{F}_z \exp\left(i\frac{\pi}{2}F_x\right), \qquad \theta = \gamma H_1 \qquad (10)$$

R represents the same rotation operator that appeared in Fig. 1c and σ_0 is the density operator of the system at the equilibrium condition. Expression (10) asserts that the spin system has nonzero coherences corresponding to the single guantum transitions, since the operator F_y has off-diagonal elements connecting the states with magnetic quantum numbers differing by 1. This forced evolution of the density operator corresponds to the classical picture where a nuclear magnetization M_z is rotated around the direction of the field H_1 (here x-direction) by an angle of θ and creates a transverse magnetization M_y.

Together with expressions (9) and (10), we know that the statistical character of the system of many spins is all involved in the longitudinal magnetization or the z component of the total spin $(\bar{F}_z)$ at the equilibrium state, if we ignore the relaxations of nuclear spins, whose quantum mechanical feature can be also included in the density operator (10) but which are not the matter to be examined in this article. Therefore the calculation of the operator formula for the magnetization $M_y(t)$ (Eqs. (8) and (9)) is able to be quite mechanically or quantum mechanically performed as shown in Fig. 2. In the next section we will see how this approach can be extended to treat the 2D NMR experiments.

3. Toward 2D Spectroscopy

A spin system must tend to the equilibrium state for a long time if the system is kept free from any external stimulations. In nuclear spin systems, two types of relaxation processes can be distinguished (6, 11): the transverse relaxation (spin-spin relaxation) which is responsible for the decay rate of the detected NMR signals and corresponds to the process of the macroscopic dephasing of the coherencies $a_i{}^*a_j$ ($\overline{a_i{}^*a_j} \Rightarrow 0$), and the longitudinal relaxation (spin-lattice relaxation) which corresponds to the returning process of the polulations of energy levels to the equilibrium populations $\overline{(a_i{}^*a_i}\Rightarrow(a_i{}^*a_i)_0)$. The longitudinal relaxation is always slower than the transverse relaxation in nuclear systems but both are relatively slow compared with other relaxation phenomena seen in electronic spin systems or optical absorption systems. Especially in solution systems, it is rare to observe relaxation rates faster than $10^3\,\mathrm{sec}^{-1}$ for nuclei with half spins. Therefore further stimulation to the spin system can be experimentally attained with ease during the time evolution of the system or the density operator $\sigma(t)$. In conventional NMR, we know such techniques as the double resonance techniques or the multiple-pulse excitations. These techniques have been quite efficiently utilized to elucidate the nonlinear properties of spin systems. For example, the connectivities of the transitions in coupled spin systems have been made visible by the spin-tickling or spin-decoupling technique (12), the relaxation times of transverse magnetizations or longitudinal magnetizations were measured with the spin-echo technique (13) or the inversion recovery technique (14), respectively, and the cross relaxation process of heteronuclear systems have been observed by the heteronuclear double resonance technique (15). All these techniques are now unified and extended to form a new comprehensive method in NMR, that is, 2D NMR spectroscopy (16), where the NMR spectra are naturally developed on the 2D frequency base. Once we obtain a frame to present the NMR spectra in this new base, we find ourselves standing before a promising opportunity. It has not only revamped the conventional methods but also made it possible to design new techniques which are not easily realized with the conventional NMR methods.

The general scheme of 2D spectroscopy is demonstrated in Fig. 3a (17) and a realization of this scheme among many possible 2D experiments is shown in Fig. 3b (18). First the spin system is prepared to an initial inequilibrium state. Normally this is done by a 90° r.f. pulse as shown in Fig. 3b where single quantum coherencies are elaborated. A new feature of this NMR experiment is seen in the newly introduced time periods, the evolution period and the mixing period. The former defines a time variable t_1 and the latter mixes or connects the evolutions of density operators during t_1 and the successive time period t_2. The data acquisition of NMR signals is only performed during the detection period t_2. By varying the delay time t_1 systematically while keeping other experimental conditions unaltered, we obtain a matrix type of data set $s(t_1,$

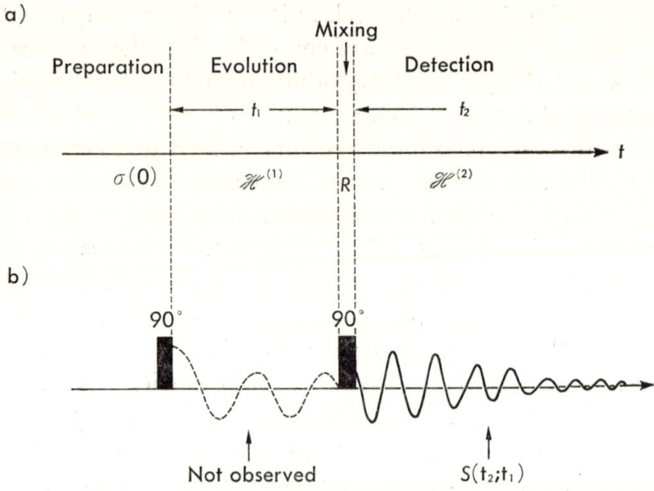

Fig. 3. Experimental scheme of 2D NMR spectroscopy. a) The general scheme of 2D spectroscopy. Here t_1 and t_2 become two variables of 2D response signals. $\mathcal{H}^{(1)}$ and $\mathcal{H}^{(2)}$ are the Hamiltonians during t_1 and t_2 periods, $\sigma(0)$ is the initial density matrix after the preparation and R represents a mixing operator. b) One realization of 2D NMR spectroscopy which elucidates the spin connectivity of coupled nuclei called "2D correlated" NMR spectroscopy (*18*). Here the first 90° pulse is used to prepare the initial magnetization (or initial density matrix) and the second pulse is applied to mix two transitions (precession frequencies) evolved during two successive time periods, the evolution period and the detection period.

t_2). After the double-Fourier transformation for the data set in the two time directions, finally the 2D spectra $S(\omega_1, \omega_2)$ result with two-dimensional spreading.

II. REALIZATION OF 2D SPECTROSCOPY

1. *Possible 2D NMR Experiments*
As shown in Fig. 3a the general scheme in 2D NMR is very flexible. Depending on one's interest of what subjects are to be examined and on what NMR parameters are to be measured, there are various possibilities for fitting the actual contents into the frame. Here we investigate possible 2D experiments and classify the numerous 2D methods proposed so far according to the implications of the general scheme.

To make the underlying principle of 2D spectroscopy clear, we will use the same argument developed in Sec. I and give the explicit formula for the density operator $\sigma(t_1, t_2)$ which determines the expression of the observable F_y (or M_y) in 2D experiments in general. As simple straightforward extension from the density operator for the case of 1D NMR experiments (Eq. (9)), the expression of the density operator for the 2D NMR experiments is given:

$$\sigma(t_1, t_2) = \exp{(-i\hat{\hat{\mathcal{H}}}^{(2)} t_2)}\hat{\hat{R}} \exp{(-i\hat{\hat{\mathcal{H}}}^{(1)} t_1)}\sigma(0) \tag{11}$$

The meaning of the above expression is immediately clear from the scheme in Fig. 3a: that the prepared initial state given by $\sigma(0)$ first evolves under the influence of the spin Hamiltonian $\mathcal{H}^{(1)}$, then suddenly the spin states are rotated (or coherencies are mixed up) by very strong and short external perturbations, for example by a 90° r.f. pulse, and finally the density operator of the system evolves obeying the Hamiltonian $\mathcal{H}^{(2)}$ during the t_2 period. Expression (11) is in principle implied also in expression (9) since $\sigma(0)$ is not specified there. Therefore if we take the term $\hat{\hat{R}} \exp{(-i\hat{\hat{\mathcal{H}}}^{(1)}t_1)}$ $\sigma(0)$ as the prepared initial state in the 1D NMR experiment, expression (9) is extended to the expression (11). With use of the same method of calculation performed in the case of the 1D NMR (Fig. 2), we obtain the expression for the magnetization component $M_y(t_1, t_2)$.

$$
\begin{aligned}
M_y(t_1, t_2) &= \mathrm{Tr}\,(F_y\sigma(t_1, t_2)) \\
&= \sum_{ij,lm} F_{yij}\exp{(\;i(\omega_{ii}^{(2)} - \omega_{jj}^{(2)})t_2)}R_{ij,lm} \\
&\qquad \exp{(-i(\omega_{ll}^{(1)} - \omega_{mm}^{(2)})t_1)}\sigma_{lm}(0) \\
&= \sum_{ij,lm} A_{ij,lm}\cos{(\omega_{ij}^{(2)}t_2 + \omega_{lm}^{(1)}t_1)}, \\
A_{ij,lm} &= F_{yij}R_{ij,lm}\sigma_{lm}(0), \qquad \omega_{ij}^{(2)} = \omega_{ii}^{(2)} - \omega_{jj}^{(2)}, \\
\omega_{lm} &= \omega_{ll}^{(1)} - \omega_{mm}^{(1)}
\end{aligned}
\tag{12}
$$

where $\omega_{ij}^{(2)}$ and $\omega_{lm}^{(1)}$ are the transition frequensies (precession frequencies) between energy levels i and j, and l and m determined by Hamiltonians $\mathcal{H}^{(2)}$ and $\mathcal{H}^{(1)}$, respectively, $R_{ij,lm}$ is the component of the mixing operator $\hat{\hat{R}}$ and connects two transitions during periods t_1 and t_2, F_{yij} is the component of F_y which has nonzero value only for the transition of one spin flip (SQT) and $\sigma_{lm}(0)$ is the component of the density operator of the system at the initial state. In order to see the remarkable characteristics appearing in the 2D NMR response $M_y(t_1, t_2)$, the irrelevant terms which are essentially equal to the 1D NMR response and the relaxation terms are neglected in expression (12).

From expression (12) the resulting 2D spectrum is known to have a peak at the position $(\omega_{lm}^{(1)}, \omega_{ij}^{(2)})$ after the double-Fourier transformation of the NMR signal $s(t_1, t_2) \propto M_y(t_1, t_2)$, when the amplitude factor $A_{ij,lm}$ has a nonzero value. By comparing the 1D NMR response $M_y(t)$ in Fig. 2 and the 2D NMR response $M_y(t_1, t_2)$ (Eq. (12)), we can observe two characteristics resulting in 2D NMR signals. First, while the transition frequencies $\omega_{ij}^{(2)}$ during the t_2 period is only determined as those of single quantum transitions due to the nature of the observable F_y, no such restriction is demanded for the setting of the transition frequencies $\omega_{lm}^{(1)}$, because what type of initial states are to be prepared and which kind of external perturbations are to be applied are the matter of choice for our design of 2D experiments. Second, the way to connect the two transition frequencies $\omega_{ij}^{(2)}$ and $\omega_{lm}^{(1)}$ has a lot of freedom depending on the choice of realizing the mixing period. It means that the experimental design critically depends on the type of the mixing pulse(s) given in the mixing period by which the connection of various transition frequencies evolved during the two time periods t_1 and t_2 is realized to give zero or finite intensities for the amplitude factor $A_{ij,lm}$. These two properties make 2D spectroscopy very flexible amd moreover enable us to open a new direction in putting useful information into NMR spectra.

To classify the possible 2D NMR methods proposed up to now, then, we base it on the following criteria: 1) how the initial state at $t_1 = 0$ is prepared $(\sigma_{lm}(0))$, 2) which kind of Hamiltonians $\mathcal{H}^{(1)}$ and $\mathcal{H}^{(2)}$ are assumed during the two time periods t_1 and t_2 which determine the precession frequencies $\omega_{lm}^{(1)}$ and $\omega_{ij}^{(2)}$, and 3) how the two transition frequencies are connected in the mixing period by the mixing pulse(s) $(R_{ij,lm})$. Roughly speaking we have four classes of 2D NMR families. Here works about 2D NMR spectroscopy, say, experimental methods, theoretical calculations and applications, are listed according to this classification.

I.   Resolved NMR $(\mathcal{H}^{(1)} \neq \mathcal{H}^{(2)}, R_{ij,lm} = \delta_{(ij)(lm)}, \sigma_{lm}(0) \neq 0 \text{ (for SQT)})$

Homonuclear J-resolved NMR (19–29); heteronuclear J- or chemical shift resolved NMR (30–43); dipolar resolved NMR in solids (44–49).

II.   Correlated NMR $(\mathcal{H}^{(1)} = \mathcal{H}^{(2)}, R_{ij,lm} \neq \delta_{(ij)(lm)}, \sigma_{lm}(0) \neq 0$ (for SQT))

Autocorrelation through J-coupling (18, 50); autocorrelation through spin-exchange (51, 52).

III.   Indirect detection $(\sigma_{lm}(0) \neq 0$ (for MQT or coherencies from other nuclei))

MQT (multiple quantum transition) detection (*18, 53–56*); indirect detection of rare nuclei (*57, 58*).

IV. Combinations of I, II, and III, and others

Heteronuclear cross-correlation (*59–65*); phase separation (*66–69*); double resonance in 2D NMR (*70–73*); *J*-scaling (*25, 74–77*); MQT detection (1D presentation) (*78–81*); zeugmatography (*82*); sensitivity calculation (*83*); miscellaneous (*84, 85*).

From the point of biological application, 2D spectroscopy has been found to hold the following virtues. It simplifies the complicated spectra of biological substances like amino acids and proteins (*20, 21, 25, 70*), saccharides (*26, 27, 29*), lipids (*39*), and organic phosphates (*40, 42*) and makes the spectral analysis of the complicated spectra an easy task without losing measurement accuracy (*50*). In the next subsection, we will examine the elaborated characteristics of 2D NMR spectroscopy in detail by taking typical 2D NMR methods, namely, homonuclear *J*-resolved and homonuclear correlated NMR spectroscopies.

2. Homonuclear 2D Correlated NMR Spectroscopy
We will explain here a simple but very fruitful 2D NMR technique which was originally suggested by Jeener and later realized by the Ernst group (*18*). This 2D NMR elucidates the spin coupling of interacted spins with a scalar interaction of the type $JI_1 \cdot I_2$ by visualizing the connectivity of different transitions on the 2D spectral domain. Two different versions (*18, 50*) have been proposed for the realization of 2D correlated NMR as shown in Fig. 4. The resulting two 2D spectra (c and c′ in Fig. 4) are mutually related by a conformal mapping and therefore they include completely the same information (*86*). In relation to the 2D *J*-resolved NMR which will be discussed in the next subsection, the version in Fig. 4a′ called spin-echo correlated spectroscopy (SECSY) is easy to understand. But here we investigate the characteristics manifested in the original version of this method.

As shown in Fig. 4a, this experiment relies on the performance of the pulse sequence of the type $90° - kt_1 - 90° -$ data acquisition, here $k =$ 0, 1, 2⋯2^n. Experiments are repeated by varying the parameter k one by one. Acquired signals (free induction decays) then make a set of data with a function form $s(t_1, t_2)$ where t_1 and t_2 are independent time variables corresponding to the sampling in two time periods and normally taken in a digital manner for the convenience of the fast Fourier transformation algorithm ($s(t_1, t_2) = s(nt_1, mt_2)$, n, m: integers.). With the double-

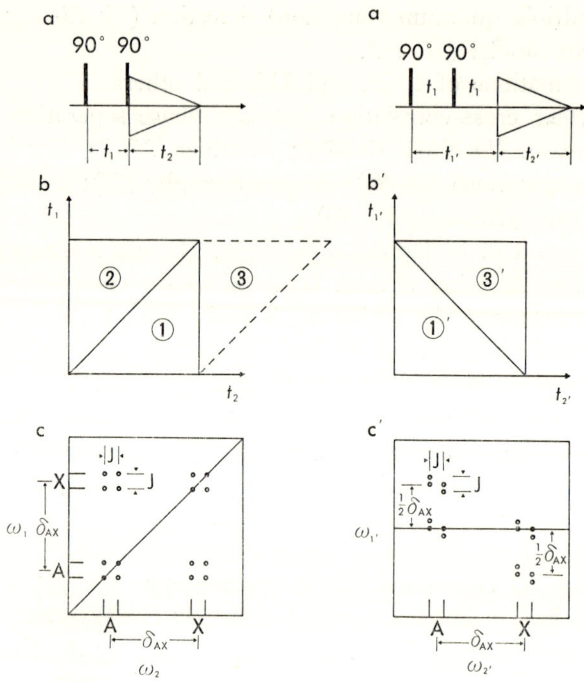

Fig. 4.   Comparison of 2D correlated (a, b, c) and 2D spin-echo correlated spectroscopy (SECSY) (a', b', c'). The experimental schemes are indicated by diagrams a and a'. The relation between the recorded data sets $s(t_1, t_2)$ and $s'(t_1', t_2')$ are visualized in diagrams b and b'. The resulting spectra are shown in diagrams c and c'. They are related by conformal mapping (from Nagayama *et al.* (*50*)).

Fourier transformation of $s(t_1, t_2)$ in both dimensions, we obtain the 2D spectra $S(\omega_1, \omega_2)$. An example of 2D spectra obtained for a weakly coupled spin system is shown in Fig. 5. Two types of 2D peaks spread with non-vanishing intensities can be seen. Eight peaks around the diagonal line (the broken straight line in Fig. 5) on the spectra result from the evolution of the magnetization which has 2D transition frequencies $(\omega_{lm}^{(1)}, \omega_{ij}^{(2)})$ originating from spin flips of the same spins in both transitions. The other eight peaks off the diagonal line correspond to the spin flips belonging to two different coupled spins in the two transitions. The latter eight peaks, which are called cross peaks or correlated peaks are very important in this 2D NMR method since the appearance of these peaks just indicates that there exists a coupling between two nuclei through the scalar spin coupling

$(JI_1 \cdot I_2)$. We will see later that this cross peaks are profoundly utilized in protein NMR to elucidate the coupling of various proton species even in a complicated spectrum where many resonance lines severely overlap. For the weakly coupled two-spin case, a total of $4 \times 4 = 16$ lines have resulted in the 2D spectrum. In general, then, we can expect n^2 number of peaks to appear in the correlated 2D spectra (this argument is not strict for the spin system such as AX_2 or A_2X_3), if we observe n resonance lines in the corresponding 1D NMR spectra. Judging from the number of peaks developed on the spectra, the 2D spectra seem to be more complicated than 1D spectra. But this argument is not taken up since the drawback of the increase in peak numbers can be compensated by the widely spreaded cross peaks which keep the same spectral informations of the original 1D resonance lines and add new informations about the connectivity of coupled spins. Actually the position and the fine structure of a group of cross peaks (we call a group of cross peaks a correlated peak from now on) carry information about the chemical shift and coupling constant of the coupled nuclei as shown in Figs. 4c and 4c′.

From the point of view of an accurate determination of coupling constants J, however, simpler spectra with a limited number of peaks are more desirable. By varying the flip angle for the second pulse from $90°$ to $180°$ we have a 2D NMR spectrum of that sort. This version of 2D NMR will be described in the next subsection.

3. *Homonuclear 2D J-resolved NMR Spectroscopy*

Let us think of what is brought about if the input power for the second pulse in 2D correlated NMR is changed. This consideration leads us to investigate the problem of how the amplitude of the magnetization components changes depending upon the flip angle θ of the second pulse. For the weakly coupled spin systems, a closed expression for the input-output relation has been given (*18*). The results have revealed that the dependence of the amplitude factor $A_{ij,lm}$ (Eq. (12)) on the flip angle θ is very complicated as compared with the nonlinear but simple expression for the case of 1D NMR response given by Eq. (10). But it has been found that there exists an extreme case where things are made simple, that is, the response for the second pulse with a flip angle of $180°$. In this case the amplitude factors $A_{ij,lm}$ corresponding to the cross peaks and the diagonal peaks aligned just on the diagonal line (Fig. 5) become zero. On the 2D spectrum of Fig. 5, nonvanishing components are correspondingly designated by arrows. The 2D NMR spectroscopy thus in-

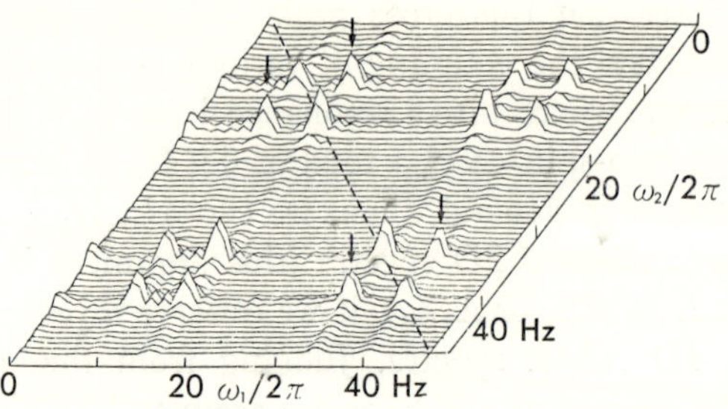

Fig. 5.   2D correlated spectrum in absolute-value mode of 2, 3-dibromo-thiophene (a AX spin system). Two 90° pulses were applied with the experimental scheme of Fig. 4a. Four spectral components which correspond to the peaks observed in the J-resolved spectrum are indicated by arrows. The diagonal line in the square representation is drawn by a straight broken line for convenience of comparison with Fig. 4c (from Aue *et al.* (*18*)).

troduced gives a new type of 2D NMR spectroscopy called homonuclear 2D J-resolved spectroscopy (*19–21*). The experimental procedure for 2D J-resolved NMR normally relies on the performance of pulse sequences of the spin-echo type $90° - k \cdot t_1 - 180° - k \cdot t_1 -$ data acquisition. Therefore the straightforward modification of 2D correlated NMR is instead attained from the spin-echo version of the correlated NMR (*50*) shown in Fig. 4a'. At any rate, the essential point in the relation between 2D correlated and 2D J-resolved NMR is the variation in the flip angle θ. As mentioned before (Fig. 4), the alternation in the start of data acquisition from the normal type to the spin-echo type results in a conformal mapping of 2D spectra so as to transfer the diagonal line in the square presentation to the horizontal line parallel to the ω_2-axis. Practically this new representation of 2D spectra is advantageous since a big data storage area as seen for 2D correlated spectra is not called for in this representation.

In 2D J-resolved NMR spectra, the chemical shifts and the fine structures of coupled nuclei can be separately displayed in different dimensions. This is schematically shown in Fig. 6. To focus on the high resolution feature retained in the spin-multiplicity, here, the frequency scale in the ω_1-direction is more expanded than that in the ω_2-direction. The spread of the 2D peaks developed in the ω_1-direction is seen to be confined in a narrow frequency range with splittings (coupling constants) of the order of 10 Hz. This is caused by the fine splitting of the multiplets. For the case of proton nuclei the scalar interaction is very small

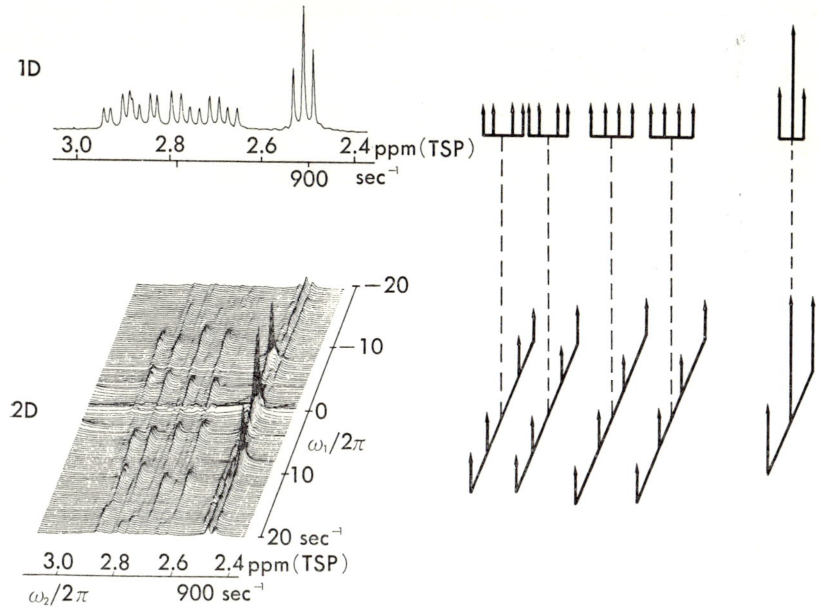

Fig. 6.   Presentation of 2D J-resolved spectrum. The multiplets of the one-dimensional (1D) spectrum (upper trace) are effectively rotated by 45° to a 2D frequency plane. The schemes on the right side indicate the spin multiplets seen in the experimental spectra on the left. The spectra were recorded at 360 MHz in a ^2H_2O solution containing 0.1 M of each of the five amino acids Ala, Ile, Met, Tyr, and His, pH−10.5, $T=25°C$. The spectral region contains a two-proton triplet of the β-methelene of Met and four one-proton multiplets (doublet-doublet) of the β-methelene protons of Tyr and His. The frequency axes are calibrated in Herz; for the chemical shifts, a calibration in ppm is also indicated (from Wüthrich *et al.* (*72*)).

(JI_1I_2). The overlapping of many spin multiplets is completely avoided in the 2D spectrum in Fig. 6. This enables us to identify fine structures belonging to the same spins with less ambiguity and determine the coupling constant J very accurately. With such a 2D J-resolved spectrum, we can study the conformations of molecules, even for macromolecules like proteins, in detail.

In comparison with 1D NMR spectra, 2D J-resolved spectra have the good property that well-resolved 2D peaks appear to hold a one-to-one correspondence to the resonance lines observed in 1D NMR (strictly, this is only true for the weakly coupled spin systems (*22*)). This feature is clearly illustrated in the schematic drawing in Fig. 6. In general, 2D resolved NMR of various types has this remarkable property. In order to see the origin of this in resolved 2D NMR, another explanation for the experimental principle can be approached on the basis of the general experimental scheme of 2D NMR spectroscopy. First, the evolution period associated with the time variable t_1 is defined as that which includes the defocussing and the refocussing periods and the 180° r.f. pulse of the

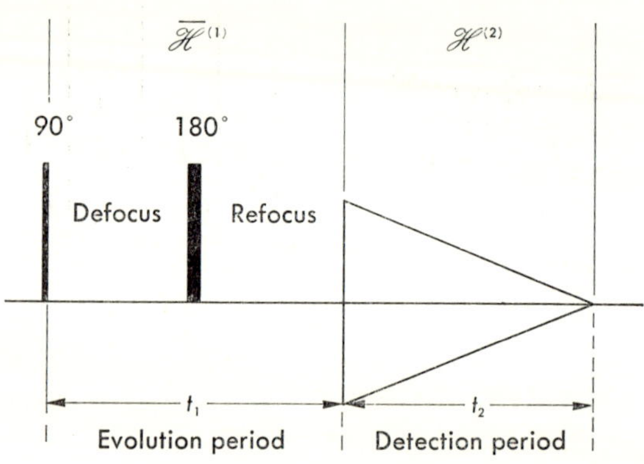

Fig. 7.   Scheme of 2D J-resolved ^{1}H NMR spectroscopy. The Hamiltonian $\mathcal{H}^{(1)}$ is averaged during the evolution period by a 180° spin-echo pulse so as to eliminate the Zeeman interaction.

$$\overline{\mathcal{H}}^{(1)} = \sum_{a,b} J_{ab} I_{za} I_{zb}, \qquad \mathcal{H}^{(2)} = \gamma H_0 \sum_{a} (1-\delta_a) I_{za} + \sum_{a,b} J_{ab} I_{za} I_{zb}.$$

spin-echo type experiment (Fig. 7). Second, the mixing pulse can be considered as null or a unit operator. No coherences are then transferred from the evolution period to the detection period except for the smooth passage of the same precession (transition) of magnetization components. Though the Hamiltonians $\mathcal{H}^{(1)}$ and $\mathcal{H}^{(2)}$ are originally the same, the chemical shift term in the Hamiltonian $\mathcal{H}^{(1)}$ is averaged out during the evolution period due to the spin-echo effect. We then have $\overline{\mathcal{H}}^{(1)} \neq \mathcal{H}^{(2)}$, $\overline{\mathcal{H}}^{(1)} = \sum_{a,b} J_{ab} I_{za} I_{zb}$ and $\mathcal{H}^{(2)} = \gamma H_0 \sum_{a} (1-\delta_a) I_{za} + \sum_{a,b} J_{ab} I_{za} I_{zb}$ where δ_a represents the chemical shift of each nucleus and J_{ab} indicates the coupling constant of coupled nuclei a and b. In weakly coupled spin systems the magnetic quantum number is a good quantum number, and then we finally obtain $A_{ij,ij} \cos (\omega_{ij}^{(1)} t_1 + \omega_{ij}^{(2)} t_2)$ for the evolution of the magnetization component $M_y(t)$ (refer to expression (12)), where i and j represent sets of magnetic quantum numbers of individual spins. The above expression should be compared with the response from one pulse excitation of the same spin system with the same spin Hamiltonian $\mathcal{H} = \mathcal{H}^{(1)} = \mathcal{H}^{(2)}$, $B_{ij} \cos(\omega_{ij}t)$, $B_{ij} = F_{ij}\sigma_{ij}(0)$ (refer to expression in Fig. 2). In consequence the following correspondence between 1D and 2D NMR holds: $A_{ij,ij} \leftrightarrow B_{ij}$ and $(\omega^{(1)}{}_{ij}, \omega^{(2)}{}_{ij}) \leftrightarrow \omega_{ij}$.

III. DATA HANDLING OF 2D SPECTRA

1. Phase Problem and Causality Principle

The double-Fourier transformation generates four types of transformed components

$$S(\omega_1, \omega_2) = \int_0^\infty dt_1 \exp(-i\omega_1 t_1) \int_0^\infty dt_2 [s(t_1, t_2)] \exp(-i\omega_2 t_2)$$

$$= S^{cc}(\omega_1, \omega_2) - S^{ss}(\omega_1, \omega_2) - iS^{cs}(\omega_1, \omega_2) - iS^{sc}(\omega_1, \omega_2), \qquad (13)$$

where superscripts c and s mean the cosine and the sine transformation, respectively. With the single-sided Fourier transformation in time in Eq. (13), positive and negative frequencies in the ω_1 and ω_2 dimensions are to be superposed in all four components. In general a remedy to retrieve the discrimination of the sign of frequencies is naturally obtained by applying a double-sided Fourier transformation in time. But this is prohibited by the causality principle. For the discrimination of the sign of frequencies in the ω_2-direction, quadrature detection (*87*), which is a standard technique utilized in conventional 1D NMR, can be successfully applied. To distinguish positive and negative frequencies in the ω_1-dimension, however, the quadrature detection does not help and therefore another alternative method should be pursued.

Linear combinations of the four components are employed in 2D NMR spectroscopy (*34*) to retrieve the sign of the ω_1-frequencies.

$$S_{\mathrm{real}}(\omega_1, \omega_2) = \begin{cases} S^{cc}(\omega_1, \omega_2) - S^{ss}(\omega_1, \omega_2), & (\omega_1 > 0, \ \omega_2 > 0) \\ S^{cc}(\omega_1, \omega_2) + S^{ss}(\omega_1, \omega_2), & (\omega_1 < 0, \ \omega_2 > 0) \end{cases} \qquad (14)$$

$$S_{\mathrm{imag}}(\omega_1, \omega_2) = \begin{cases} S^{cs}(\omega_1, \omega_2) + S^{sc}(\omega_1, \omega_2), & (\omega_1 > 0, \ \omega_2 > 0) \\ S^{cs}(\omega_1, \omega_2) - S^{sc}(\omega_1, \omega_2), & (\omega_1 < 0, \ \omega_2 > 0) \end{cases} \qquad (15)$$

Here $S_{\mathrm{real}}(\omega_1, \omega_2)$ and $S_{\mathrm{imag}}(\omega_1, \omega_2)$ correspond to the real and imaginary parts in the right side of Eq. (13) but the positive and negative frequencies are now distinguished in the ω_1-direction. In order to display phase sensitive 2D spectra, a combination of $S_{\mathrm{real}}(\omega_1, \omega_2)$ and $S_{\mathrm{imag}}(\omega_1, \omega_2)$ is applied (*68*) in the same way as employed in the 1D NMR to correct the actual setting of the instrument's phase-sensitive detector. A much simpler presentation of 2D spectra is obtained by calculating the absolute-value spectra.

$$|S|(\omega_1, \omega_2) = [S_{\text{real}}^2(\omega_1, \omega_2) + S_{\text{imag}}^2(\omega_1, \omega_2)]^{1/2} \tag{16}$$

No phase adjustment is of course required in Eq. (16) but we have some urdesirable drawbacks in the presentation because the line shape is no more Lorentzian in the absorption mode and has a $1/\omega$ tailing (25). The long tailing is manifested for the peaks in the absolute-value spectra of Figs. 5 and 6.

From the points of view of display and analysis of high resolution NMR data an ideal 2D spectrum would be as follows: all the 2D peaks developed on an experimental spectrum have the same identical line shape with the same sign in intensity. This is actually realized in conventional 1D NMR spectra. Except for some particular cases, this is not feasible in the phase sensitive 2D NMR spectra. When we perform a double-Fourier transformation for a time function of the form given by Eq. (12) and distinguish positive and negative frequencies in the ω_1-dimension, phase-sensitive 2D spectra $S_{\text{real}}(\omega_1, \omega_2)$ present peaks having mixed line shapes with the absorption mode and the dispersion mode (25). Next we will consider this problem further from the point of view of the causality principle.

2. Projection Cross-section Theorem

The projection cross-section theorem which may be applied to any double-Fourier transformation in general (88) claims that any central cross-section $c(t) = s(t \cos \phi, t \sin \phi)$ in the time domain relates the corresponding projection $P(\omega)$ in the frequency domain in the form of a Fourier transformation pair (25).

$$c(t) \xleftrightarrow[\mathscr{F}^{-1}]{\mathscr{F}^1} P(\omega) \tag{17}$$

This relation is visualized in Fig. 8. This theorem implies that the projection of a 2D spectrum $S(\omega_1, \omega_2)$ obtained from the linear operation on the original time signal $s(t_1, t_2)$ contains only limited information on the function $s(t_1, t_2)$. It is dependent on the function values of $s(t_1, t_2)$ lying on a straight line with slope ϕ through the point $t_1 = t_2 = 0$. Particularly interesting conclusions are drawn for $s(t_1, t_2)$ by taking into account the fact that the causality principle must be fulfilled. It requires that

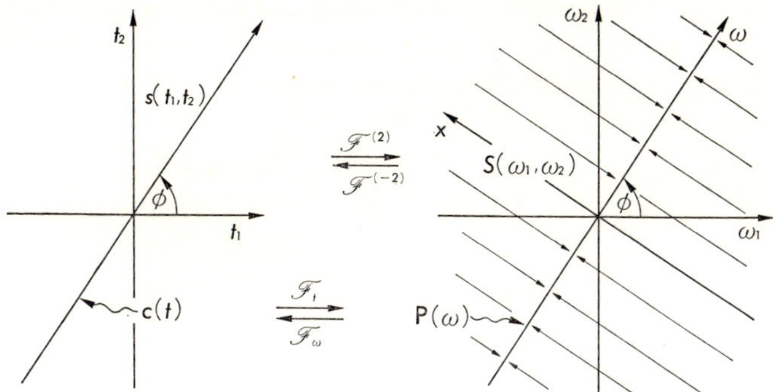

Fig. 8.   Illustration of the projection cross-section theorem. The terms $s(t_1, t_2)$ and $S(\omega_1, \omega_2)$ represent the signals in the time domain and the frequency domain, respectively; $c(t)$ is a cross-section in the time domain ; and $P(\omega)$ is the corresponding projection in the frequency domain. The direction of projection, x, is indicated by the arrows perpendicular to $P(\omega)$ (from Nagayama *et al.* (25)).

$$s(t_1, t_2) = 0 \quad \text{for} \quad t_1 < 0 \quad \text{and/or} \quad t_2 < 0,$$

i.e., $s(t_1, t_2)$ has nonzero values exclusively in the first quadrant. The Fig. 8 shows that for $\pi/2 < \phi < \pi$ the cross-section $c(t)$ is equal to zero except for $t = 0$ at which point $c(t)$ has a finite value. The Fourier transformation is then also equal to zero for $\pi/2 < \phi < \pi$. This shows that the projections for $\pi/2 < \phi < \pi$ of the spectrum of any 2D physical response must be equal to zero. This also suggests that any phase-sensitive 2D spectrum must not be positively constrained, otherwise the projection $P(\omega)$ in any direction of the 2D spectrum should have a spectrum with finite values. The appearance of phase-sensitive spectra containing positive and negative peak values, as pointed out in the previous subsection, is therefore a natural consequence. To violate this restriction which is strictly caused from the causality principle, we now have three remedies. 1) to apply a nonlinear operation to circumvent the projection cross-section theorem itself; 2) to extend the time axis of t_1 or t_2 to the negative direction; or 3) to give up distinguishing the positive and negative frequencies and superpose them to retrieve the positive constrained sepectrum. The last remedy is equal to presenting only a Fourier transformed component $S^{cc}(\omega_1, \omega_2)$ in Eq. (13).

The absolute value 2D spectrum given as an alternative presentation for the phase-sensitive spectrum shows an example in the sense of remedy 1 (*18*, *25*). The positively constrained 2D spectrum calculated by Eq. (16) enables us to pick up the projection and cross-section spectra as shown in Fig. 9. Note that the broad but simplified projection spectrum

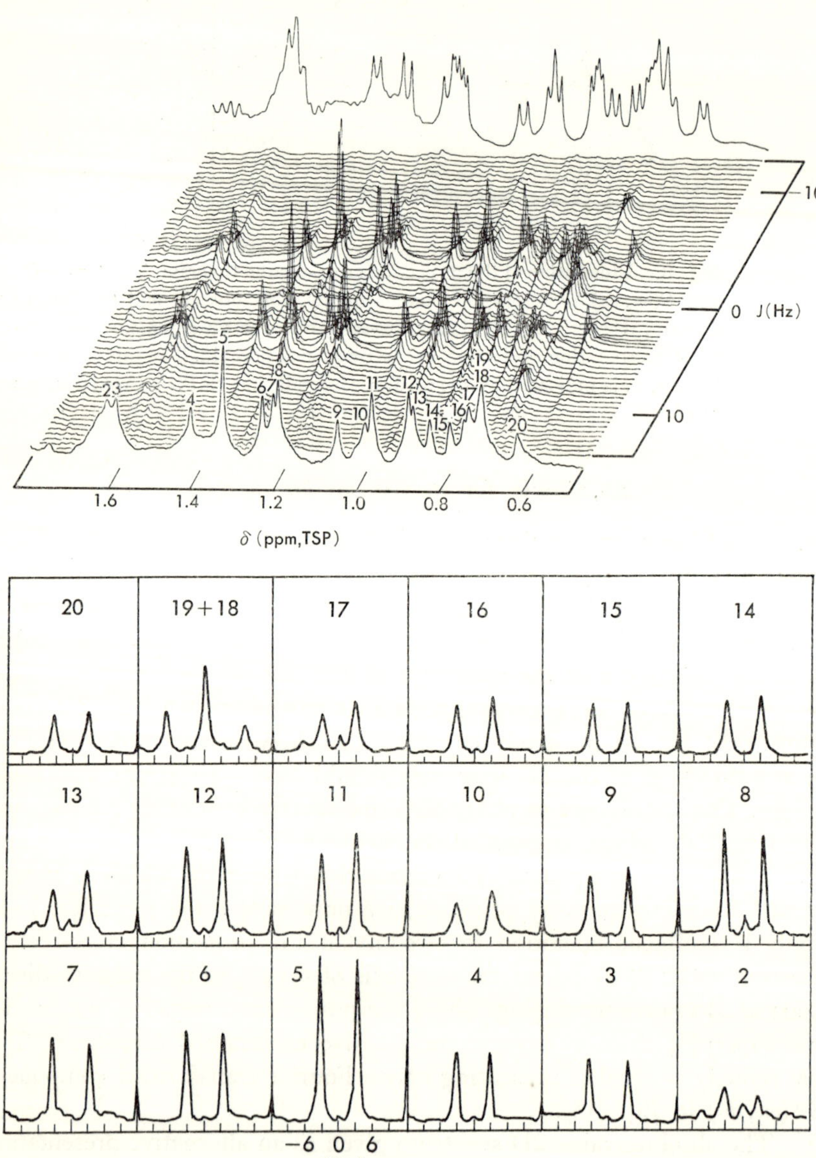

(the bottom trace in Fig. 9a) corresponds to a broad-band decoupled spectrum containing only the information about chemical shifts (19). In contrast to the projection spectrum, the cross-sections of the 2D spectrum are found to show relatively narrow multiplet lines (Fig. 9b). It has verified (25) that the line shape of the cross-section spectrum becomes rigorously Lorentzian, if the line widths of 2D peaks in the ω_1- and the ω_2-directions are equal. Despite some drawbacks the representation in the absolute-value mode is convenient owing to the plausible nature mentioned above together with the ease of data handling. In heteronuclear 2D NMR, a selective 180° pulse applied just before detection acts so as to reverse the direction of time for the t_1 period (66). By the combination of the normal and the reversed precession 2D experiments, we reach to the remedy 2. Another method for obtaining pure absorption 2D spectra in the heteronuclear case is given by the " gated decoupler " technique which is known to be successful if heteronuclei of coupled counter-parts are intrinsically weakly-coupled (35). A technique of the type of remedy 3 could be extended to the homonuclear case with the use of the decoupling technique applied in the detection period (71). In 2D correlated NMR of the original version (Fig. 4a) phase-sensitive spectra $S^{cc}(\omega_1, \omega_2)$ are usually presented, since the discrimination of the positive and negative frequencies are not called for. Although peaks with positive or negative intensities simultaneously spread on the 2D frequency base, individual peaks have pure phase (18). The drawback in remedy 3 that it cannot distinguish positive and negative frequencies has been recently saved by relying also on two different experiments with gated decoupling for the heteronuclear 2D J-resolved NMR (69). If we can

← Fig. 9.   2D J-resolved spectrum in absolute-value mode of basic pancreatic trypsin inhibitor. Upper : high-field region from 0.4 to 1.8 ppm, which contains the resonances of nineteen methyl groups (97), of the 360 MHz NMR spectrum of a 0.01 M solution of basic pancreatic trypsin inhibitor (BPTI) in ^2H_2O, pH=4.5, $T=60°C$. Prior to the Fourier transformation, the 2D data set was weighted in the t_1 and t_2 directions by window functions cos $(t_x/2T_x)$ exp $(t_x/0.4T_x)$, with $x=1,2$; $T_1=2.46$ sec, and $T_2=1.23$ sec are the maximum acquisition times in the t_1 and t_2 domains. The top trace shows the conventional 1D spectrum; the 2D J-resolved spectrum was computed from 64× 8,192 data points and is presented as a (J, δ) spectrum (refer to Fig. 10); the bottom trace shows the projection of the 2D spectrum with $\phi=45°$. Lower : presentation of cross-sections of 2D J-resolved spectrum of Fig. 9a. Methyl resonances of BPTI. The numeration of the multiplets corresponds to that in Fig. 9a (from Nagayama et $al.$ (25)).

perform the double-sided Fourier transformation for the frequency variable of t_2, that is equivalent to extend the experimental t_2 time-axis in the negative direction, 2D spectra in the pure absorption mode can be developed. This type of 2D experiment (remedy 2) has been realized for homonuclear J-resolved spectroscopy by extending the time variable of t_2 to the refocusing region in the evolution period (Fig. 7) (*69*). Although this method has a severe experimental restriction that the sampling resolution $\Delta\omega_2$ in the ω_2 dimension cannot be arbitrarity set and must always be larger than the whole spectral range of the ω_1-axis, it can be applied to the small molecules where the resonance lines appear well-separated with narrow line widths.

3. Conformal Mapping and Similarity Theorem

In relation to the presentation of 2D spectra, it is known that any class of 2D NMR experiments must have two types of experimental procedures (*86*). As an example two experimental schemes for 2D correlated NMR are shown in Fig. 4. As seen there, the normal and spin-echo type schemes are similar and the presented two 2D spectra are mutually transfered by a conformal mapping.

$$S(\omega_1, \omega_2) \longleftrightarrow S'\left(\frac{\omega_1+\omega_2}{2}, \omega_2\right) \tag{18}$$

Transformation (18) can be also performed by a mathematical manipulation imposed on the 2D spectra. However it is important to notice that the above mapping transformation (18) purely results from an alternation of the physical realization of the general experimental schemes. That physical mapping process is interpreted by a theorem in the Fourier transformation in general; *i.e.*, the similarity theorem (*89*). It claims that in a multi-dimensional Fourier transformation a similarity transformation applied to the time variables corresponds to a similarity transformation in the frequency variables (*86*):
if

$$f(t) \xleftrightarrow{\quad\mathscr{F}^n\quad} F(\omega)$$

then

$$f(At) \xleftrightarrow{\quad\mathscr{F}^n\quad} \frac{1}{|A|}F(\omega A^{-1}) \tag{19}$$

When we put

$$A = \begin{bmatrix} 2 & 0 \\ -1 & 1 \end{bmatrix}, \qquad A^{-1} = \begin{bmatrix} \dfrac{1}{2} & 0 \\ \dfrac{1}{2} & 1 \end{bmatrix}, \qquad t = \begin{pmatrix} t_1 \\ t_2 \end{pmatrix}$$

and $\boldsymbol{\omega} = (\omega_1, \omega_2)$ into relation (19), we obtain the mapping procedure indicated by expression (18) (Figs. 4c and 4c'). Strictly speaking, the desired conformal mapping should have a form that it does not give any skewness in the mapped spectra.

$$S(\omega_1, \omega_2) \longleftrightarrow S'\left(\frac{\omega_1 + \omega_2}{2}, \frac{\omega_2 - \omega_1}{2}\right) \tag{20}$$

Using the relation (19) conversely, we obtain the associated transformation in the time domain.

$$s(t_1, t_2) \longleftrightarrow s'\left(\frac{t_1 - t_2}{2}, \frac{t_1 + t_2}{2}\right) \tag{21}$$

This type of transformation, however, has no meaning again in the physical experiments because the signal $s(t_1, t_2)$ has nonzero values only in the first quadrant as mentioned before. Therefore another consideration should be made. A purely mathematical transformation (spectral tilting)

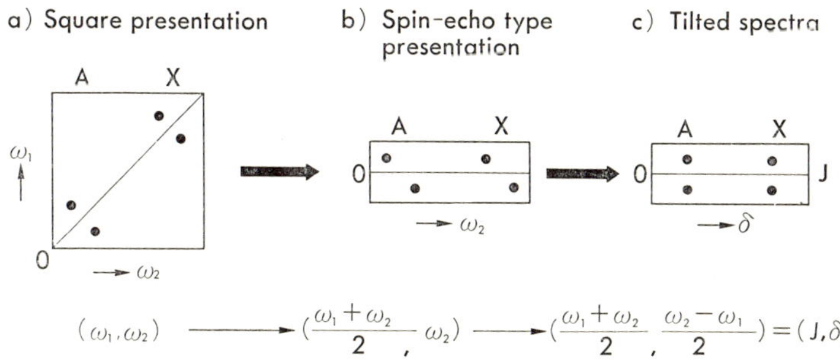

Fig. 10. Conformal mapping of three types of presentations of 2D J-spectra. The similarity transformations associated with the mappings are shown at the bottom of the figure. a) Square representation obtained with the scheme as drawn in Fig. 4a by alterating the second pulse from 90° to 180°; b) spin-echo type presentation obtained with use of the scheme shown in Fig. 7; c) presentation of the tilted spectrum obtained from the presentation (b) by shifting the ω_2 coordinate according to the similarity transformation shown at the bottom. The ω_1- and ω_2-axes are renamed the J- and δ-axes, respectively.

of 2D spectra has given a solution for this problem (25). The conformal mapping process applied to the 2D J-resolved NMR is illustrated in Fig. 10. The mapping in the time domain of the type (21) recovers its physical sense if the response signals have values in two quadrants of the 2D time space. The double-sided Fourier transformation introduced in t_1 or t_2 time-axes also solves the above mapping problem. Such methods (66, 69) directly present 2D spectra in the form in Fig. 10c without the aid of mathematical procedures.

If spectral tilting is imperfectly performed, a partially decoupled spectrum results in the projection of the 2D spectrum. This so-called J-scaling (25, 74–77) puts another virtue on the resolved type of 2D NMR. For example useful information could be gained by this technique for spin identification (74). Practically, however, the J-scaled spectrum can be obtained easier without relying on 2D NMR but with a multi-pulse 1D NMR method (75, 77).

IV. METHODS OF ANALYSIS IN 2D CORRELATED NMR AND 2D J-RE-SOLVED NMR APPLIED TO PROTEINS

Two dementional correlated NMR of the spin-echo type and the 2D J-resolved NMR are mutually connected only by changing the flip angle of the second pulse from 90° to 180° or *vice versa* (compare Fig. 4a' and Fig. 7). Therefore both techniques are easily combined in an experiment. Here methods of data analysis in this combined 2D experiment are ex-plained which serve to identify spin systems, to assign the individual resonance lines and to determine chemical shifts and coupling constants of individual spin groups in proteins. A small protein with a molecular weight of 6,500, basic trypsin inhibitor of bovine pancreas (BPTI) and its derivative reduced at a disulfide bond were used in NMR experiments.

1. Spin Identification

In the conventional NMR, a coupled spin group is identified in a spec-trum by applying selective spin decoupling to a particular part of reson-ance lines belonging to the spin group. Contrary to such a step-by-step method, 2D correlated NMR has a remarkable character that the connec-tivities of transitions for various coupled nuclei are visualized all at once with one experiment. An example for the case of the AX spin system is schematically illustrated in Fig. 4c'. There two sets of cross-peaks (two correlated peaks) aligned on a line intersecting the $\omega_1 = 0$ axis at an angle of 135° indicate the presence of coupling between the A and X spins. The chemical shift positions of the two correlated peaks are

$$A: \quad \omega_1 = \frac{\delta_A - \delta_X}{2}, \quad \omega_2 = \delta_A,$$

$$X: \quad \omega_1 = \frac{\delta_X - \delta_A}{2}, \quad \omega_2 = \delta_X, \tag{22}$$

where δ_A and δ_X are the chemical shifts of A and X spins. Equation (22) suggests that correlated peaks always appear with original chemical shifts in the ω_2 dimension and one half of the chemical shift difference of coupled nuclei in the ω_1 dimension. From now on, therefore, we rename the ω_1-axis the chemical shift difference axis ($\Delta\delta$) and the ω_2-axis the chemical shift axis (δ) in 2D correlated spectra of the spin-echo type.

The presence of correlated peaks in the rigid form mentioned is quite convenient for analyzing and identifying a 2D spectrum systematically. Neglecting the fine structures arising from spin multiplicities, connectivity patterns of 2D correlated spectra for the aliphatic proton groups of 20 amino acid residues can be classified as shown in Table I. The glycine and alanine amino acid residues are seen to depict the sim-

TABLE I

Connectivity Diagrams of 2D Correlated Peaks for Aliphatic Groups of 20 Amino Acids

No. of spins	Amino acids	Weakly coupled spins
2	Gly Ala[a]	
3	Asn, Asp, Cys, His, Phe, Ser, Trp, Tyr	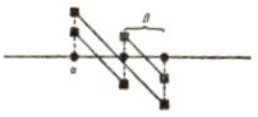
3	Thr[a]	
4	Val[a]	
5	Gln, Glu, Met	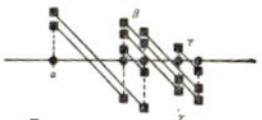
6	Ile[a]	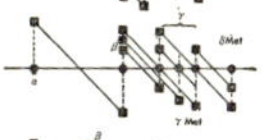
6	Leu[a]	
7	Arg, Pro	Given by extension of the pattern for the
9	Lys	5 spin system but not shown here.

The fine structure for each correlated peak is neglected.

[a] Methyl groups are counted as by one spin.

plest connectivity pattern. For spin systems bigger than two spins, more complicated patterns result. From the classification of 2D correlated patterns (2D connectivity diagrams), it is found that there exist some particular groups which permit us to assign the type of amino acids without doing any further experiments. Those are amino acid residues of threonine, valine, isoleucine, leucine, and lysine. For the residues of glycine and alanine, we have to utilize the difference of the spin systems (AX or AX$_3$) reflected in the difference of the fine structures that is also visible in the correlated spectra. With use of the identification of 2D patterns and fine structures, as a whole 10 groupings of the 2D correlated pat-

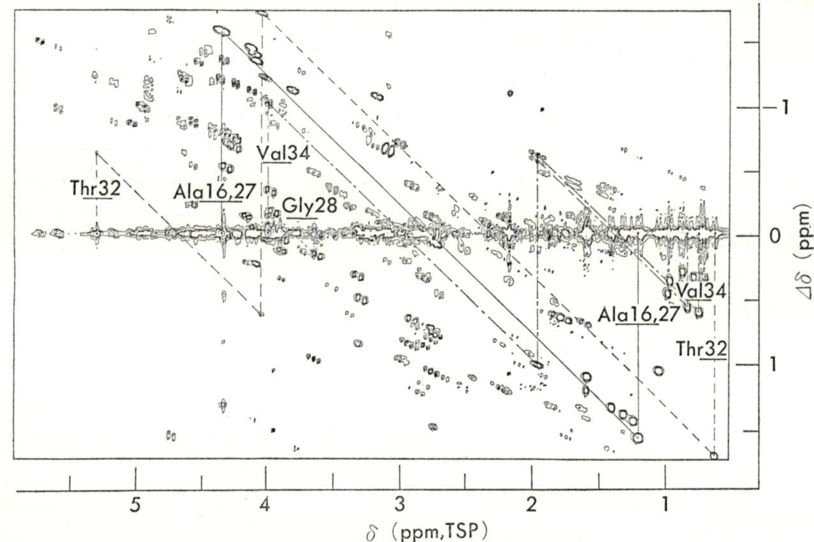

Fig. 11. Contour plot for a 2D correlated spectrum of the aliphatic region of BPTI. The measurement was performed in 6 hr for a 0.01 M BPTI solution in ^{2}H$_2$O, pH=7.0, T=68°C. The original spectrum was computed from 256 × 1,024 data points by carrying out zero-supplemented Fourier transformation (94) in both time directions. A sampling resolution of 2.441 Hz was applied in both frequency dimensions. To enhance the spectral resolution, a window function of the shifted sine bell (95) was multiplicated to the time domain signals in both time directions prior to each of the Fourier transformations. The δ-axis shows the chemical shift of proton resonances and the Δδ-axis indicates the difference in the chemical shifts of coupled nuclei. Connectivities between the components of the coupled nuclei in the following spin systems are indicated: — — — Thr 32; — - — Val 34; —— Ala 16 and Ala 27; - - - - - Gly28 (from Nagayama et al. (86)).

terns are distinguished easily as far as the aliphatic spin groups of amino acid residues are concerned.

A few examples of identified connectivity diagrams in an actual protein (BPTI) NMR are shown in Fig. 11, where a contour plot of the 2D spectrum is favorably presented. Five glycines out of six glycine constituents have been identified, while only one glycine is indicated by a connection in the figure. Furthermore five alanines which have the same patterns as glycine but are distinguishable from its fine structure are seen on the spectrum. Eventual overlapping of Ala 16 and 27 in the 2D spectra was only due to the low resolution of the digital representation inherent in the 2D correlated NMR which deals with a very large two dimensional spectral region. With the use of 2D J-resolved spectroscopy which requires less data area to spread spectra, both were discriminated as will be shown later. One connectivity diagram indicated by Val 34 made us give it a unique assignment since the BPTI includes only one valine residue. Threonine 32 has also shown an example of unique pattern diagrams, although we cannot distinguish and assign among 3 threonines involved in the BPTI simply from the spectrum.

We consider, here again, the characteristics brought about in the correlated spectroscopy. Among the various virtues inherent in this type of spectroscopy of great importance from the point of view of application to biological molecules is the high identification capability to attribute a few resonance lines to a spin group as in a 2D diagram. This enables us to trace out a particular spin group of coupled nuclei from one experiment to another with less ambiguity as compared with 1D NMR, even if the resonances are overlapped with resonances arising from other spin groups. To examine this situation more clearly, let us pay attention to the heavily overlapped peaks in the region of 3.5 to 4.5 ppm in Fig. 11. As already mentioned, the center spectrum around $\Delta\delta=0$ corresponds to the 1D spectrum of this protein. In such a 1D spectrum, the tracing of changes in the positions of resonance lines from one experiment to the other is rather difficult due to the simultaneous movement of many peaks. This situation takes place very often in protein NMR associated with changes in the experimental conditions, say, temperature or pH. This uncertainty of tracing the identical lines can be avoided in 2D NMR by using well resolved correlated peaks and comparing the total change burdended on 2D diagrams under various experimental conditions. Conversely speaking, even for the case of one ro two correlated peaks overlapped in 2D spectra, small perturbations in the chemical shift positions

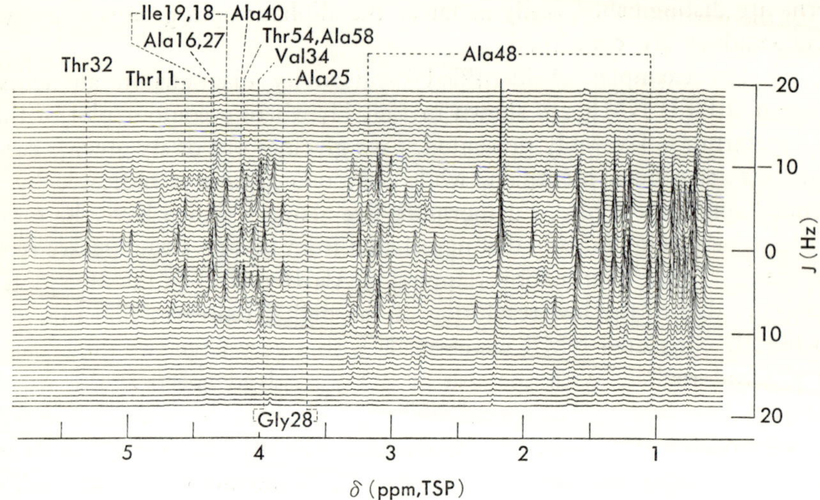

Fig. 12.   Three-dimensional plot for a 2D *J*-resolved spectrum of the aliphatic region of BPTI. The measurement was performed in 16.5 hr with use of the same solution and the same experimental conditions (pH=7.0, $T=$ 68°C), as of the 2D correlated spectrum shown in Fig. 11. The spectrum was computed from $64 \times 4{,}096$ data points by carrying out zero-supplemented Fourier transformations in both time-directions. A sampling resolution of 0.61 Hz was applied in both frequency dimensions. A window function of the increasing exponential multiplied by a cosine function was used to enhance the spectral resolution. The δ-axis shows the chemical shift of proton resonances and the *J*-axis indicates the coupling constants of coupled spin systems. The assignments for the α-proton resonances which show doublets or quartets are designated in the figure (from Nagayama *et al.* (*86*)).

introduced by changing the experimental conditions allows us to recover the unambiguous decomposition of connected 2D pattern diagrams. These characteristic features have to be fully utilized to recognize, identify, and assign proton resonances in proteins.

A *J*-resolved spectrum for the aliphatic protons of the BPTI is shown in Fig. 12. It gives close ups of the fine structures of multiplets with higher resolution, which are also manifested in 2D correlated spectra (Fig. 11) but not immediately clear because of the low resolution and larger number of peaks. Note that in an overlapped spectral region from 3.5 to 4.5 ppm, plausible peak identifications are only feasible by referring to the corresponding 2D correlated spectra taken under the same experimental conditions, otherwise positioning of chemical shifts cannot be fixed.

With the use of the 2D *J*-spectra taken under the same experimental condition, we are able to confirm the identification and assignment of proton resonances performed in the previous step by investigating the fine structures of multiplets resulting in the *J*-resolved spectroscopy. For example, the α-proton of alanine and nonequivalent α-protons of glycine could be distinguished by finding the difference in the coupling constants and spin multiplets (doublet or quartet). The doublet nature of the fine structure for the α-proton of valine, isoleucine, and threonine also gives a clue for discriminating them from the rest of the aliphatic protons of amino acid residues. Examples of these are shown in Fig. 13 together with their assignments. Now we have a reasonable answer to why the close combination of correlated and *J*-resolved spectroscopy is important in protein NMR.

2. Determination of Coupling Constants

In order to determine coupling constants accurately, a cross-section representation was recommended as suitable (*25*). The corresponding spin

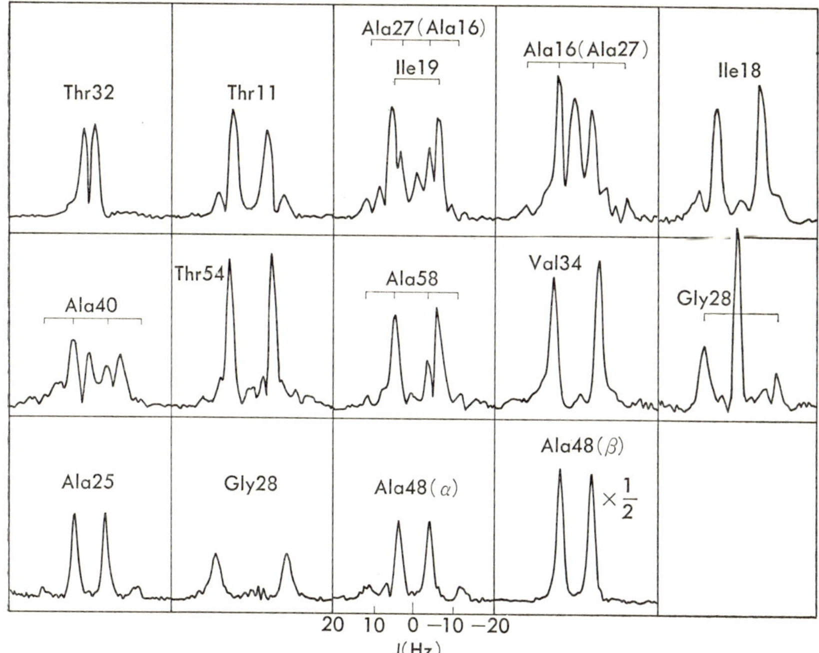

Fig. 13. Presentation of cross-sections of the 2D *J*-spectrum shown in Fig. 12. The chemical shifts and the coupling constants for these spin multiplets are listed in Table V in the Appendix (from Nagayama and Wüthrich (*96*)).

multiplets shown in Fig. 13 are collected from Fig. 12. Note that they are fine structures of spin multiplets for the α-protons from the heavily overlapped spectral region of a protein (BPTI) NMR spectrum. The obtained coupling constants $^3J_{\alpha\beta}$ thus determined give a basis for discussing the conformational aspects of proteins. In Fig. 13 eventual overlapping is seen between the Ile 19 doublet and Ala 27 quartet, which are resolved under different conditions (85°C), but doublets of alanines 16 and 27 are well separated in contrast to the 2D correlated spectrum (Fig. 11) that has a lower resolution of the sampling points. The other multiplets like singlets, doublets, triplets, and doublet-doublets are visible from 3.5 to 5.7 ppm in Fig. 13. Many of them were finally assigned to α-protons of particular amino acid residues. Around 5.6 ppm, for example, a recognized triplet and doublet separated from each other by 0.07 ppm were attributed to the α-protons of Tyr 21 and Cys 30, respectively. Coupling constants $^3J_{\alpha\beta}$ obtained for the BPTI molecule are collected in the Appendix.

If 2D J-peaks appear to be well separated from the overlapped spectral region and assignments for those peaks have already been proposed with conventional NMR, we can omit the experimental step with 2D correlated NMR. Direct comparison of the 1D and 2D spectra affords enough information to identify and assign the 2D J-peaks. A good example for protein NMR is seen for the resonances of peptide NH protons in BPTI. Many reports have been published on the resonance assignments for well separated NH protons in the protein (*90–93*). Relying on the results of such 1D NMR studies, resonance assignments for NH protons have been carried out in 2D J-spectra. Figure 14 shows a comparison of a normal 1D spectrum and a 1D-like reconstructed spectrum which was obtained from cross-sections of a 2D J-spectrum of this molecule by giving the chemical shift positions determined from a projection spectrum. In the figure the correspondence between the 1D and 2D spectra is immediately clear.

The apparent coupling constants $^3J_{\text{NH}\alpha\text{H}}$ are given in Fig. 14 as readings of separation of double peaks at the peak maxima. It is necessary to interpret these values with caution as the apparent values are always larger in the 2D J-spectrum than in the 1D spectrum. We could not say *a priori* which is closer to the intrinsic values for those coupling constants. By taking into account the nonlinear operation (Eq. (16)) performed to obtain the absolute value 2D spectra, we have calculated how the apparent value for the coupling constant deviates from the

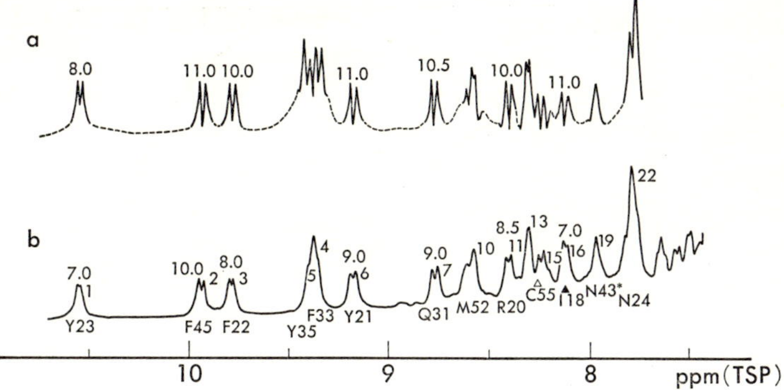

Fig. 14. Comparison of 2D and 1D spectra for the low-field NH proton resonances of BPTI. The measurement was performed in 14.5 hr for a 0.02 M BPTI solution in ^2H_2O, pH=4.2, T=36°C. The 2D spectrum was computed from 64×4,096 data points by carrying out zero-supplemented Fourier transformations in both time directions. A sampling resolution of 0.305 Hz for the ω_1-dimension and a sampling resolution of 1.22 Hz for the ω_2-dimension were applied. a: 1D spectrum reconstructed from the cross-sections of the original 2D spectrum. b: 1D spectrum computed from the first spin-echo free induction decay in the 2D data set with t_1=0. Note that the 2D time domain signal for t_1=0 corresponds to the normal 1D free induction decay. Reading of the apparent double-peak separation for several NH doublets are designated by figures. The numeration and assignment of the NH resonances are also indicated (93).

intrinsic value in the 2D spectrum. For the doublet case the result is compared with the apparent separation of the doublet peaks experienced in the 1D NMR (Fig. 15). A Lorentzian curve was assumed for the simulation of a 1D doublet and the same exponential decay time, which leads to the Lorentzian curve in the 1D NMR, was applied to both time directions for the simulation of a 2D doublet. From Fig. 15 two remarkable trends for the 2D cross-section spectrum in the absolute-value mode can be made out: first, making a good contrast to the case of the 1D NMR, two peaks of a 2D doublet repel each other in the overlapped case with the Γ/A ratio of values up to 2.5, and second, that the overlapped two peaks of the doublet show double-peak feature even for a very large Γ/A ratio up to about 3.0. Particularly, the latter trend should be compared to the trend seen in 1D NMR where the double-peak begins to coalesce already with a Γ/A ratio of about 2.0. Both effects originate from the special treatment to calculate the absolute value spectra (Eq. (16)), that is, the nonlinear operation, together with the function

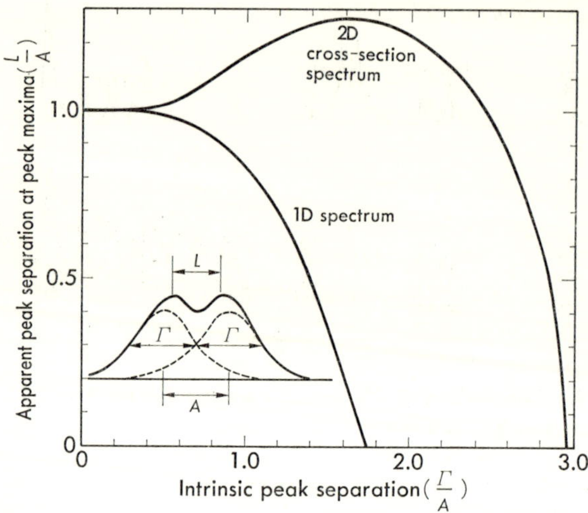

Fig. 15. Relation between the intrinsic peak separation and the apparent peak separation. The effect of overlapping of two peaks for a doublet was evaluated by assuming the same line width $\Gamma/2$ and intensity in each of the two peaks. A Lorentzian line shape was assumed for the 1D doublet simulation. For the simulation of the 2D doublet, which corresponds to the cross-section spectrum of 2D doublet peaks, a relaxation rate $(\Gamma/2)^{-1}$ is assumed for each of two time periods, t_1 and t_2.

form of 2D response given by Eq. (12). As already mentioned, the function form of the type of " phase twist " (*34*) inevitably results in a mixing of the absorption mode and the dispersion mode in a spectrum after the double-Fourier transform of the time signal. The apparent high ability to separate each peak in spin multiplets impresses upon us that 2D NMR gives much better resolution to the spectrum, besides the 2D separation. But we must be careful to treat the obtained apparent values for coupling constants, because we might make a dangerous overestimation for those values. In this context, digital filtering to enhance the resolution of resonance peaks, in other words, to reduce the ratio of Γ/A by narrowing the line widths of individual peaks, is a necessary manipulation to get reliable values for the coupling constants (*25*). From Fig. 15 we can see that the apparent values have small deviation (less than 5%) from the intrinsic value if the ratio Γ/A is supressed to a value smaller than 0.7. Examples of cross-sections raised in Fig. 13 all satisfy this condition except for the α-protons of Thr 32, and the overlapped Ile 19 and Ala 27.

3. *Individual Assignments of 2D Resonance Peaks of Coupled Aliphatic Protons*

We will quickly survey the method of resonance assignments using the combination of the 2D correlated spectrum and the 2D J-spectrum. The detailed procedure for resonance assignment and the determination of NMR parameters in actual systems, here, chemical shifts and coupling constants $^3J_{\alpha\beta}$ for α- and β-protons of BPTI, will be published elsewhere (*96, 96a*). Only the results are given in Table V in the Appendix.

The tactics employed to assign observed resonances, which are thought to open an approach to the systematic assignment of NMR signals in proteins, are summarized as follows.

1) Identification and classification of 2D correlated spectral patterns to the particular spin systems according to the diagrams drawn in Table I. 2) Spin multiplicities and coupling constants of the α-β proton coupling for identified spin systems are determined with 2D J-resolved spectra. 3) For the nuclei recognized and assigned by other workers, we have taken their results. For the nuclei of which coupled counterpart have been assigned, assignments of the nuclei are automatically drawn in the first step. 4) For spin groups for which assignments are not attainable from other reports; a) specify the titratable groups by pH titration experiments and b) modify the residues with the use of the established chemical modification techniques. 5) Then possible individual assignments are limited by comparing a set of observed chemical shifts and coupling constants for a spin system with shifts predicted from the ring current effect and coupling constants calculated with the use of a Karplus curve on the basis of the X-ray structure.

For our systematic assignment, the first step is very crucial. To obtain the unambiguous connectivity of 2D diagrams for particular spin groups, information accessible from other steps was also referred to. Spin multiplicities and coupling constants of α- and β-protons, and the slight change in shift positions introduced by pH titration were found to be informative in this task. Moreover, several experiments with different temperatures have been carried out to give small perturbations to chemical shifts. This in many cases offered key information to decompose overlapped 2D patterns. Individual assignments for the connectivity diagrams shown in Fig. 11 have been accomplished with the aid of published assignments about methyl groups of these amino acids (*97*). The same procedure has been applied to the assignments of α-protons shown in Fig. 12. On the other hand, an assignment for the unequivalent α-protons of Gly 28 in Figs. 11 and 12 was obtained at the step 4 with the systematic method described above. The method for the resonance asignments of peptide NH protons (Fig. 14) has been already explained, and it corresponds here to the assignment step 3.

V. APPLICATIONS TO THE STUDY OF CONFORMATION OF BASIC TRYPSIN INHIBITOR OF BOVINE PANCREAS

As a result of many theoretical considerations (*98–104*) and crystallographic studies (*105–106*) proteins give a picture of semi-liquid particles rather than rigid molecules. This view suggests that the protein is loosened in the native state and fluctuates in its conformation to a large degree. In this context, NMR gives an alternative tool to investigate protein flexibility. Recently a series of NMR studies performed with BPTI and its derivatives by the group headed by Wüthrich have revealed many aspects of flexibility of this protein. They have led to propose a model about the internal mobility of proteins (*107–117*). Here we approach this general problem by viewing it through the relatively slow motions reflected on chemical shifts and coupling constants of proton resonances. A small protein basic trypsin inhibitor of bovine pancraeas (BPTI, MW =6,500) (Transylol®, Bayer Leverkusen) and its derivative (RCAM BPTI), which was obtained by reducing a disufide-bond 14-38 and blocking mercapto groups with carboxyaminomethyl (*118*), were utilized in this study.

The amino acid sequence of the inhibitor (*118a*) is:

```
1              5              10             15
Arg Pro Asp Phe Cys Leu Glu Pro Pro Tyr Thr Gly Pro Cys Lys
16             20             25             30
Ala Arg Ile Ile Arg Tyr Phe Tyr Asn Ala Lys Ala Gly Leu Cys
31             35             40             45
Gln Thr Phe Val Tyr Gly Gly Cys Arg Ala Lys Arg Asn Asn Phe
46             50             55
Lys Ser Ala Glu Asp Cys Met Arg Thr Cys Gly Gly Ala;
```

S–S bond: Cys 5-Cys 55, Cys 14-Cys 38, and Cys 30-Cys 51.

1. *Rotation Freedom about C^α-C^β Bonds*

In the 1D spectrum of protein ^{1}H-NMR we would usually see well separated resonance peaks only in the higher or lower field region (both ends of the spectral region of protein NMR spectra). Many NMR studies of proteins, therefore, have concentrated their efforts only on studying well separated peaks like those of methyl groups, aromatic groups, and peptide and other NH's. Judging from the 1D spectra of free amino acids, relatively simple spectra are manifested for α-protons with fine singlet, doublet, triplet, and doublet-doublet structures *(119)*. But they have long been abondoned because of the overlapping problem in protein ^{1}H-NMR despite their useful structural information. As shown in Figs. 11 to 14 and explained therein, this obstacle has now been overcome by 2D NMR spectroscopy.

1) Temperature dependence of coupling constants $^3J_{\alpha\beta}$

An accurate and unambiguous determination of coupling constants $^3J_{\alpha\beta}$ can be made for the residues isoleucine, threonine, and valine owing to the strong and well resolved doublet of their α-protons (Fig. 12). They therefore become good probes to discuss the protein flexibility at different locations in the protein. To interpret the obtained constants in structural term, of course, we need a pertinent model to the analysis together with a Karplus curve which connects the observed values of coupling constants and the dihedral angles, here, χ^1. From the view point of side chain flexibility, the rotamer population analysis as made in low molecular-weight molecules using Pachler equation *(120)* would be suitable. But we cannot rule out the fixed angle analysis for the protein which is expected to have locally rigid structures in it. This problem is just a matter to be examined in the NMR study. In the next subsection we discuss the problem in comparison with the X-ray structure. In this subsection, we investigate the observed coupling constants for above three types of amino acids under the assumption that they reflect the thermal fluctuation or libration of side chains.

Temperature dependence of coupling constants $^3J_{\alpha\beta}$ and their discrepancies with the crystallographically predicted values for six amino acid residues are collected in Table II. A bigger difference in the constants between those determined by NMR and those predicted by X-ray was evidenced closer toward the N- and C-terminii. On the other hand, at the α-proton of Thr 32 the accordance of NMR and X-ray values is fairly good for both BPTI and RCAM BPTI. We can expect about 7 Hz for the averaged coupling constants in the case of free rotation about the C^α-C^β bond. Except for the case of Thr 32, therefore, the negative sign of the discrepancies has just shown that changes in the direction of higher mobility are favored at the region near the two terminii. This supports the general idea that the protein is more flexible in its structure near the end of polypeptide chains. In big contrast to the big discrepancy with the X-ray predicted values, changes in coupling constants brought about by

TABLE II

Changes in Coupling Constants $^3J_{\alpha\beta}$ with Temperature and the Chemical Modification for BPTI

Amino acids	BPTI		RCAM BPTI	
	Discrepancy[a] with the X-ray predicted values (Hz)	Dependence[b] with temperature (Hz)	Discrepancy[a] with the X-ray predicted values (Hz)	Dependence[c] with temperature (Hz)
Thr 11	−3.2	−0.3	−4.1	−0.6
Ile 18	−1.8	−0.6	−2.7	−1.5
Ile 19	−1.9	0.0	−1.9	−0.6
Thr 32	−0.2	0.3	−0.2	0.0
Val 34	−1.7	−0.6	−1.7	−1.2
Thr 54	−3.1	0.0	−3.1	0.0

The values of the observed and the X-ray predicted coupling constants $^3J_{\alpha\beta}$ are listed in Table V in the Appendix.

[a] $^3J_{\alpha\beta}$ (NMR)$-^3J_{\alpha\beta}$ (X-ray) at 30°C, pH 4.2.

[b] $^3J_{\alpha\beta}$ (85°C)$-^3J_{\alpha\beta}$ (30°C) at pH 4.2.

[c] $^3J_{\alpha\beta}$ (65°C)$-^3J_{\alpha\beta}$ (30°C) at pH 4.1.

raising the temperature and reducing the disulfide bond Cys 14-Cys 38 are rather small. As far as viewing this protein through the C^α-C^β bond rotations of hydrophobic side chains, it seems to maintain a similar dynamic structure over a wide range of temperature and hold almost the same averaged conformation in the reduced molecule at low temperature. However, as for the flexibility of side chains some difference in the nature of local environments are seen between the intact and reduced protein. In the RCAM BPTI Ile 18 and Val 34, which are close to the modified center have shown a sizable dependence of their coupling constants on temperature as compared with the original jump in those from the crystallographically predicted values. But the effect of the modification seems not to propagate distantly. The RCAM BPTI is known to reduce its denaturation temperature from more than 100°C for the intact protein to 72°C at pH 4.2 (*121*). When a kind of rescaling of temperature is done according to this finding, we reach the following interpretation of protein flexibility. That is, protein flexibility is much stressed near the denaturation region and the fluctuation of local environments would become larger depending on the instability of the local structure prior to the large amplitude fluctuation.

2) *Comparison with the X-ray structure*

About 80 coupling constants have been obtained, to discuss the conformational aspects of the protein. Though the coupling constants for all of the assigned residues were not necessarily fully determined because of the overlapping of α-proton resonances still present in the 2D J-resolved spectra.

Except for alanyl residues of which methyl group rotate very rapidly in proteins (*116*), the coupling constants $^3J_{\alpha\beta}$ of amino acids are expected to quantitatively determine the χ^1 angles of the constituents. To go in this direction, however, first we have to overcome two hazardous problems as mentioned before: first, which Karplus curves empirically obtained should be taken as the basis of the calculation of the dihedral angle, and second, what kind of a model should be employed to interpret the observed constants? The rotamer population analysis or the fixed dihedral angle analysis could be used.

Among the several options reported so far, we have employed the most recent proposition for the Karplus curve by DeMarco *et al.* (*122*) for the reason that it was determined by NMR experiments using a rigid cyclopeptide ferrichrome and its derivatives of which the crystallographic structures have been established. The so-called "ferrichrome curve" was found to essentially coincide with the Karplus curve frequently utilized for peptide conformation analyses given by Kopple *et al.* (*123*) within an rms deviation of 0.8 Hz.

No *a priori* model can be assumed for the analysis of coupling constants. In order to examine the side chain flexibility, therefore, we have applied three criteria and classified the amino acid residues on this basis. 1) Group I: residues whose two coupling constants are compatible within 1 Hz deviation with $^3J_{\alpha\beta_2}$ and $^3J_{\alpha\beta_3}$ calculated from the crystallographic structure (*124*). 2) Group II: no agreement between the observed and the calculated, but explainable by assuming one unique χ^1 angle with allowance of 1 Hz ambiguity. 3) Group III: residues not belonging to either Group I or Group II. Residues belonging to the Group I are thought to maintain the local structures about C^α-C^β bonds similar to those of the X-ray structure. Residues of Group II are the candidates which differ in their local coordinates from the X-ray structure. For the residues of Group III, we could safely claim large fluctuations or rotation about the C^α-C^β bonds with a rate faster than the NMR time scale, here, about 10^{-2} sec. The results of such a categorization are summed up in Table III. Note discrimination between Group II and Group III based on the obtained coupling constants is only applicable to the amino acid residues which have an α-proton coupled to two β-protons (*120*). Therefore, amino acids residues

TABLE III

Classification of Amino Acid Residues in BPTI Based on the Observed Coupling Constants $^3J_{\alpha\beta2}$ and $^3J_{\alpha\beta3}$[a]

Group I	Group II	Group III
Asn 24	Lys 26	Arg 17, Arg 20
		Asn 43, (Asn 44)[b]
		Asp 3, Asp 50
		Glu 7, Glu 49, Gln 31
	Ser 47	Met 52
Phe 33, Phe 45	Phe 4	Phe 22
	Tyr 10	Tyr 21, (Tyr 35)[b]
Leu 6	Leu 29	
		Pro 9
Cys 5, Cys 14, Cys 30		Cys 55
Cys 38, Cys 51		
Thr 32		⎧Thr 11, Thr 54⎫[c] ⎨Ile 18, Ile 19 ⎬ ⎩Val 34 ⎭

[a]   The categorization is explained in the text.
[b]   Most probable classification. See the footnote to Table V.
[c]   These residues cannot be classified in either Group II or Group III.

isoleucine, threonine, and valine cannot be classified into Group II or Group III.

From the table we can see several aspects of the solution conformation of BPTI. First, hydrophilic amino acids Asn, Asp, Glu, Gln, and Arg are likely to be mobile in solution with one exception (Asn 24). Second, there are some difference in the mobility of hydrophobic groups; the C^α-C^β bond rotation of phenylalanines looks like more rigid than that of tyrosines. Third, χ^1 angles about C^α-C^β bonds for cysteines have the same values in solution structure as in the X-ray structure (Cys 55 is able to be classified into Group I by making an allowance in criterion 1 to within 1.5 Hz.)

A crystallographic study of BPTI protein (124) has told us that Arg 1, Thr 11, Arg 20, Asn 24, Gln 31, Tyr 35, Asn 43, Asn 44, Ser 47, Asp 50, and Arg 53 make intramolecular hydrogen bonds at the end part of side chains. As for these H-bonded side chains which are expected to be forced to have fixed local structures, however, we could not obtain any positive evidence that side chain motions have been severely restricted over a wide temperature range. Two possible examples of the restricted motion by the H-bond have been observed in Asn 24 and Asn 43 at the low temperature of 30°C (refer to Table V). This finding makes a contrast to the results obtained for cysteines where the disulfide bond seems to anchor the side chain to a fixed position.

Residues Phe 4, Tyr 10, Lys 26, Leu 29, and Ser 47 in Group II suggest that there exist some differences between the protein structures in solution and in crystal. To confirm this, the interpretation given to Group II residues should be further tested, because it is probable that it could replaced by another explanation emphasizing the fluctuations in the rotamer populations about the C^α-C^β bond. Especially rapid rearrangement of the side chain between a few rotamers could be rather postulated for residues Try 10 and Lys 26, if the chemical shift equivalence of two β-protons is seriously taken. Conclusively, the recently prevailing new picture for the protein structure has been also confirmed in this study by investigating the structural aspects of BPTI with 2D NMR. In detail, however, there are big variations in the amount of fluctuations depending on the local environments. The hydrophobic surroundings around phenylalanines seem to be more rigid than those around tyrosines. The order of magnitude of the proton vicinal coupling constants is 10 Hz. Therefore, relatively slow motions in the structural fluctuations (10^{-2} sec) are reflected to suffice averaging the coupling values.

2. Structural Implications of NMR Parameters

1) Comparison with numerically calculated structural parameters

The comparison of the structure obtained from NMR results with the X-ray structure in terms of rotation about the C^α-C^β bond leads us to the idea that the structure in solution is more loosened than the structure in crystal. In crystal, protein molecules are thought to interact with neighboring proteins. Particularly, side chains on the protein surface cannot avoid being influenced by the side chains of the proteins nearby. For the BPTI protein, the importance of surrounding protein molecules in the crystal has been reported (125). Table IV shows the results which were evaluated based on the real-space refined cyrystal structure (124). The amino acid residues listed in the table are thought to be anchored at the end parts of their side chains and are then expected to strengthen or tighten the structure of the overall protein conformation in crystal. This nearest neighbor interactions among proteins in crystal could give answers to the two severe questions about protein conformations: 1) Why is the X-ray crystallography successful in determining the three-dimensional structure of proteins including the orientation of individual side chains?; 2) What is the

TABLE IV

Importance of Surroundings in the Crystal Structure of the BPTI Molecule (from Gelin and Karplus (*125*))

Residue	Interaction[a]	Distance (A)	vdW[b]	elec[b]	H-bond[b]
A. Side chains significantly improved by inclusion of interactions with surroundings					
Arg-17	$N_{\eta 1}\cdots$solvent	3.61	−0.17	0	0
	$N_{\eta 2}\cdots C_{\alpha}58$	2.65	3.5	0.68	0
	$N_{\eta 2}\cdots$solvent	2.94	0.17	0	−2.4
Lys-26	$C_{\delta}\cdots C_{\beta}(58)$	3.67	−0.13	0.03	0
	$C_{\epsilon}\cdots C_{\beta}(58)$	3.42	0.01	0.12	0
	$N_{\zeta}\cdots C_{\beta}(58)$	3.12	0.51	0.45	0
Gln-31	$N_{\epsilon 2}\cdots$solvent	3.25	−0.17	0	0
Arg-39	$N_{\epsilon}\cdots O_{\epsilon 1}(49)$	3.11	−0.28	0.57	−3.0
	$C_{\zeta}\cdots C_{\beta}(48)$	3.67	−0.15	0.23	0
	$N_{\eta 1}\cdots O_{\eta}(21)$	2.78	0.87	0.99	−2.3
	$N_{\eta 2}\cdots$solvent	2.94	0.19	0	−2.4
	$N_{\eta 2}\cdots O_{\epsilon 2}(49)$	3.06	2.3	−3.9	0
Glu-49	$O_{\epsilon 1}\cdots N(3)$	2.95	−0.12	2.41	0
	$O_{\epsilon 1}\cdots$solvent	3.02	−0.12	0	3 0
	$O_{\epsilon 1}\cdots N_{\epsilon}(39)$	3.11	−0.28	0.57	−3.0
	$O_{\epsilon 1}\cdots C_{\alpha}(2)$	3.34	−0.20	−1.24	0
	$O_{\epsilon 2}\cdots N_{\epsilon 2}(39)$	3.06	−0.23	−0.39	0
Asp-40	$O_{\delta 1}\cdots$solvent	2.84	0.30	0	−3.5
	$O_{\delta 1}\cdots$solvent	2.85	0.26	0	−3.5
	$O_{\delta 2}\cdots$solvent	2.96	−0.04	0	−3.3
Met-52	$C_{\gamma}\cdots C_{\gamma}(14)$	3.77	−0.04	−0.04	0
	$C_{\epsilon}\cdots C_{\beta}(16)$	3.50	−0.03	0.08	0

B. Side chains not significantly improved, in poor agreement with X-ray (χ_1, χ_2)

Asp-3:   many exterior interactions, all$>$3.7 Å; no hydrogen bonds

Glu-7:   very few interactions; one hydrogen bond ($O_{\epsilon 1}\cdots$solvent, 3.08 Å, −2.76 kcal/mol)

Lys-15: shortest interaction ($N_{\zeta}\cdots$solvent) 4.37 Å; no hydrogen bonds

Lys-41: few interactions$<$4 Å; no hydrogen bonds

Arg-42: very few interactions, few shorter than 5 Å

Lys-46: many interactions, but very few$<$4 Å; no hydrogen bonds

[a]   Residue atom, atom in adjacent protein unit or solvent atom.

[b]   Energies in kilocalories per mole calculated from distances in X-ray coordinates.

origin of its discrimination ability about the flexibility of side chains between the crystal structure determined by X-ray and the solution structure suggested by NMR? Starting from this viewpoint and taking into account some structural resemblances still retained between the crystal and solution structures for the rigid parts in the BPTI protein as suggested by

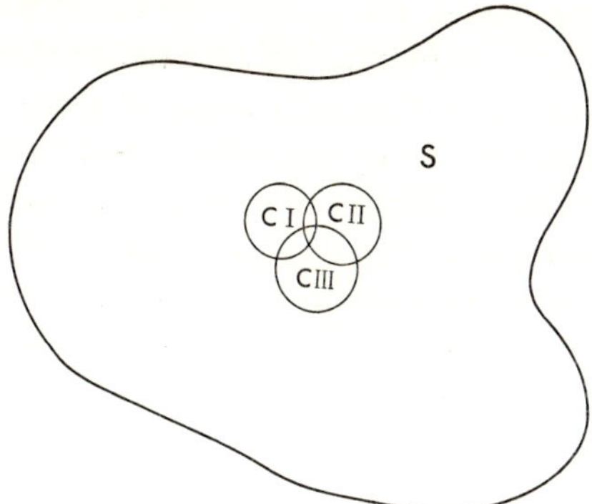

Fig. 16. Schematics of the conformational space of proteins that depict the relation between the solution structures and the crystal structures of proteins. S, conformational space covered by the solution structures; C, conformational space covered by the crystal structures. I, II, and III indicate the difference in crystal structures resulting from the difference in the crystal forms.

Group I in Table II, we offer a schematic idea explaining the relation between the two structures (Fig. 16). On this understanding we will compare the various NMR parameters with the structural parameters calculated on the basis of the X-ray structure.

In order to examine the stuructural aspects of the BPTI molecule in solution, we have postulated two spectral measures to evaluate conformational fluctuation of side chains around the C^α-C^β bond: 1) flexibility factor $|{}^3J_{\mathrm{NMR}} - {}^3J_{\mathrm{X\text{-}ray}}|$ that is a measure of the difference between the crystal structure and the solution structure and calculated by taking an average of the absolute values of subtractions between NMR-observed and X-ray predicted coupling constants ${}^3J_{\alpha\beta_2}$ and ${}^3J_{\alpha\beta_3}$ for each of amino acid residues; and 2) the temperature coefficient of chemical shift $\overline{|\Delta\delta_\beta/\Delta T|}$, that, is a measure of structural change dependent on temperature and calculated by taking an averaged temperature coefficient of chemical shifts of one or two β-protons observed for each of amino acid residues. Since the X-ray "rigid structure" reveals (124) that χ^1 angles about C^α-C^β bonds fall close to the angles corresponding to the three rotamer conformations about the C^α-C^β bond, *trans*, *gaushe⁺*, and *gaushe⁻*, the flexibility factor is expected to reflect the flexible nature of side chains in terms of rotamer fluctuation about the C^α-C^β bond. As to the calculation of this factor there are two choices caused by the ambiguity about stereospecific assignments for two

β-protons. Here we have always taken the smaller one as the measure on the understanding that the solution structure includes the crystal structure as one of the possible fluctuations in the conformational space (Fig. 16).

As for the structural parameters about the conformation of the BPTI, M. Gō and N. Gō evaluated the following theoretical parameters (*126*): residue accessibility (*126a*), accessible surface area of individual α- and β-carbons (*127*), volume of Voronoi polyhedron (*128*) and the numbers of near atoms. Among those calculated parameters, residue accessibility, which is a structural measure of protein conformations to express how a side chain is extended to the surrounding water solvent, has given an interesting correlation with two measures obtained from the NMR parameters.

A correlation diagram of the flexibility factor and residue accessibility is shown in Fig. 17a. With a few exceptions which imply some importance in the specificity of local structures as seen later, a broad correlation is evidenced between these two quantities. The implication is that flexibility of side chains as far as rotation about C^{α}-C^{β} bonds concerned is primarily determined by the accessibility of the surrounding water molecules to the side chains. This result seems to agree with our general understanding about the solution structure of proteins. However it is rather surprising to find such a correlation between two parameters, if we notice that the residue accessibility was calculated on the basis of the "rigid structure" obtained by the X-ray crystallography (*124, 129*). As a future problem it is necessary to refine the structural parameter of residue accessibility by taking into consideration the dynamical properties of the protein. Nevertheless we may safely claim such a broad correlation between the solution and crystal structure, provided that we hold the view schematically shown in Fig. 16.

Concerning the amino acid residues which do not hold correlation between the two parameters, we can see some interesting trends. The residues belonging to the hydrophobic group, isoleucine, leucine, and valine, have shown smaller flexibility factors than expected for the group having a medium size of residue accessibility. This might be reasoned as being due to the rigid structure of water surrounding this hydrophobic group, say, an iceberg structure (*130*). Cysteines 14 and 38, which mutually make a disulfide bond look less flexible in spite of their accessibility to water. A very small flexibility factor characterizes all the six cysteine groups. The other characteristic trend visible in the correlation diagram is the large flexibility factor resulting in two tyrosine residues. Tyrosines 10 and 21 have shown fine structures of triplet in the cross-sections of the 2D

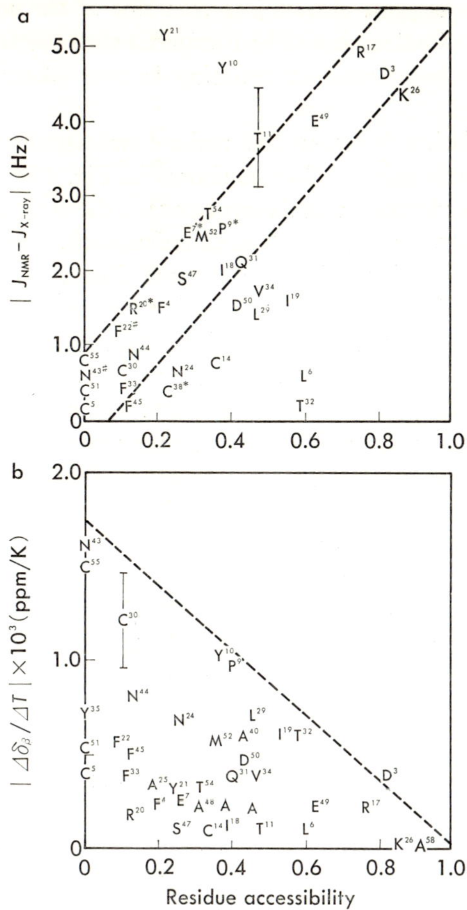

Fig. 17. Correlation between the NMR parameters and a structural parameter, residue accessibility. The residue accessibility of individual amino acid residues was calculated from a summation of the main-chain accessible surface area and the side-chain accessible surface area by normalizing it with the accessible surface area of an extended tripeptide (Gly-X-Gly; $X=20$ amino acids) (126a). a: flexibility factor vs. residue accessibility. Coupling constants $^3J_{\alpha\beta}$ were determined with a 2D J-resolved spectrum obtained for a BPTI solution in ^2H_2O pH$=$4.2, $T=$68°C. $^3J_{\alpha\beta}$ for E7, P9, R20, and C38 were collected from the data measured at $T=$85°C, pH$=$4.2. $^3J_{\alpha\beta}$ for N43 and F22 is that measured at $T=$30°C, pH$=$4.2. b: temperature coefficient of the chemical shift of the β-proton vs. residue accessibility. The temperature coefficients were evaluated from the data of 2D J-spectra obtained for a BPTI solution at $T=$30°C, 68°C, and 85°C, pH$=$4.2.

spectra of BPTI, which suggest almost free rotations of the side chains about C^α-C^β bonds. This finding should be compared with the less flexible nature about the C^α-C^β bond rotation resulting in four phenylalanines, 4, 22, 33, and 45.

The second correlation between the residue accessibility and the structural measure determined by NMR is shown in Fig. 17b. Although we could not see here a vigorous correlation between the residue accessibility and the temperature coefficient of the chemical shift, a sort of anticorrelation seems to be manifested in the diagram. That is, the residues which have high accessibility to water give small temperature coefficients. But the residues with lower residue accessibility are forced to have a wide variety of temperature coefficients. Together with the result shown in Fig. 17a, this trend may be interpreted as the side chains that are well extended to solvent water show relatively small changes in the averaged local structure in solution, in other words, the dynamic structure of the local environment around such side chains is well averaged by the jump form the crystal phase to the solution phase. Similar trends have already been mentioned concerning the rotational properties of the C^α-C^β bonds for aliphatic hydrophobic residues (Table II). On the other hand, the β-protons around the protein core have a wide variety in their coefficient as suggested by the second trend. No matter what type of source causes the structural dependent shift resulting for the β-protons, either the ring current shift (131–134) or the back-bone carbonyl magnetic anisotropy (135–137), such a variety should only be interpreted on the basis of the structural specificity realized in each local region in the actual protein. At least two residues, Pro 9 and Cys 55, have good reason to show a rather big temperature coefficient of chemical shift despite the low residue accessibility, since the β-protons of those residues are very close to the phenylalanyl or tysosyl ring, within a distance from 3 Å to 4 Å (Pro 9-Phe 22 and Phe 33, Cys 55-Tyr 23). For a proton which is influenced by the ring current of neighboring aromatic rings, very little change in the mutual distance explains the relatively large change in shift (132, 134). Instead of going into a detailed investigation of this problem, we will look at the problem from another side, placing more emphasis on the overall properties of the protein structure. Next what kinds of structural changes occur with raising the temperature are investigated.

2) *Thermal expansion of the protein molecule*
The proton chemical shift in proteins is primarily determined by the electronic structure of individual amino residues (119, 138). But it also

carries information about the three dimensional structure of the molecule. In proton NMR the structural information appears as a discreapancy in the observed shift as compared with the primary shift (random coil value of the chemical shift). In this context a correlation between the structural dependent shift, which is here evaluated by the difference in the random coil value (*139*) and the observed value of proton chemical shifts ($\Delta\delta_\mathrm{I}$), and the dependence of the observed shift on temperature ($\Delta\delta_\mathrm{II}$) would give an interesting information on the thermal properties of the protein structure. In the previous subsection we have made clear that change in the averaged structure of proteins is more vividly evidenced in the chemical shifts of protons inside the protein. Therefore it might be reasonable to take such a correlation for a type of proton directly attached to the protein backbone here the α-proton. A correlation diagram for the identified α-protons of BPTI is given in Fig. 18.

Before giving a detailed discussion about the thermal aspects of proteins we point out two important trends seen in the diagram. In NMR study of organic molecules the observed change in shifts dependent upon the experimental conditions, say, temperature or pH, is very often inter-

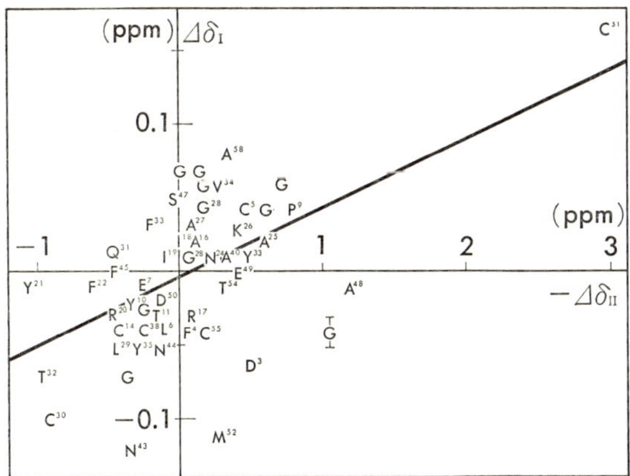

Fig. 18. Correlation between the temperature dependence of chemical shift ($\Delta\delta_\mathrm{I}$) and the structural dependence shift ($\Delta\delta_\mathrm{II}$) for the α-proton. $\Delta\delta_\mathrm{I}=\delta(30°\mathrm{C})-\delta_\mathrm{random\,coil}$, $\Delta\delta_\mathrm{II}=\delta(85°\mathrm{C})-\delta(30°\mathrm{C})$. Where $\delta(85°\mathrm{C})$ and $\delta(30°\mathrm{C})$ represent the observed chemical shifts for α-proton resonances of BPTI obtained at $T=85°\mathrm{C}$ and $30°\mathrm{C}$, pH$=4.2$. $\delta_\mathrm{random\,coil}$ is the chemical shift of the α-protons of individual amino acids in the random coil state (*139*). The straight line indicates a regression line with the values: slope$=4.7\times10^{-2}$ and regression coefficient $r=0.50$.

preted as the result of a rapid interconversion of a few dynamically equilibrated states (*140*). The chemical shift is then connected to the populations of a few states, for example populations of the random coil and helix states in the helix-coil transitions of polypeptides (*141–145*). This so-called chemical exchange mechanism is also a good candidate to explain the temperature change in the shift values of BPTI. Actually this is a prerequisite to interpreting protein NMR spectra, because we have only picked up proton resonances which have shown one-to-one correspondence to the constituent protons with observable peak heights. By advancing such an argument further ahead, one might imagine that the observed change in shift in our study reflects a rapid interconversion of two states, namely, the native and denatured states. But such a claim is completely against the result shown in Fig. 18. We have not seen any strictly linear relationship between two quantities that would be always expected for the two-state model. The origin of the small change in shift observed prior to the denaturation (folding) is then thought to reflect some structural change in the native state of the protein, though we cannot yet specify the nature of the change. This is our implicit stand point employed through this study.

Second, we pay attention just to the broad correlation itself manifested in the diagram. As seen in Fig. 18 the data points in the diagram are scattered over almost all directions. The number of data points scattered in the first and the third quadrants, however, fairly exceeds that appeared in the second and the fourth quadrants. Then as a general tendency we observe that the value of the chemical shift increases or decreases associated with raising temperature so as to loosen the structure, it means that the chemical shift observed at higher temperature favors the value at the random coil state of the protein. An extremely loosened structure of proteins is, of course, that at the random coil state of proteins. Of importance here is that the very small temperature-dependent shifts observed in the native state as a whole have shown a similar characteristic trend to be elaborated in the large amplitude fluctuation of denaturation.

To evaluate the amount of loosening in the protein structure at high temperature, we made an attempt to estimate the thermal expansion coefficient of the protein molecule in solution. As mentioned before the structural dependent shift in proton NMR is caused by diamagnetic fields originating from the magnetically anisotropic centers near protons. No matter what kinds of sources give this effect the influencing fields are well approximated by a type of dipole field (*146, 147*).

$$\sigma(\boldsymbol{r}_i) \propto \frac{1}{r_i^3} f(\theta_i) \tag{23}$$

Here $\boldsymbol{r}_i$ is the interaction vector to a proton measured from the anisotropic center, and r_i and θ_i are the polar coordinates of the vector $\boldsymbol{r}_i$ that represent the distance and orientation of a proton. For a very small change in the interaction vector, the variation of the structural dependent shift $\sigma(\boldsymbol{r}_i)$ is given by

$$\sigma(\boldsymbol{r}_i+\Delta\boldsymbol{r}_i) = \sigma(\boldsymbol{r}_i)\left(1-3\frac{\Delta r_i}{r_i}+\frac{f'(\theta_i)}{f(\theta_i)}\Delta\theta_i-3\frac{f'(\theta_i)}{f(\theta_i)}\frac{\Delta\theta_i\Delta r_i}{r_i}\right) \tag{24}$$

Then the dependence of the relative change in shift on the small variation in the vector becomes.

$$\frac{\sigma(\boldsymbol{r}_i+\Delta\boldsymbol{r}_i)-\sigma(\boldsymbol{r}_i)}{\sigma(\boldsymbol{r}_i)} = -3\frac{\Delta r_i}{r_i}+\frac{f'(\theta_i)}{f(\theta_i)}\Delta\theta_i-3\frac{f'(\theta_i)}{f(\theta_i)}\frac{\Delta\theta_i\Delta r_i}{r_i} \tag{25}$$

Equation (25) asserts that if a molecule uniformly expands ($\Delta\theta_i=0$), the shift $\sigma(\boldsymbol{r}_i)$ also changes uniformly depending only on $\Delta r_i/r_i=$const., regardless of the magnitude of the original value of shift $\sigma(\boldsymbol{r}_i)$. Judging from the diagram in Fig. 18, this is not the case for the BPTI molecule for the same reason given before to reject the two-state model. The heterogeneity of local structures in proteins must be carefully considered. As implied in Eq. (25), the relative change in shift will show great variation in various local protons for nonuniform changes in the overall structure of the molecule. To extract some overall properties about the stuructural aspect of the protein conformation, then, an average of the expression given by the left side in Eq. (25) should be taken. Assuming the averaging-out of the cross-term between the angle and distance variations in Eq. (25), we obtain.

$$\overline{\frac{\Delta\sigma(\boldsymbol{r}_i)}{\sigma(\boldsymbol{r}_i)}} = -3\overline{\frac{\Delta r_i}{r_i}}+\overline{\frac{f'(\theta_i)}{f(\theta_i)}\Delta\theta_i} \tag{26}$$

In our study the first term on the right side of Eq. (26), is thought to reflect the volume expansion of the molecule. The second term, which is rather difficult to interpret with a simple structural term, can be thought of as corresponding to a sort of skewness of the protein structure associated with alternation of external parameters, here, temperature.

Linear regression analysis has been performed on the diagram in Fig. 18 to evaluate the average of relative change in shift $\overline{\Delta\sigma(\boldsymbol{r}_i)/\sigma(\boldsymbol{r}_i)}$, where more than 50 data points are presented to make such an analysis practical. The slope of the regression line designated by a straight line in

Fig. 18 corresponds to the left side of Eq. (26) and finally yields 8.5×10^{-4} K^{-1} as the temperature coefficient of that quantity. If the second term can be neglected as compared with the first term in Eq. (26), the temperature coefficient of the relative change in shift $\overline{\Delta\sigma(r_i)/\sigma(r_i)\Delta T}$ is directly related to the thermal expansion coefficient in volume owing to the factor 3 multiplying to the linear expansion ratio $\overline{\Delta r_i/r_i}$ in the first term of Eq. (26). We thus obtain an expansion coefficient for the BPTI molecule.

$$\frac{\Delta V}{V\Delta T} = 8.5\times10^{-4}\mathrm{K}^{-1} \qquad (30°\mathrm{C} \rightarrow 85°\mathrm{C})$$

It might be thought that it is too daring to assign a macroscopic view of thermal expansion to a very small system with an atomic scale and dangerous to define a expansion coefficient for a heterogeneous system like protein molecules. In fact we have not yet clarified the meaning of protein thermal expansion with the use of molecular terms. We do not take it to be like a mechanical expansion of a bulk matter but rather think of it as representing a measure of change in the protein structure averaged at dynamic equilibrium in the conformational space (Fig. 16). With that understanding, we can attempt to compare the obtained value of the expansion coefficient for the protein to the typical values of volume expansion coefficient reported for organic liquids and solids (148).

organic liquids: $8–17 \times 10^{-4}$ K^{-1} ($\sim 20°$C, acetone, benzene, chloroform, diethylether, ethanol, ethylchloride, toluene, phenol),

protein (BPTI): 8.5×10^{-4} K^{-1} ($30 \rightarrow 85°$C),

organic solids: $2–5 \times 10^{-4}$ K^{-4} ($\sim 20°$C, anthracene, benzoinic acid, camphor, naphthalene, phenol, potassium acetate, urea).

The obtained value for the protein is seen to fall between the values of organic liquids and solids being slightly closer to the values for organic liquids. This result gives another example supporting the notion that the protein molecule is a semi-liquid particle, as was explained in a previous section.

Finally we would like to make some comments about the relation between our findings and the recently proposed structural model of proteins, "the hydrophobic cluster model" (109, 110, 117). This protein model claims that there are a few stable hydrophobic domains which are loosely bound to each other by intramolecular interactions. Of course we must define what is implied in "stable" domains but it is very interesting to assume that there exist a few cooperative domains in such a small protein (MW=6,500). Provided that we accept such a model, we have to pursue further problems concerning protein flexibility, focussing on the special nature of each local environment. The analysis carried out in this section has primarily been aimed at that purpose. Unfortunately, however, we have failed to specify domain structures clearly in the protein in our study. For example, from Fig. 18 we could pick out would-be rigid trends around Tyr 21, Phe 22, Phe 45, Ala 48, and Glu 49, and those seem to be candidates for hydrophobic domains. But they are also explained by stressing the nonuniform change in the protein structure together with the implication of Eq. (26). To do this kind of task more extensive experiments are needed to observe proton resonances for γ-, δ-, ε-, NH and aromatic protons with higher resolution. Also we must consider what the relation between small amplitude fluctuations in the native state and the large amplitude fluctuations manifested in protein denaturation is, since "the hydrophobic cluster model" is thought to reflect the comformational property of the latter.

VI. GENERAL DISCUSSIONS AND CONCLUDING REMARKS

The new NMR measurement technique, 2D spectroscopy, is now opening a new field in investigation of the structural aspects of protein molecules in solution in molecular terms. As an example, a systematic assignment and the determination of NMR parameters, which are developed to discuss the structural properties of a protein (BPTI), are described. Experiments have been performed with particular classes of 2D spectroscopy, 2D J-resolved and 2D correlated NMR spectroscopies. As described in Sec. II, there are tremendous varieties in 2D NMR realization. Therefore the application of 2D NMR to biological systems may have more variety and flexibility, and is not confiend to the two types of 2D NMR utilized herein. In this context, for example, 2D NOE NMR (*52*) offers a very powerful and promising option in systematic resonance assignments (*149, 150*). Thus various types of 2D NMR could be combined in a more effective manner (*96a*).

Two-dimensional spectroscopy in general does not require any special NMR techniques or electric devices. If a good software and a big data storage capacity are available to handle the experiment and the data obtained, any NMR spectrometers with pulse FT equipment can be used for 2D experiments. Of course much longer experimental and data processing time is required to carry out these procedures but that drawback is compensated by the high degree of resolution and accuracy of information realized in 2D specrta. At any rate, 2D NMR is an extension of multi-pulse NMR measurements. Consequently the necessary care does not exceed that usually taken in such measurements.

Next we will make clear the applicable range of this new technique to actual protein systems within our experience of using the two types of 2D NMR. Since both methods rely on the appearance of well-separated fine structures in the NMR spectra, molecules with low molecular weight are always favored. For the proton nuclei at 360 MHz, protein molecules with a molecular weight lower than 10^4 (glucagon, melittin, cardiotoxin, bacterial cytochrome c_{551}, BPTI and HPI) result in good 2D spectra that fit to the spin analysis for methyl, aromatic and α-protons. For protein molecules with a molecular weight higher than 10^4 (cytochrome c horse and procollagen) it is more difficult to get good 2D spectra due to the broadening and overlapping of many resonance peaks invevitable in the ^{1}H-NMR of big molecules. But this problem is always related to more general problems concerning the sampling resolution and the signal-to-noise ratio in NMR measurements. In general 2D NMR measurements need the longer experimental time or larger numbers of accumulations to get the same signal-to-noise ratio of the spectra as obtained in 1D NMR (83). This is caused by the delay time t_1 given before starting the data acqui-sition. When a spin system has a very rapid relaxation rate, which is the case in large molecules, the level of detected signals in the detection period naturally becomes low after the delay t_1. To avoid this difficulty one frequently applies a short acquisition time of data sampling in both the t_1 and t_2 periods. This remedy, however, impairs the virtues in-herent in 2D spectroscopy because of the resulting low resolution in the sampling points. The signal-to-noise ratio and the sampling resolution of a spectrum are mutually complementary (151). We cannot shorten the ex-perimental time of NMR measurements without ruining the sampling resolution. But if we can afford low sampling resolution in the spectra, we can shorten it without ruining the signal-to-noise ratio. In our study shorter experimental times were utilized to carry out the 2D correlated NMR (Fig. 11) as compared with those applied in the 2D J-resolved NMR (Fig. 12). The resulting signal-to-noise ratio in the 2D correlated spectra seemed to be even better than that resulting in the 2D J-spectra. This illustrates just the effect mentioned above, because a sampling resolu-tion (2.44 Hz) of four times of the resolution (0.61 Hz) of the latter was given to the 2D correlated NMR measurements. A low resolution of 2.44 Hz is, however, high enough to determine the chemical shift of proton resonances with an accuracy of 0.01 ppm at 360 MHz NMR. Therefore to overcome the experimental difficulty of obtaining good spectra in proteins bigger than 10^4, we must be very careful to compromise the two requirements of a good signal-to-noise ratio and good sampling resolution, and to find out the optimal conditions accroding to the purpose of each particular NMR experiment.

One interesting conclusion has been drawn from this 2D NMR study about the structural aspects of proteins in solution. It has been asserted that proteins in solution would be more flexible than proteins in crystal. This idea is schematically illustrated in Fig. 16. This proposition, however, holds some contradictions, since we know that the protein molecule itself should be identical either in crystal or in solution. This apparent contradiction has been resolved by asking the origin of the difference in structures bringing about the difference in interactions between proteins in the two phases.

Nearest neighbour interactions of protein molecules in the crystal lattice have been thought to be one possibility to explain this. The schematic idea proposed here (Fig. 16) also implies that the nearest neighbour interaction restricts in general the possible conformations of proteins and in some cases fixes them in very rigid conformation selected from the ensemble of many dynamically equilibrated conformations in solution as experienced in the crystal structure. Our model for protein structures in solution, which was drawn from only one protein sample, should be further tested. For example, a comparison of X-ray structures of a protein in various crystal forms would reveal how the local structures of the protein are dependent on the crystal forms (*153*). Concerning NMR study, 2D NMR approaches similar to this work could be applied to other protein samples with higher molecular weight. On the other hand, for a theoretical approach to the protein conformations, molecular dynamics of the protein structure in a long time scale (up to 10^{-3} sec) could in the future be compared with the results obtained by NMR. Lastly we repeat our conclusion again. The structure of protein in solution is not completely the same as that realized in crystal. The crystal structure is a "rigid" extreme and the solution structure is a "flexible" extreme.

SUMMARY

This paper deals with a few aspects of recently developed new NMR technique, two-dimensional (2D) NMR, ranging from its principle to its biological applications. First the historical background of the new technique is surveyed and its underlying principle is explained. To cover the versatile applicability and flexibility of the technique in the NMR studies, then, various experimental techniques proposed up to now are classified from the general point of view of this new experimental frame. In Sec. III spectroscopic characteristics experienced in the practical data handling are described. We observe there that the causality principle and several theorems of multidimensional Fourier transformation play a key role to costrain the appearance of 2D spectra in general. Applications of 2D NMR to the protein study are described in the following sections. For the successful use of NMR techniques in the elucidation of protein structures in solution, the individual resonance assignments for each of the amino acid residues are a crucial first step. The combination of two classes of 2D NMR experiments, J-resolved and correlated 2D NMR, was found to make it feasible to obtain this critical data in a systematic way.

Based on the individual assignments of $C^\alpha H$ and $C^\beta H$ proton resonances in the basic pancreatic trypsin inhibitor (BPTI), coupling constants, $^3J_{\alpha\beta}$, between $C^\alpha H$ and $C^\beta H$ and their chemical shifts were determined with 2D J-resolved NMR. The results were compared with the values of $^3J_{\alpha\beta}$ predicted from the crystallographically determined structure with a Karplus-type curve. Flexibility about C^α-C^β bonds in BPTI was then discussed for the individual residues. A model to explain the difference in internal mobility of the protein recognized between the NMR results and the prediction from the X-ray structure was also proposed. On this model NMR parameters were compared with a theoretical parameter, residue accessibility, calculated from the X-ray structure. With a few exceptions, most of amino acid side chains in the interior of the protein were found to be locked into unique spatial orientations, with the mobility restricted to rapid torsional fluctuations about a unique χ^1 value. For the residues on the protein surface structual rearrangement was found which includes rapid averaging between two or several distinct populated values of χ^1. As an extension of the picture for the protein flexibility in solution, the thermal expansion coefficient of the protein molecule was estimated with the NMR chemical shift data. The result has characterized the structural aspect of the protein molecule, that is the protein is not strictly rigid in the native state and has a liquid-like property in the solution phase.

Acknowledgments

The author wishes to dedicate this article to Professor K. Wüthrich and Professor R. R. Ernst to commemorate of his stay with their groups in ETH-Zürich from 1976 to 1979. He expresses his thanks to all group members of Prof. K. Wüthrich's and Prof. R. R. Ernst's laboratories for their discussions and wishes to express his particular thanks to Dr. P. Bachmann for his support in writing a software program for 2D NMR. He is also grateful to Dr. M. Gō and Dr. N. Gō for the results of the calculations of residue accessibility and other structural parameters. He gratefully acknowledges illuminating discussions and critical reading of the manuscript to Dr. E. Bartholdi. This study was supported by research grants provided by the Swiss Federal Institute of Technology and by the Swiss National Science Foundation.

APPENDIX

Listed here are chemical shifts of α-, β-, and γ-protons and coupling constants $^3J_{\alpha\beta}$ for the assigned amino acid residue in BPTI. Evidences utilized for the resonance assignments are also shown. For the ring current calculation, we have followed the procedure proposed by Perkins & Wüthrich (*133*). That is: 1) the Johnson-Bovey shift equation (Johnson & Bovey, (*152*)) was employed: 2) the center of mass and the mean square plane of the aromatic ring were calculated to put the cylindrical coordinate system onto each ring; 3) assumed electron density of π-electrons was 6.0 and 5.64 for the phenylalanine and tyrosine ring current, respectively; 4) tetragonal geometry was assumed to put hydrogens to carbon atoms with bond lengths of 1.070 Å for methine groups, 1.073 Å for methlene groups and 1.101 Å for methyl groups. Three dimensional atomic coordinates were taken from the real space refined structure of BPTI (Deisenhofer & Steigemann, (*124*)).

To calculate the $^3J_{\alpha\beta}$ coupling constants the χ^1 angles about C^α-C^β bonds were taken from the refined X-ray structure and the "ferrichrome curve" proposed by DeMarco *et al.* (*122*) was employed as a Karplus curve.

TABLE V

Collections of Individual Resonance Assignments and Evidence for Assignments in the Proton NMR Spectrum of the Basic Pancreatic Trypsin Inhibitor

Resonance assignment		30°C pH 4.2	Chemical shift from TSP (ppm)						Coupling constant $^3J_{\alpha\beta}$ (Hz)		
			Shift difference between pH 7.0 and pH 3.1 at 68°C[a]	Shift difference between reduced BPTI and BPTI at 30°C, pH 4.2[a]	Shift difference between 85°C and 30°C at pH 4.2[a]	Shift difference between the observed and random coil values at 30°C, pH 4.2[b]	Ring current shift[c]	Assigned shift in other works[d]	30°C pH 4.2[e]	85°C pH 4.2[e]	Predicted from the X-ray structure[f]
Asp 3	α-CH	4.26	−0.07		−0.06	−0.51	−0.07				
	β-CH	2.79	−0.16			−0.05 or 0.04	0.01		6.1	6.1	10.9
	β-CH	2.79	−0.16			0.04 or −0.05	0.02		6.1	6.1	1.7
Phe 4	α-CH	4.58	0.03			−0.08	0.23				
	β-CH	2.97	−0.06		−0.02	−0.02 or −0.25	0.00		{16.0}*§	5.5*§	4.1
	β-CH	3.33	0.04			0.11 or 0.34	0.00			4.3*§	2.7
Cys 5	α-CH	4.33	0.03		0.04	−0.36	0.26				
	β-CH	2.74				−0.21 or −0.53	−0.17		12.2	12.8	12.9
	β-CH	2.88			−0.02	−0.39 or −0.07	−0.05		3.7	3.1	3.9
Leu 6	α-CH	4.51			−0.04	0.12	0.16				
	β-CH	1.83				0.18	0.05		12.8	11.6	12.8
	β-CH	1.83				0.18	0.06		3.1	4.3	4.1
	γ-CH	1.68			−0.02	0.03	0.03	1.73			
Glu 7	α-CH	4.55				0.25	0.05				
	β-CH	2.19	0.04			0.10 or 0.23	0.04			6.4	2.6
	β-CH	2.19	0.04			0.23 or 0.10	0.04			8.5	9.0
	γ-CH	2.64	−0.15		−0.05	0.36 or 0.30	0.16				
	γ-CH	2.64	−0.15		−0.05	0.36 or 0.30	0.14				

Pro 9	α-CH	3.68	0.04	0.06	0.04	−0.79	−0.55				
	β-CH	1.33			−0.07	−0.97 or −0.65	−1.37			3.7	3.6
	β-CH	0.11			0.04	−1.87 or −2.19	−2.07	0.09 or 0.21		7.9	12.9
Tyr 10	α-CH	4.93		−0.08	−0.02	0.33	0.17				
	β-CH	2.97	0.02	0.02	−0.06	−0.16 or 0.05	−0.04		8.5	8.5	3.6
	β-CH	2.97	0.02	0.02	−0.06	0.05 or −0.16	0.01		8.5	8.5	12.9
Thr 11	α-CH	4.53		−0.03	−0.03	0.18	0.25				
	β-CH	4.04				−0.18	−0.12	4.04	8.8	8.5	12.0
Cys 14	α-CH	4.55		−0.12	−0.04	−0.14	−0.07				
	β-CH	2.79		0.06 or 0.19		−0.17 or 0.20	−0.17			12.2	12.4
	β-CH	3.48		−0.50 or −0.64		0.20 or 0.52	−0.12			3.1	2.1
Ala 16 or	α-CH	4.28				−0.07	−0.08	4.3			
Ala 27	β-CH$_3$	1.18				−0.22	−0.12	1.19	7.3	7.3	
Arg 17	α-CH	4.31		−0.03	−0.03	−0.09	−0.09	4.29			
	β-CH	1.60		−0.06		−0.20 or −0.32	−0.04			7.3	12.9
	β-CH	1.60		−0.06		−0.32 or −0.20	−0.05			7.3	2.1
Ile 18	α-CH	4.19		0.09	0.02	−0.03	−0.02	4.25			
	β-CH	1.88	0.02	−0.03		−0.01	−0.02	1.87	11.0	10.4	12.8
Ile 19	α-CH	4.31		0.07		0.09	−0.14	4.30			
	β-CH	1.96		−0.04	−0.03	0.07	−0.01	1.96	11.3	11.0	12.9
Arg 20	α-CH	4.69			−0.03	0.29	0.13	4.70			
	β-CH	0.86	0.02			−0.94 or −1.06	−0.37	1.62		11.6	12.9
	β-CH	1.80				0.12 or 0.00	0.17			3.7	2.9
Tyr 21	α-CH	5.68				1.08	0.32	5.70			
	β-CH	2.70				−0.22 or −0.43	0.24	2.70	7.0	6.7	12.5
	β-CH	2.70				−0.43 or −0.20	0.16	2.70	7.0	6.7	2.1
Phe 22	α-CH	5.26	0.02			0.60	0.58	5.29			
	β-CH	2.80				−0.19 or −0.42	−0.04	2.81	4.9	(7.9)	5.7
	β-CH	2.92				−0.30 or −0.07	−0.21	2.87	3.7		2.0
Tyr 23	α-CH	4.29	0.03			−0.31	0.11	4.31			
	β-CH	2.73			0.03	−0.19 or −0.40	0.07	2.72		(14.7)	2.5

Resonance assignment	30°C pH 4.2	Shift difference between pH 7.0 and pH 3.1 at 68°C[a]	Shift difference between reduced BPTI and BPTI at 30°C pH 4.2[a]	Shift difference between 85°C and 30°C at pH 4.2[a]	Shift difference between the observed and random coil values at 30°C, pH 4.2[b]	Ring current shift[c]	Assigned shift in other works[d]	30°C pH 4.2[e]	85°C pH 4.2[e]	Predicted from the X-ray structure[f]
					Chemical shift from TSP (ppm)			Coupling constant $^3J_{\alpha\beta}$ (Hz)		
Tyr 23 β-CH	3.46				0.33 or 0.54	0.14	3.45			12.7
Asn 24 α-CH	4.60				−0.16	0.01	4.57			
β-CH	2.85			−0.04	0.02 or 0.10	−0.34		3.7	3.7	4.3
β-CH	2.18			0.03	−0.57 or −0.65	−0.35		12.8	12.2	12.8
Ala 25 α-CH	3.76				−0.59	−0.19	3.74			
β-CH$_3$	1.55				0.15	0.10	1.57	7.3	7.3	
Lys 26 α-CH	4.06				−0.34	−0.02	4.08			
β-CH	1.86				−0.01 or 0.11	−0.04		8.2	7.9	10.6
β-CH	1.86				0.11 or −0.11	−0.03		8.2	7.9	1.8
Ala 27 α-CH	4.30			0.03	−0.06	−0.07	4.3			
or										
Ala 16 β-CH$_3$	1.16				−0.14	−0.12	1.19	7.3	7.3	
Gly 28 α-CH	3.91				−0.06	−0.01		{−16.5}	{−17.1}	
α-CH	3.61				−0.36	−0.02				
Leu 29 α-CH	4.73			−0.05	0.34	0.08				
β-CH	1.40				−0.20	−0.03			3.7	2.0
β-CH	1.71			−0.06	0.06	−0.01			6.7	5.6
γ-CH	1.4				−0.2	−0.09	1.47			
Cys 30 α-CH	5.61			−0.10	0.92	−0.18	5.60			
β-CH	2.65		−0.03	0.03	−0.31 or −0.63	0.06	2.67	12.2	11.6	12.7
β-CH	3.68		−0.02	−0.10	0.40 or 0.72	0.18	3.69	3.4	3.7	2.5
Gln 31 α-CH	4.82				0.45	−0.01	4.87			

	β-CH	1.72			−0.02	−0.41 or −0.29	−0.51		10.4	10.7	12.8
	β-CH	2.15				0.14 or 0.02	−0.02		3.7	3.7	2.6
	γ-CH	2.02	−0.06	0.03		−0.36 or −0.55	−0.07				
	γ-CH	1.83	−0.03		0.02	−0.55 or −0.36	−0.23				
Thr 32	α-CH	5.30			−0.07	0.95	−0.21	5.28			
	β-CH	4.04			−0.04	−0.18	−0.22	4.01	2.4	2.7	2.6
Phe 33	α-CH	4.8		0.03	0.03	0.19	0.25	4.89			
	β-CH	2.98		−0.04	−0.02	−0.01 or −0.24	−0.01		5.8	5.2	5.3
	β-CH	3.11				−0.11 or 0.12	0.18		2.7	2.4	2.1
Val 34	α-CH	3.91			0.06	−0.27	−0.19	3.92			
	β-CH	1.95			−0.02	−0.18	−0.02	1.96	11.0	10.4	12.7
Tyr 35	α-CH	4.89	0.02	−0.05	−0.05	0.29	0.25	4.89			
	β-CH	2.50		0.02		−0.63 or −0.42	−0.56	2.52		3.7+	6.0
	β-CH	2.67			−0.03	−0.25 or −0.46	−0.31	2.69		11.6+	11.8
Gly 12	α-CH	2.92		0.15	−0.04	−1.05	−0.69		{−18.3}	{−18.9}	
or	α-CH	4.21		−0.35	−0.02	0.24	−0.08				
Gly 36	α-CH	3.23		0.38	0.04	−0.54	−0.27			{−18.3}	
or	α-CH	4.32		0.13	−0.07	0.35	−0.29				
Gly 37	α-CH	3.25		0.29	0.07	−0.72	−0.58		{−18.9}	{−18.9}	
	α-CH	3.89		0.12	0.06	−0.08	−0.34				
Cys 38	α-CH	4.96		−0.13	−0.06	0.27	0.12				
	β-CH	3.80	−0.97 or −1.12			0.52 or 0.84	0.04			6.3	5.3
	β-CH	3.15	−0.47 or −0.38			0.19 or −0.13	−0.02			1.5	2.1
Ala 40	α-CH	4.09		0.66		−0.26	0.27	4.09			
	β-CH₃	1.19		0.08	0.04	−0.25	−0.19	1.21	7.0	7.0	
Asn 43	α-CH	5.07		−0.03	−0.11	0.31	0.21				
	β-CH	3.39		−0.07	−0.14	0.56 or 0.63	0.25		11.6	7.6	11.8
	β-CH	3.31			−0.05	0.55 or 0.48	0.21		3.7	7.6	1.8
Asn 44	α-CH	4.89	0.02	0.03	−0.05	0.13	0.24	4.88			
	β-CH	2.51	−0.03	0.05	−0.02	−0.04 or 0.03	−0.03			3.7+	3.9
	β-CH	2.79	−0.02	−0.03	−0.07	−0.25 or −0.32	−0.26	2.78		11.6+	12.9

Resonance assignment	30°C pH 4.2	Chemical shift from TSP (ppm)						Coupling constant $^3J_{\alpha\beta}$ (Hz)		
		Shift difference between pH 7.0 and pH 3.1 at 68°C[a]	Shift difference between reduced BPTI and BPTI at 30°C, pH 4.2[a]	Shift difference between 85°C and 30°C at pH 4.2[a]	Shift difference between the observed and random coil values at 30°C, pH 4.2[b]	Ring current shift[c]	Assigned shift in other works[d]	30°C pH 4.2[e]	85°C pH 4.2[e]	Predicted from the X-ray structure[f]
Phe 45 α-CH	5.12				0.46	0.22	5.12			
β-CH	3.42	0.02		−0.04	0.20 or 0.43	0.19		12.0	12.0	12.9
β-CH	2.79				−0.20 or −0.43	0.09	2.79	4.9	3.7	3.9
Ser 47 α-CH	4.52			0.05	0.02	−0.16				
β-CH	3.84				−0.05 or 0.19	−0.16		3.4	3.4	5.6
β-CH	4.08				0.19 or −0.05	−0.02		3.4	3.4	2.0
Ala 48 α-CH	3.15	0.04			−1.20	−0.60	3.14			
β-CH$_3$	1.03				−0.37	−0.38	1.04	7.3	7.3	
Glu 49 α-CH	3.88	−0.08			−0.42	−0.13				
β-CH	1.87	−0.03			−0.10 or −0.22	−0.10		(16.0)	8.8	12.4
β-CH	2.03	−0.05			−0.06 or 0.06	−0.08			6.4	2.1
γ-CH	2.41	−0.16			0.10 or 0.00	−0.06				
γ-CH	2.28	−0.18		−0.02	0.00 or 0.10	−0.08				
Asp 50 α-CH	4.29			−0.02	0.11	−0.07				
β-CH	2.73	−0.06		−0.03	−0.02 or −0.11	0.02			10.4	9.1
β-CH	2.89	−0.07		−0.02	0.05 or 0.14	0.13			4.3	2.1
Cys 51 α-CH	1.68		0.03	0.16	−3.01	−2.28				
β-CH	3.18				−0.10 or 0.22	−0.09			4.9	4.0
β-CH	2.89			−0.05	−0.07 or −0.39	−0.19			12.8	12.8
Met 52 α-CH	4.17	0.03		−0.12	−0.34	−0.12				
β-CH	2.00			−0.03	−0.01 or −0.13	−0.10			9.8#	12.6

	β-CH	2.07		0.04	0.03	−0.06 or 0.06	−0.06			4.3#	2.4
	γ-CH	2.70			−0.05	0.07	−0.07				
	γ-CH	2.70			−0.05	0.07	−0.04				
Thr 54	α-CH	4.07				−0.28	−0.03				
	β-CH	3.99			−0.02	−0.23	−0.11	3.95	9.8	9.8	12.9
Cys 55	α-CH	4.62			−0.04	−0.07	−0.38	4.31			
	β-CH	2.26	−0.03		−0.07	−0.70 or −1.02	−0.40		(14.3)	11.6	12.7
	β-CH	2.03		−0.03	−0.09	−1.25 or −0.96	−1.42			3.1	2.5
Gly 56	α-CH	3.96	0.06			−0.01	−0.07		{−17.1}*	{−17.1}*	
or											
Gly 57	α-CH	3.80	0.02		0.04	−0.17	−0.08				
Ala 58	α-CH	4.01	−0.14		−0.02	−0.34	−0.02	3.99			
	β-CH$_3$	1.31	−0.05			−0.09	−0.03	1.31	7.3	7.3	

[a] Changes in chemical shifts bigger than 0.02 ppm are listed.

[b] Most recent data by Bundi and Wüthrich (139) were assumed for the amino acid chemical shifts in the random coil state. Due to the ambiguity for the stereospecific assignments of two geminal β-protons, two choices are included in the list.

[c] For the β-protons of phenylalanines and tyrosines, the contribution of their own ring current shift was subtracted. The negative sign of shifts in b and c corresponds to up-field shift.

[d] Values were collected from the reports by Wüthrich et al. (97), Dubs et al. (93), Wüthrich and Wagner (113), Perkins and Wüthrich (133), and Wagner et al. (150).

[e] Coupling constants in parentheses ( ) indicate the summation of $^3J_{\alpha\beta2}$ and $^3J_{\alpha\beta3}$. Coupling constants in parentheses { } are those of the proton-proton geminal coupling. Values given here are the readings of the distance of peak maxima of α-proton multiplets without corrections for overlapping. A resolution of 0.3 Hz was employed for the reading by interpolating one point to successive sampling points. But that accuracy is only applicable to values for the strong and well resolved doublets of the α-protons of isoleucine, valine and threonine, and methyls of alanine. *, values measured at 68°C; §, values obtained from β-proton resonances; +, values from the reading of severely overlapped multiplets of two α-protons; #, values obtained at 41°C for RCAM BPTI.

[f] The Karplus curve used is $9.5 \cos^2 \theta - 1.6 \cos \theta + 1.8$ (Hz) for $^3J_{\alpha\beta}$ (DeMarco et al. (122)). Here θ is either $\theta^{\alpha\beta2}$ or $\theta^{\alpha\beta3}$ which is related to the torsional angle χ^1 with $\theta^{\alpha\beta2} = \chi^1 - 120°$ and $\theta^{\alpha\beta3} = \chi^1$.

REFERENCES

1 E. M. Purcell, H. C. Torrey, and R. V. Pound, *Phys. Rev.*, **69**, 37 (1946).

2 F. Bloch, W. W. Hansen, and M. Packard, *Phys. Rev.*, **70**, 474 (1946).

3 R. R. Ernst and W. A. Anderson, *Rev. Sci. Instrum.*, **37**, 93 (1966).

4 I. I. Rabi, J. R. Zacharias, S. Millman, and P. Kush, *Phys. Rev.*, **53**, 318 (1938).

5 W. E. Lamb and R. C. Ratherford, *Phys. Rev.*, **72**, 241 (1947).

6 F. Bloch, *Phys. Rev.*, **70**, 460 (1946).

7 R. Kubo and N. Hashitsume, *Prog. Theor. Phys.* Suppl. **46**, 210 (1970).

8 A. Abragam, " The Principles of Nuclear Magnetism," Clarendon Press, Oxford (1961).

9 H. Primas, *Helv. Phys. Acta*, **34**, 331 (1961).

10 A. G. Redfield, *IBM J. Res. Develop.*, **1**, 19 (1957).

11 N. Bloembergen, E. M. Purcell, and R. V. Pound, *Phys. Rev.*, **73**, 679 (1948).

12 A. L. Bloom and J. N. Shoolery, *Phys. Rev.*, **97**, 1261 (1955).

13 E. L. Hahn, *Phys. Rev.*, **80**, 580 (1950).

14 H. Y. Carr and E. M. Purcell, *Phys. Rev.*, **94**, 630 (1954).

15 I. Solomon, *Phys. Rev.*, **99**, 559 (1955).

16 The idea of 2D spectroscopy was proposed by J. Jeener in 1971 and experimentally realized by R. R. Ernst in 1974. The first description of the 2D NMR experiment appeared in 1975 (R. R. Ernst, *Chimia*, **29**, 179 (1975)).

17 R. R. Ernst, W. P. Aue, P. Bachmann, A. Höhener, M. Linder, B. H. Meier, L. Müller, A. Wokaun, K. Nagayama, and K. Wüthrich, Proceedings of the XXth Congress AMPERE Tallinn, Springer-Verlag, Berlin, p. 15 (1979).

18 W. P. Aue, E. Bartholdi, and R. R. Ernst, *J. Chem. Phys.*, **64**, 2229 (1976).

19 W. P. Aue, J. Karhan, and R. R. Ernst, *J. Chem. Phys.*, **64**, 4226 (1976).

20 K. Nagayama, K. Wüthrich, P. Bachmann, and R. R. Ernst, *Biochem. Biophys. Res. Commun.*, **78**, 99 (1977).

21 K. Nagayama, K. Wüthrich, P. Bachmann, and R. R. Ernst, *Naturwissenschaften*, **64**, 581 (1977).

22 A. Kumar, *J. Magn. Reson.*, **30**, 227 (1978).

23 R. Niedermeyer and R. Freeman, *J. Magn. Reson.*, **30**, 617 (1978).

24 G. Bodenhausen, R. Freeman, G. A. Morris, and D. L. Turner, *J. Magn. Reson.*, **31**, 75 (1978).

25 K. Nagayama, P. Bachmann, K. Wüthrich, and R. R. Ernst, *J. Magn. Reson.*, **31**, 133 (1978).

26 L. D. Hall and S. Sukumar, *J. Am. Chem. Soc.*, **101**, 3120 (1979).

27 L. D. Hall, S. Sukumar, and G. R. Sullivan, *J. Chem. Soc., Chem. Commun.*, 292 (1979).

28 A. Bax, A. F. Mehlkopf, and J. Smidt, *J. Magn. Reson.*, **35**, 167 (1979).

29 L. D. Hall, G. A. Morris, and S. Sukumar, *J. Am. Chem. Soc.*, **102**, 1747 (1980).

30 L. Müller, A. Kumar, and R. R. Ernst, *J. Chem. Phys.*, **63**, 5490 (1975).

31 G. Bodenhausen, R. Freeman, and D. L. Turner, *J. Chem. Phys.*, **65**, 839 (1976).

32 G. Bodenhausen, R. Freeman, R. Niedermeyer, and D. L. Turner, *J. Magn. Reson.*, **24**, 291 (1976).

33 L. Müller, A. Kumar, and R. R. Ernst, *J. Magn. Reson.*, **25**, 383 (1977).

34 G. Bodenhausen, R. Freeman, R. Niedermeyer, and D. L. Turner, *J. Magn. Reson.*, **26**, 133 (1977).

35 R. Freeman, G. A. Morris, and D. L. Turner, *J. Magn. Reson.*, **26**, 373 (1977).

36 G. Bodenhausen, R. Freeman, and D. L. Turner, *J. Magn. Reson.*, **27**, 511 (1977).

37 G. Bodenhausen, R. Freeman, G. A. Morris, and D. L. Turner, *J. Magn. Reson.*, **28**, 17 (1977).

38 G. Bodenhausen, and R. Freeman, *J. Magn. Reson.*, **28**, 303 (1977).

39 D. L. Turner and R. Freeman, *J. Magn. Reson.*, **29**, 587 (1978).

40 P. H. Bolton and G. Bodenhausen, *J. Am. Chem. Soc.*, **101**, 1080 (1979).

41 R. Freeman, S. P. Kempsell, and M. H. Levitt, *J. Magn. Reson.*, **35**, 447 (1979).

42 J. R. Everett, D. W. Hughes, A. D. Bain, and R. A. Bell, *J. Am. Chem. Soc.*, **101**, 6776 (1979).

43 L. Müller, *J. Magn. Reson.*, **36**, 301 (1979).

44 R. H. Hester, J. L. Ackerman, V. R. Cross, and J. S. Waugh, *Phys. Rev. Lett.*, **34**, 993 (1975).

45 J. S. Waugh, *Proc. Natl. Acad. Sci. U.S.*, **73**, 1394 (1976).

46 M. E. Stoll, A. J. Vega, and R. W. Vaughan, *J. Chem. Phys.*, **65**, 4093 (1976).

47 R. K. Hester, J. L. Ackerman, B. L. Neff, and J. S. Waugh, *Phys. Rev. Lett.*, **36**, 1081 (1976).

48 S. J. Opella and J. S. Waugh, *J. Chem. Phys.*, **66**, 4919 (1977).

49 E. F. Rybaczewski, B. L. Neff, and J. S. Waugh, *J. Chem. Phys.*, **67**, 1231 (1977).

50 K. Nagayama, K. Wüthrich, and R. R. Ernst, *Biochem. Biophys. Res. Commun.*, **90**, 305 (1979).

51 B. H. Meier and R. R. Ernst, *J. Am. Chem. Soc.*, **101**, 6441 (1979).

52 J. Jeener, B. H. Meier, P. Bachmann, and R. R. Ernst, *J. Chem. Phys.*, **71**, 4546 (1979).

53 A. Wokaun and R. R. Ernst, *Chem. Phys. Lett.*, **52**, 407 (1977).

54 A. Wokaun and R. R. Ernst, *Mol. Phys.*, **36**, 317 (1978).

55 G. Bodenhausen, N. M. Szeverenyi, R. L. Vold, and R. R. Vold, *J. Am. Chem. Soc.*, **100**, 6265 (1978).

56 G. Bodenhausen, R. L. Vold, and R. R. Vold, *J. Magn. Reson.*, **37**, 93 (1980).

57 A. A. Maudsley and R. R. Ernst, *Chem. Phys. Lett.*, **50**, 368 (1977).

58 A. A. Maudsley, A. Wokaun, and R. R. Ernst, *Chem. Phys. Lett.*, **55**, 9 (1978).

59 A. A. Maudsley, L. Müller, and R. R. Ernst, *J. Magn. Reson.*, **28**, 463 (1977).

60 G. Bodenhausen and R. Freeman, *J. Magn. Reson.*, **28**, 471 (1977).

61 R. Freeman and G. A. Morris, *J. Chem. Soc., Chem. Commun.*, 684 (1978).

62 G. Bodenhausen and R. Freeman, *J. Am. Chem. Soc.*, **100**, 320 (1978).

63 L. Müller and R. R. Ernst, *Mol. Phys.*, **38**, 963 (1979).

64 D. C. Finster, W. C. Hutton, and R. N. Grimes, *J. Am. Chem. Soc.*, **102**, 400 (1980).

65 L. Müller, *J. Magn. Reson.*, **38**, 79 (1980).

66 P. Bachman, W. P. Aue, L. Müller, and R. R. Ernst, *J. Magn. Reson.*, **28**, 29 (1977).

67 R. Freeman, S. P. Kempsell, and M. H. Levitt, *J. Magn. Reson.*, **34**, 663 (1979).

68 M. H. Levitt and R. Freeman, *J. Magn. Reson.*, **34**, 675 (1979).

69 A. Bax, A. F. Mehlkopf, and J. Smidt, *J. Magn. Reson.*, **35**, 373 (1979).

70 K. Nagayama, P. Bachmann, R. R. Ernst, and K. Wüthrich, *Biochem. Biophys. Res. Commun.*, **86**, 218 (1979).

71 K. Nagayama, *J. Chem. Phys.*, **71**, 4404 (1979).

72 K. Wüthrich, K. Nagayama, and R. R. Ernst, *Trends Biochem. Sci.*, **4**, N178 (1979).

73 A. Wokaun and R. R. Ernst, *Mol. Phys.*, **38**, 1579 (1979).

74 R. Freeman and G. A. Morris, *J. Magn. Reson.*, **29**, 173 (1978).

75 W. P. Aue and R. R. Ernst, *J. Magn. Reson.*, **31**, 533 (1978).

76 L. D. Hall, G. A. Morris, and S. Sukumar, *Carbohydr. Res.*, **76**, C7 (1979).

77 W. P. Aue and R. R. Ernst, *J. Magn. Reson.*, **38**, 375 (1980).

78 H. Hatanaka, T. Terao, and T. Hashi, *J. Phys. Soc. Japan*, **39**, 835 (1975).

79 H. Hatanaka and T. Hashi, *J. Phys. Soc. Japan*, **39**, 1139 (1975).

80 S. Vega, T. W. Shattuck, and A. Pines, *Phys. Rev. Lett.*, **37**, 43 (1976).

81 S. Vega and A. Pines, *J. Chem. Phys.*, **66**, 5624 (1977).

82 A. Kumar, D. Welti, and R. R. Ernst, *J. Magn. Reson.*, **18**, 69 (1975).

83 W. P. Aue, P. Bachmann, A. Wokaun, and R. R. Ernst, *J. Magn. Reson.*, **29**, 523 (1978).

84 A. Kumar and R. R. Ernst, *J. Magn. Reson.*, **24**, 425 (1976).

85 G. Bodenhausen, S. P. Kempsell, R. Freeman, and H. D. W. Hill, *J. Magn. Reson.*, **35**, 337 (1979).

86 K. Nagayama, A. Kumar, K. Wüthrich, and R. R. Ernst, *J. Magn. Reson.*, **40**, 321 (1980).

87 J. D. Ellétt, M. G. Gibby, U. Haeberlen, L. M. Huber, M. Mehring, A. Pines, and J. S. Waugh, *Adv. Magn. Reson.*, **5**, 117 (1971).

88 R. N. Bracewll, *Aust. J. Phys.*, **9**, 198 (1956).

89 D. C. Champeny, "Fourier Transformation and Their Physical Applications," Academic Press, London (1973).

90 S. Karplus, G. H. Synder, and B. D. Sykes, *Biochemistry*, **12**, 1323 (1973).

91 T. D. Marinetti, G. H. Synder, and B. D. Sykes, *Biochemistry*, **15**, 4600 (1976).

92 T. D. Marinetti, G. H. Synder, and B. D. Sykes, *Biochemistry*, **16**, 647 (1977).

93 A. Dubs, G. Wagner, and K. Wüthrich, *Biochim. Biophys. Acta*, **577**, 177 (1979).

94 E. Bartholdi and R. R. Ernst, *J. Magn. Reson.*, **11**, 9 (1973).

95 A. DeMarco and K. Wüthrich, *J. Magn. Reson.*, **24**, 201 (1976).

96 K. Nagayama and K. Wüthrich, *Eur. J. Biochem.*, in press.

96a K. Nagayama and K. Wüthrich, *Eur. J. Biochem.*, in press.

97 K. Wüthrich, G. Wagner, R. Richarz, and S. J. Perkins, *Biochemistry*, **17**, 2253 (1978).

98 N. Gō, *Adv. Biophys.*, **9**, 65 (1976).

99 N. Gō, *Biopolymers*, **17**, 1373 (1978).

100 T. Ooi, K. Nishikawa, M. Oobatake, and H. A. Scheraga, *Biochim. Biophys. Acta*, **536**, 390 (1978).

101 B. R. Gelin and M. Karplus, *Proc. Natl. Acad. Sci. U.S.*, **72**, 2002 (1975).

102 J. A. McCammon, B. R. Gelin, and M. Karplus, *Nature*, **267**, 585 (1977).

103 J. A. McCammon and M. Karplus, *Proc. Natl. Acad. Sci. U.S.*, **76**, 3585 (1979).

104 J. A. McCammon, P. G. Wolynes, and M. Karplus, *Biochemistry*, **18**, 927 (1979).

105 H. Frauenfelder, G. A. Petsko, and D. Tsernoglou, *Nature*, **280**, 558 (1979).

106 D. J. Artymiuk, C. C. F. Blake, D. E. P. Grace, S. J. Oatley, D. C. Phillips, and M. J. E. Sternberg, *Nature*, **280**, 563 (1979).

107 G. Wagner, A. DeMarco, and K. Wüthrich, *Biophys. Struct. Mech.*, **2**, 139 (1976).

108 R. Hetzel, K. Wüthrich, J. Deisenhofer, and R. Huber, *Biophys. Struct. Mech.*, **2**, 159 (1976).

109 G. Wagner and K. Wüthrich, *Nature*, **275**, 247 (1978).

110 K. Wüthrich and G. Wagner, *Trends Biochem. Sci.*, **3**, 227 (1978).

111 R. Richarz and K. Wüthrich, *Biochemistry*, **17**, 2263 (1978).

112 G. Wagner, H. Tschesche, and K. Wüthrich, *Eur. J. Biochem.*, **95**, 239 (1979).

113 K. Wüthrich and G. Wagner, *J. Mol. Biol.*, **130**, 1 (1979).

114 R. Richarz, D. Sher, G. Wagner, and K. Wüthrich, *J. Mol. Biol.*, **130**, 19 (1979).

115 G. Wagner and K. Wüthrich, *J. Mol. Biol.*, **130**, 31 (1979).

116 R. Richarz, K. Nagayama, and K. Wüthrich, *Biochemistry*, in press.

117 K. Wüthrich, G. Wagner, R. Richarz, and W. Braun, *Biophys. J.*, in press.

118 H. Jering and H. Tschesche, *Eur. J. Biochem.*, **61**, 443 (1976).

118a B. Kassell and M. Laskowski, *Biochem. Biophys. Res. Commun.*, **20**, 463 (1965).

119 K. Wüthrich, "NMR in Biological Research: Peptides and Proteins," North-Holland/American Elsevier, Amsterdam (1976).

120 K. G. R. Pachler, *Spectrochim. Acta*, **20**, 581 (1964).

121 J. P. Vincent, R. Chicheportiche and M. Lazdunski, *Eur. J. Biochem.*, **23**, 401 (1971).

122 A. DeMarco, M. Llinas, and K. Wüthrich, *Biopolymers*, **17**, 617 (1978).

123 K. D. Kopple, G. R. Wiley, and R. Tauke, *Biopolymers*, **12**, 627 (1973).

124 J. Deisenhofer and W. Steigeman, *Acta Chrystallogr.*, **B31**, 238 (1975).

125 B. R. Gelin and M. Karplus, *Biochemistry*, **18**, 1256 (1979).

126 M. Gō and N. Gō., personal communication.

126a M. Gō and S. Miyazawa, *Int. J. Peptide Protein Res.*, **15**, 211 (1980).

127 A. Shrake and J. A. Rupley, *J. Mol. Biol.*, **79**, 351 (1973).

128 J. D. Bernal and J. L. Finney, *Discuss. Faraday Soc.*, **43**, 62 (1967).

129 R. Huber, D. Kukla, A. Rühlmann, O. Epp, and H. Formnek, *Naturwissenschaften*, **57**, 389 (1970).

130 H. S. Frank and M. W. Evans, *J. Chem. Phys.*, **13**, 507 (1945).

131 A. Kowalsky, *J. Biol. Chem.*, **237**, 1807 (1962).

132 H. Sternlicht and D. Wilson, *Biochemistry*, **6**, 2881 (1967).

133 S. J. Perkins and K. Wüthrich, *Biochim. Biophys. Acta*, **576**, 409 (1979).

134 S. J. Perkins and R. A. Dwek, *Biochemistry*, **19**, 245 (1980).

135 J. A. Pople, *J. Chem. Phys.*, **37**, 60 (1962).

136 J. W. ApSimon, P. V. DeMarco, D. W. Mathieson, W. G. Craig, A. Karim, L. Saunders, and W. B. Whalley, *Tetrahedron*, **26**, 119 (1970).

137 H. L. Tiegelaar and W. H. Flygare, *J. Am. Chem. Soc.*, **94**, 343 (1972).

138 R. A. Dwek, "NMR in Biochemistry," Oxford Univ. Press, Oxford (1973).

139 A. Bundi and K. Wüthrich, *Biopolymers*, **18**, 285 (1979).

140 R. Kubo, *Nuovo Cimento* **6** (Suppl. No. 3), 1063 (1957).

141 R. Ullman, *Biopolymers*, **9**, 471 (1970).

142 K. Nagayama and A. Wada, *Chem. Phys. Lett.*, **16**, 50 (1972).

143 K. Nagayama and A. Wada, *Biopolymers*, **12**, 2443 (1973).

144 E. M. Bradbury, C. Crane-Robinson, and P. G. Hartman, *Polymer*, **14**, 543 (1973).

145 K. Nagayama and A. Wada, *Biopolymers*, **14**, 2489 (1975).

146 J. A. Pople, *J. Chem. Phys.*, **24**, 1111 (1956).

147 H. M. McConell, *J. Chem. Phys.*, **27**, 226 (1957).

148 "Landort-Börnstein Tabellen," 1927, 1931, 1933, Springer-Verlag, Berlin.

149 A. Kumar, R. R. Ernst, and K. Wüthrich, *Biochem. Biophys. Res. Commun.*, **95**, 1 (1980).

150 G. Wagner, A. Kumar, and K. Wüthrich, *Eur. J. Biochem.*, in press.

151 R. R. Ernst, *Adv. Magn. Reson.*, **2**, 1 (1966).

152 C. E. Johnson and F. A. Bovey, *J. Chem. Phys.*, **29**, 1012 (1958).

153 G. E. Shulz and R. H. Schirmer, "Principles of Protein Structures," Springer-Verlag, New York (1979).

Received for publication July 3, 1980.

Adv. Biophys., Vol. 14, pp. 205–238 (1981)

THE CAP STRUCTURE IN EUKARYOTIC MESSENGER RNA AS A MARK OF A STRAND CARRYING PROTEIN INFORMATION

KIN-ICHIRO MIURA

National Institute of Genetics, Mishima, Japan

Usually, genome nucleic acids consist of double-stranded filaments. Even the single-stranded viral genome assumes a double-stranded state during replication, which is called a replicative form or a replicative intermediate. Since two strands are complementary to each other in the base pairs, their replication proceeds without errors. Each strand functions as a template for each opposite strand. However, the genetic information resides in one strand of the double-stranded filament for each information. The strand containing protein information is transcribed as a messenger RNA (mRNA) by RNA polymerase. Is there any sign in genome double strands for RNA polymerase to select the template chain?

For the purpose of solving this problem, a small genome like viral nucleic acid is favorable material. A virus containing double-stranded RNA seemed convenient for us at the starting stage of this project, since the genome is divided into some 10 segments and each segment seems to contain one cistron (information for one protein molecule) (*1*). If we separate these RNA segments, we can obtain each gene unit. This is an advantage for studies on gene structure and its expression mechanism even at present although gene engineering techniques have been developed. We were also

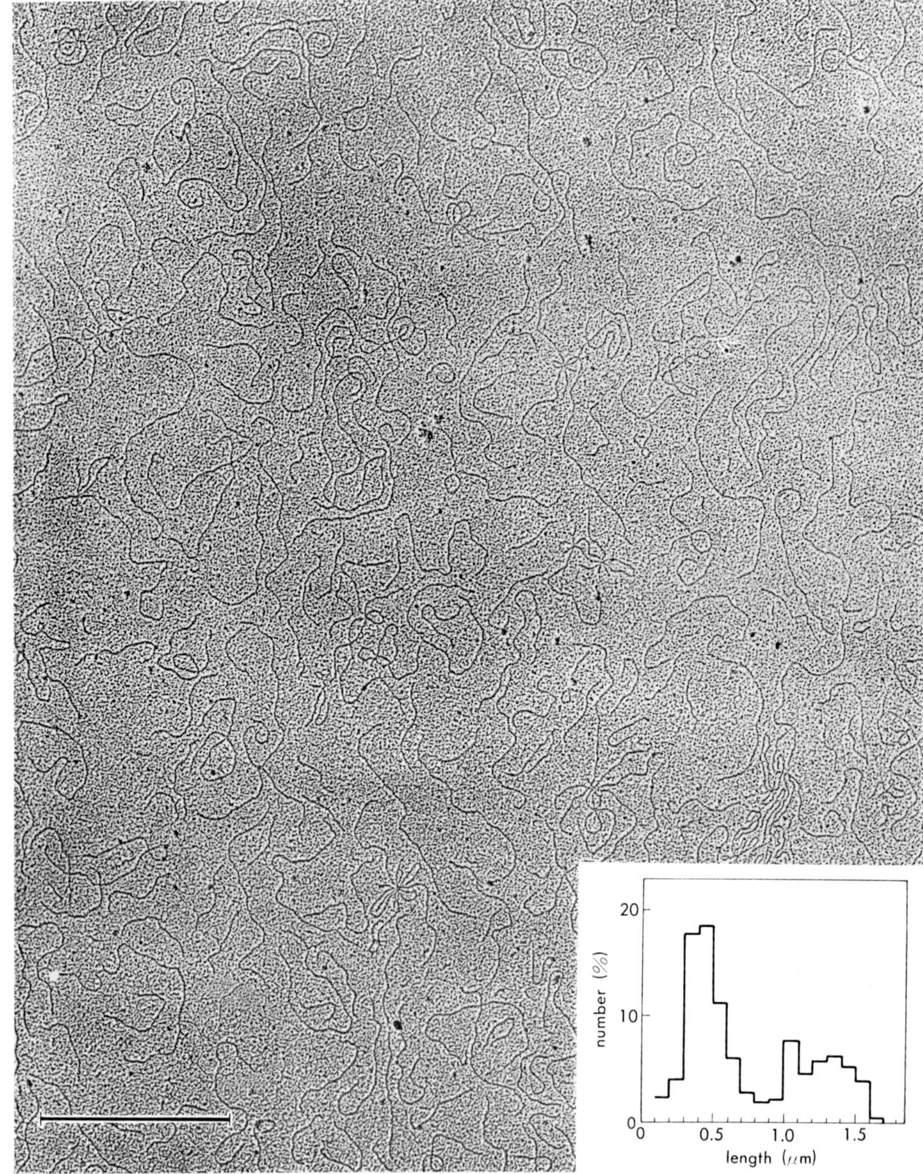

Fig. 1. Electron micrograph of CPV double-stranded RNA (*3*). The black bar indicates 1 μm. The inset graph shows the length distribution of the RNA filaments, expressed as percentages of 450 filaments.

interested in the problem of how the expression of the segmented genome is regulated. The genomes of rice dwarf virus (RDV) and cytoplasmic polyhedrosis virus (CPV) of silkworms were formerly identified as double-stranded RNA by us (2, 3); and further, the physical characters of double-stranded RNA have been studied in these viral RNAs (3–5). Even if a microanalytical technique including radioisotopes is used, it is better to prepare a large amount of genome nucleic acids. In this respect we chose CPV as the main material for this research.

I. GENOME STRUCTURE OF CPV

Electron microscopic observation shows that the double-stranded RNA preparation from silkworm CPV consists of linear filaments of heterogeneous length, as shown in Fig. 1 (3). However, CPV RNA is separated into 10 discrete bands upon electrophoresis on a polyacrylamide gel column as shown in Fig. 2. The total size of a set of 10 RNA bands reaches the molecular size of the whole genome RNA in a virion. When the CPV particle was oxidized *in situ* with periodate, the 3′-terminal nucleosides of all the RNA segments were oxidized. Therefore, if all the segments of CPV RNA link together in a definite order in the virus particle, they are not linked by phosphodiester bonds, but associate in other ways.

Fig. 2.   Electrophoretogram of CPV RNA segments (1). RNA was stained with acrydine orange after electrophoresis on a polyacrylamide gel column.

The terminal structure of the RNA segments of CPV was analyzed. The 3′-termini were analyzed by labeling with [³H]NaBH₄ after oxidation of the cis-diol at the 3′-terminal nucleoside (6, 7). Alkaline digestion of the 3′-terminal-labeled RNA should yield a [³H] nucleoside trialcohol. In fact, cytidine trialcohol (C′) and uridine trialcohol (U′) were identified by phosphocellulose column chromatography (Fig. 3) and two-dimensional thin layer chromatography. However, about 30% of the radioactivity of [³H] in the labeled RNA preparation was not identified as ordinary nucleoside trialcohols (the first peak in Fig. 3a). This unknown material was separated from nucleoside trialcohols by a charcoal column (Fig. 4). It

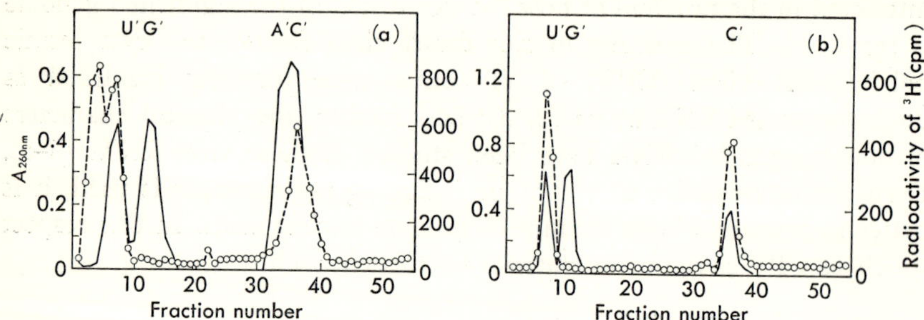

Fig. 3.   Phosphocellulose column chromatography of the alkaline and phosphatase digests of the [³H] 3′-terminally-labeled CPV RNA. Sample of (b) is the ethanol-eluted fraction in Fig. 4, which contains only nucleoside trialcohols in the sample of (a). ● absorption at 260 nm of standard nucleoside trialcohols ; ○ [³H] radioactivity.

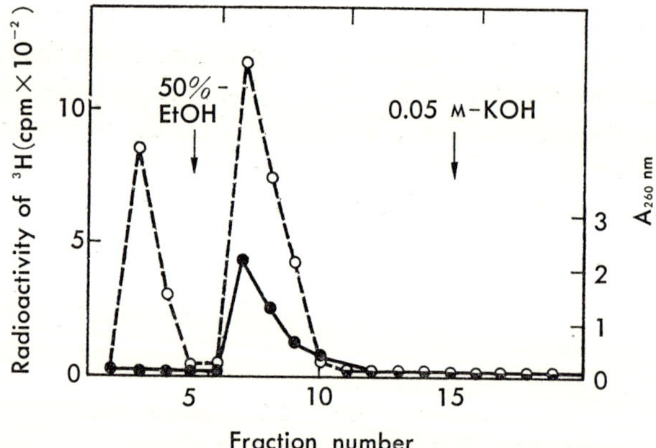

Fig. 4.   Charcoal column chromatography of alkaline digests of the [³H] 3′-terminally-labeled CPV RNA (7). [³H]-labeled nucleoside trialcohols were eluted with 50% ethanol. The first radioactive peak eluted with water was an unknown material. ○ radioactivity; ● absorption at 260 nm.

seemed to be linked to each double-stranded RNA segment by covalent bonds, but not to be contaminants or artifacts. Certainly it was identified later as an intrinsic component in one strand of double-stranded RNA.

A few nucleotide sequences from the 3′-termini of CPV RNA were analyzed by digestion with ribonucleases on the above-mentioned [cis-diol ³H]-labeled CPV RNA or by sequential removal of nucleotides from the 3′-termini. Every RNA segment gave two kinds of 3′-terminal sequence:

-G-C-C and -A-C-U. This means that one strand terminates with the former sequence and the other with the latter.

An attempt to label the 5'-termini of segmented CPV RNA with [^{32}P] phosphate was made using nucleotide kinase, as is used generally in the analysis of the 5'-termini of DNA and RNA. However, ^{32}P-labeling at the 5'-terminal was not successful for double-stranded CPV RNA, even if the RNA was treated first with phosphomonoesterase to remove any pre-existing free phosphates. As completeness of the double-standed state was considered to inhibit phosphate incorporation at the 5'-end of CPV RNA, nucleotides near the 3'-terminus of the complementary strand were deleted chemically one by one, repeating a sequential operation of oxidation, β-elimination and phosphatase treatment as shown in Fig. 5. RNA samples, from which two or three nucleotides had been split from the 3'-end, were able to incorporate [^{32}P] phosphate at the 5'-end by nucleotide kinase.

On alkaline hydrolysis of the 5'-terminal-labeled [^{32}P] RNA, appearance of ^{32}pNp ([5'^{32}P] nucleoside 3', 5'-diphosphate) is expected. How-

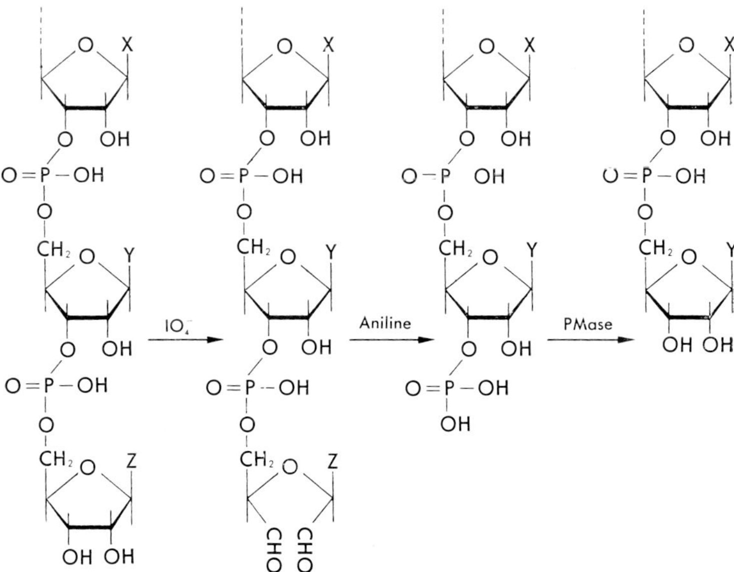

Fig. 5. Stepwise removal of a nucleotide residue from the 3'-terminus of RNA by periodate oxidation, β-elimination with aniline and phosphatase treatment.

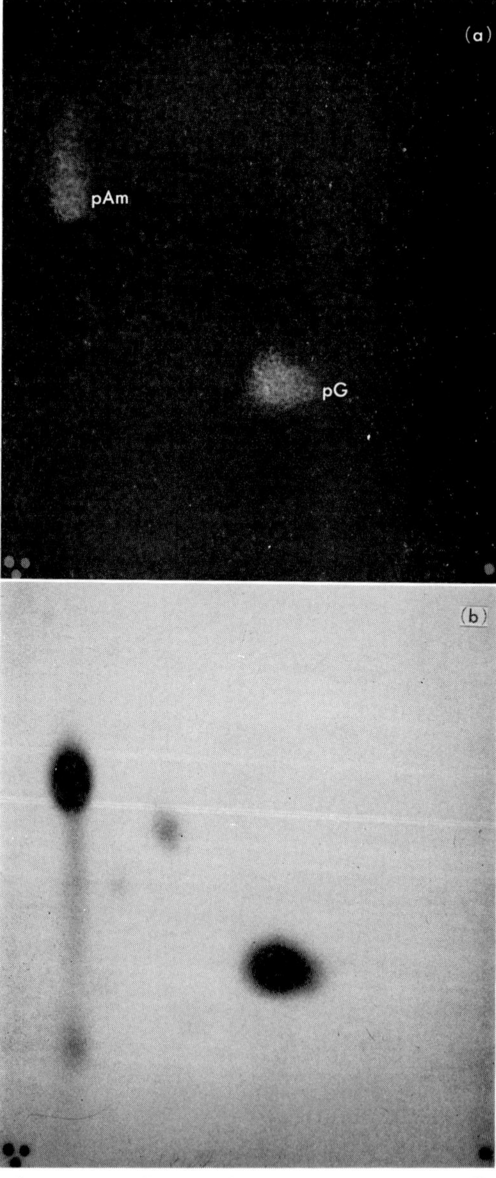

Fig. 6. Identification of pAm by fingerprinting. [^{32}P]-labeled nucleotides obtained by nuclease P_1 digest of [5′-^{32}P]-labeled CPV RNA were chromatographed two-dimensionally. a: photography by ultraviolet light. b: radioautography.

ever, dinucleotide ^{32}pNpNp appeared in addition to ^{32}pGp. The two components were present in almost the same amounts. The nucleoside on the 5′ side, that is, the 5′-terminal nucleoside in the original RNA, was considered as a nucleoside derivative modified at the 2′-position of ribose moiety since it is resistant to alkaline hydrolysis and to ribonuclease T_2 digestion. This type of nucleotide was obtained as a 5′-nucleotide ^{32}pN by the use of nuclease P_1 from *Penicillium citrinum*, which is purified by Fujimoto, Kuninaka, and Yoshino, Yamasa Shoyu Co. for production of 5′-mononucleotides from RNA (*8*). The nucleotide ^{32}pN was identified as 2′-O-methyladenylic acid ^{32}pAm by chromatographic analysis with an authentic material (Fig. 6). A few nucleotide sequences from the 5′-termini of the CPV RNA were determined by digestion with ribonucleases on the [5′^{32}P]-labeled RNA. Every RNA segment gave two kinds of 5′-terminal sequence: G-G-C- and Am-G-U-. As these were present in the same amounts, each corresponds to the 5′-terminal of each strand.

Thus, it is concluded that each segment of CPV RNA has the following structure:

$$
\begin{array}{ccc}
{}^{5'}\text{Am}-\text{G}-\text{U} & \longrightarrow & \text{G}-\text{C}-\text{C}^{3'} \\
{}_{3'}\text{U}\ -\text{C}-\text{A} & \longleftarrow & \text{C}-\text{G}-\text{G}_{5'}
\end{array}
$$

This reveals that an RNA segment takes a perfect double-stranded structure and the polarity of two strands is in antiparallel (*9*). And it is clear that one strand is distinguished from the other by the 5′-terminal methylated nucleotide.

II. CHAIN SELECTION IN TRANSCRIPTION OF DOUBLE-STRANDED RNA GENES

Transcription experiments for double-stranded RNA were carried out in order to know which strand is copied as an mRNA.

Since the replication or transcription of viral double-stranded RNA does not take place through the use of cellular RNA polymerase, the virus containing double-stranded RNA carries its own RNA polymerase in its core (capsid). In the case of CPV, pretreatment of a virus preparation with the organic solvent dichlorodifluoromethane is essential for the activation of RNA polymerase. The pretreated virus preparation synthesizes single-stranded RNA in a test tube with the addition of only the substrate NTP (nucleoside triphosphates) and bentonite, which is added for protection of the product single-stranded RNA since it is a nuclease adsorber. The product RNA is released from a virus particle, leaving genome double-stranded RNA in the particle, so that the transcript (mRNA) is easily purified by centrifugation and molecular sieving with a Sephadex. The transcript has a size and sequence unique to each genome segment (*10, 11*).

Judging from the 5′-terminal nucleotides of double strands of CPV RNA, the starting nucleotide at the 5′-terminus of the transcript should be either adenylic acid or guanylic acid. The phosphates derived from the substrate purine nucleoside triphosphate may be reserved at the 5′-termi-

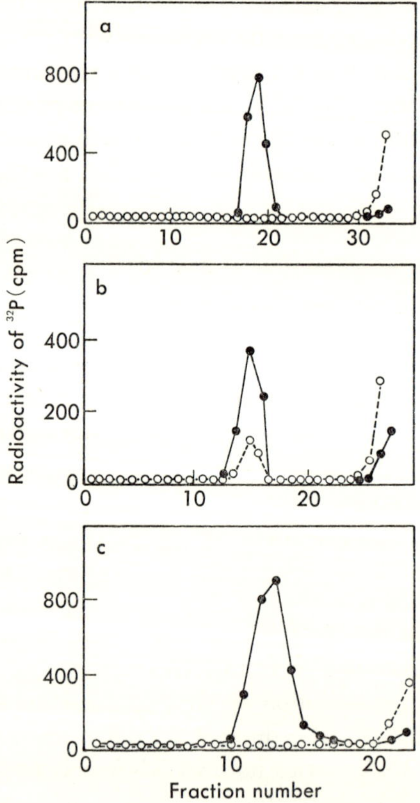

Fig. 7. Gel filtration of the CPV mRNA synthesized with β- or γ-[^{32}P]-labeled ATP or GTP (*12*). a: [β-^{32}P] GTP. b: [β-^{32}P] ATP. c: [γ-^{32}P] ATP. Every sample was chromatographed on a Sephadex G-50 column with [^{3}H] U-labeled RNA as a size marker. ● [^{3}H] radioactivity; ○ [^{32}P] radioactivity.

nus of the transcript. Thus, the incorporation of the β- or γ-[^{32}P]-labeled purine nucleoside triphosphate into the transcript was tested. Among these, only β-[^{32}P]ATP was incorporated into the transcribed RNA (Fig. 7). With this sample the terminal structure of the newly synthesized RNA was identified as ppA-G-Y (*12*). The 3'-terminus of this RNA was identified as C. These structures are common for all RNA species corresponding to the genome segments. Therefore, the mRNA of CPV has the same sequence as the 5'-terminal-modified strand of genome segments (*9*, *13*, *14*).

III. DISCOVERY OF A STRANGE MODIFIED STRUCTURE IN INFORMATIONAL RNA STRANDS

Only single-stranded RNA transcripts are released outside a virus particle to be used for protein synthesis. These single-stranded RNAs pair with complementary strands, which are synthesized later, to construct a progeny genome. One of the duplex strands of progeny genes should be methylated at the 5'-terminal nucleoside ribose. We considered that the terminal nucleoside of the mRNA may be methylated as genome RNA if a methyl-donor is added to the transcription reaction mixture. This experiment was carried out by Furuichi. When S-adenosyl-methionine (SAM) that is known to be a direct methyl-donor for nucleic acid was added to the reaction mixture of the *in vitro* transcription of CPV, the stimulation of RNA synthesis and the incorporation of methyl groups into RNA were observed (*15*).

It was then surveyed whether methylation occurred at the 5'-terminal region. Adenosine 5'-triphosphate labeled with [^{32}P] phosphate at the β-position ([β-^{32}P]ATP) and [methyl-^{3}H]SAM were added to the transcription reaction mixture. The product RNA was digested with a few different kinds of nuclease. Each digest was chromatographed on a DEAE-cellulose column with 7 M urea containing solvent to separate different-sized oligonucleotides. If both the [^{32}P] and [methyl-^{3}H] locate at the 5'-terminus of the RNA as expected, ^{32}P and ^{3}H would act together in any nuclease digest. This was found to be true, as shown in Fig. 8 (*14*). The terminal nucleotide can be obtained by *Penicillium* nuclease P$_1$ and it must be ppAm. This can be split into phosphate groups and nucleoside Am by phosphomonoesterase (phosphatase). But this was not the case as shown in chromatograms (5) and (6) in Fig. 8. Therefore, β-phosphate of the 5'-terminal Am of RNA should be blocked by some material such as XppAm (—denotes labeling with ^{32}P and =with ^{3}H). This type of com-

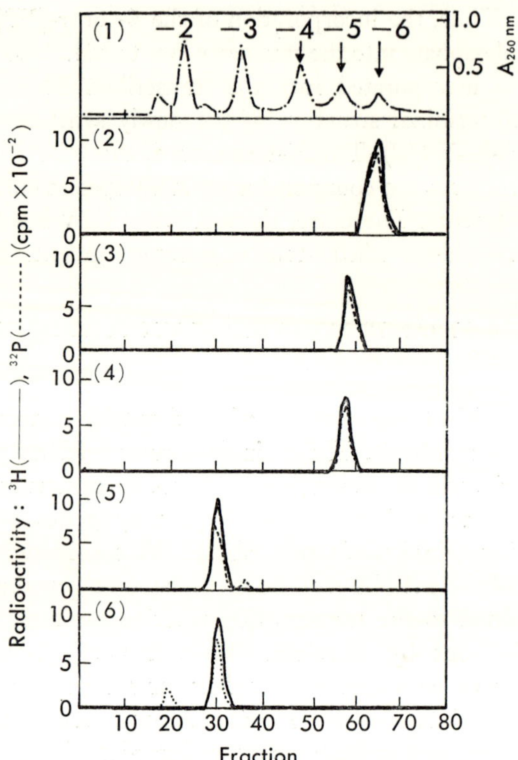

Fig. 8. DEAE-urea column chromatography of the oligonucleotides obtained from the 5′-terminal part of CPV mRNA synthesized *in vitro* in the presence of $\beta[^{32}P]$ ATP and [³H] SAM (*14*). The mRNA was digested with pancreatic ribonuclease A (2), ribonuclease T_1 (3), alkali (4), *Penicillium* nuclease P_1 (5), and P_1 plus phosphomonoesterase (6). (1) is the size marker, that is, the pancreatic ribonuclease A digest of rRNA. Numerals indicate net charge.

pound will be split by phosphodiesterase or pyrophosphatase into Xp and pAm. Certainly, venom phosphodiesterase splits the compound, but the products were three components: [³²P] inorganic phosphate, [³H] 2′-O-methyladenosine 5′-phosphate (pAm) and an unknown [methyl-³H] containing nucleotide-like material as shown in Fig. 9b. The [³H]-labeled unknown material hardly moved from its origin upon electrophoresis, so it is different from an ordinary nucleotide. It seemed to carry a plus charge in its nucleosidic part to cancel the negative charge of a phosphate group. Phosphomonoesterase deleted a phosphate group from the unknown material,

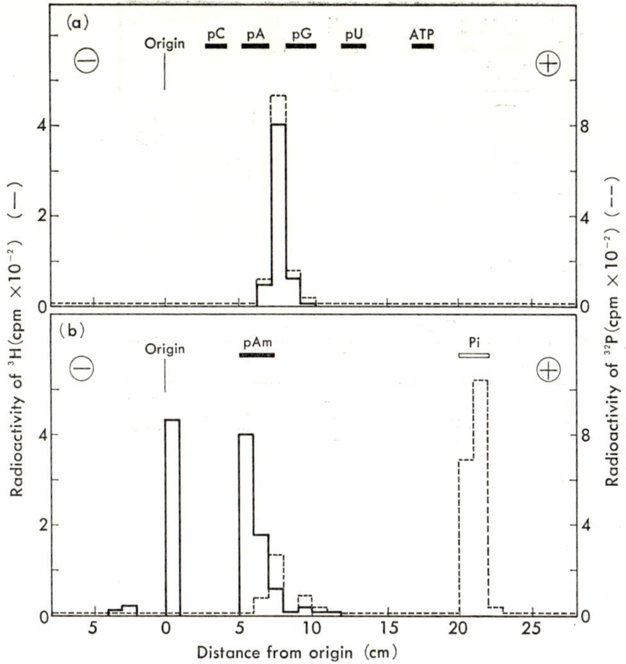

Fig. 9. Paper electrophoresis of the phosphodiesterase digest of the 5′-terminal component obtained by *Penicillium* nuclease P₁ digestion of the CPV mRNA which was labeled with [³H] SAM and [β-³²P] ATP (*16*).

and the remaining nucleosidic material carrying [³H] moved to the anode in pH 7.2 electrophoresis faster than any ordinary nucleoside (Fig. 10a). The mobility of the nucleosidic material was almost similar to that of SAM. However, the mobility in some chromatographies of the nucleosidic material is not the same as SAM or m¹A, which carries a plus charge, but it was the same as 7-methylguanosine (m⁷G) as shown in Fig. 10b. Thus the compound obtained from the 5′-terminal part of CPV mRNA must be m⁷G⁵′ppp⁵′Am. All three phosphates are protected from phosphomonoesterase attack. It is clear that the right and middle phosphates came from the α- and β-phosphates of the substrate ATP. That the left phosphate links to the 5′-position of m⁷G was confirmed by an incorporation experiment with [α-³²P]GTP (guanosine 5′-triphosphate), which was added as a substrate to the synthesis of CPV mRNA in the presence of SAM. Thus the unexpected blocking structure was discovered at the 5′-terminus of CPV mRNA (*16*). The structural formula of this part is

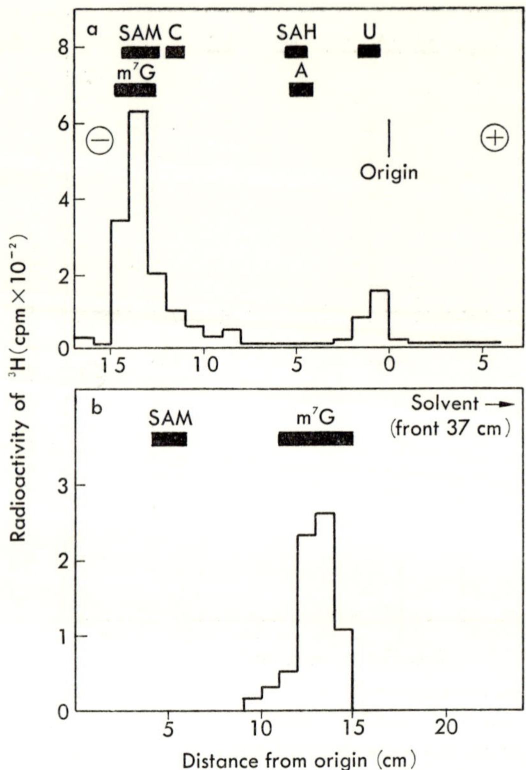

Fig. 10.   Identification of 7-methylguanosine linked to the terminal part of CPV mRNA (*16*). Fraction I in Fig. 9 was digested with phosphomonoesterase. The digest was electrophoresed on Whatman 3 MM paper at pH 3.5 (a). The digest was developed on Whatman 3 MM paper with *n*-butanol-acetic acid-H_2O (5:1:4, v/v/v) (b).

shown in Fig. 11. This structure was finally identified by chromatographic analysis of it and its degradation products with the joint use of the chemically synthesized materials. We owe the chemical synthesis of this structure to Prof. Tsujiaki Hata and his colleagues at the Tokyo Institute of Technology (*17*). This type of structure, in which two nucleosides link 5′ to 5′ in a confronting state and two pyrophosphate linkages are involved, had not been known previously in nucleic acid molecules.

This unusual structure at the 5′-terminus of RNA is synthesized in the early stage of transcription in a virus particle (*18*). All the enzymes that participate in the steps of the modified structure formation are in-

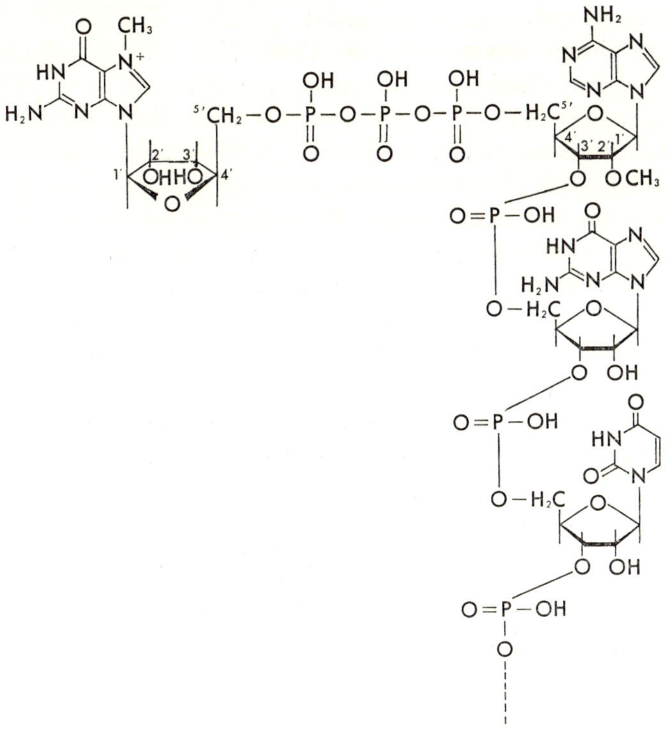

Fig. 11.   Cap structure found at the 5'-terminus of CPV mRNA.
m⁷G⁵'pppAm-G-U-⋯⋯.

volved in the virus core. Besides RNA polymerase, nucleoside triphosphate phosphohydrolase, which splits the γ-phosphate in a nucleoside
triphosphate (19), RNA guanylyl transferase, which joins pG and ppA
through pyrophosphate linkage, and methylation enzymes (RNA (guanine-
7-) methyltransferase and RNA (nucleoside-2'-) methyltransferase) were
contained in a virus particle. These enzymes would be arranged systematically in a CPV particle. By electron microscopic observations with a
technique using both negative staining and rotary shadowing, we recently
showed that the enzymes participating in RNA synthesis seem to be
located in the basal part of a projection (20).

Because an odd modified structure was detected at the 5'-terminus
of CPV mRNA, the 5'-terminal structure of genome RNA was reinvesti-
gated. If there is a similar modified structure in genome RNA, it is quite
reasonable that the phosphorylation at the 5'-terminus by nucleotide kinase

was achieved only after a sequential operation of oxidation, β-elimination, and phosphomonoesterase treatment (*14*). This possibility was examined in the following experiment. If the m⁷G residue in CPV dsRNA contains cis-diol at the 2′- and 3′ positions in the ribose moiety as expected from the results on mRNA (see Fig. 11), it ought to be labeled by a reduction reaction with [³H]NaBH₄ after periodate oxidation. The [³H]-labeled RNA was digested with nulease P₁ and phosphomonoesterase, and the digest was chromatographed on DEAE cellulose with 7 M urea-containing solution. As shown in Fig. 12 (upper), a −2-charged material was detected in addition to nucleoside trialcohols which were derived from the 3′-termini. These were analyzed further by phosphocellulose chromatography (Fig. 12 (lower)). Here again an unknown material appeared in addition to nucleoside trialcohols (C′ and U′). This situation is the same as the experiments shown in Fig. 3. The −2-charged material in Fig. 12

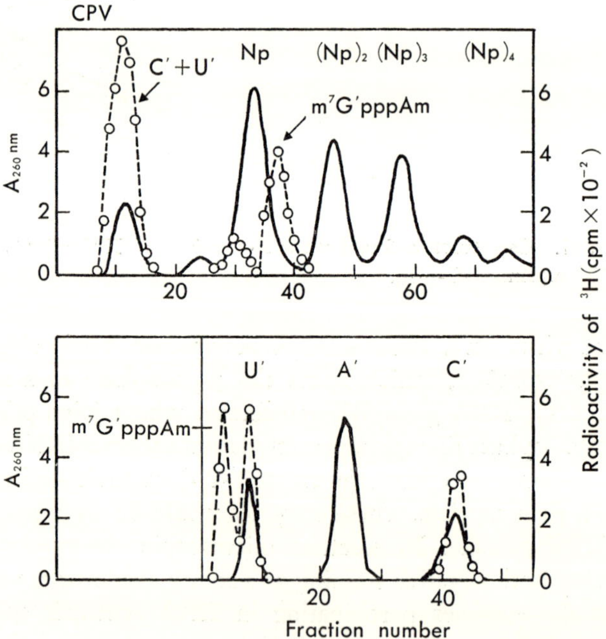

Fig. 12. Column chromatography of the *Penicillium* nuclease P₁ and phosphomonoesterase digest of [³H]-cis-diol-labeled CPV double-stranded RNA (*14*). Upper: DEAE / urea column. Lower: phosphocellulose column (elution condition was a little different from that in Fig. 3). —— absorption at 260 nm of standard materials; ◯ [³H] radioactivity.

(upper) was further analyzed by phosphodiesterase digestion, and it was identified as $m^7G'^{5'}pppAm$. Here, m^7G' represents a trialcohol of 7-methylguanosine, and its amount was just half of the sum of U' and C', trialcohols of uridine and cytidine from the 3'-termini. These results led us to the conclusion that the 5'-termini of one of the dsRNA is methylated and blocked by m^7G as found in mRNA which was synthesized *in vitro* in the presence of SAM. If double-stranded CPV RNA was phophorylated at the 5'-terminus with $[\gamma^{32}P]$ ATP by nucleotide kinase after only phophatase treatment but without removal of preexisting m^7G blockade, only an unmodified 5'-terminus was phosphorylated as ^{32}pG (*69*).

Thus, we can write the relationship between the structures of genome dsRNA and of the single-stranded transcript as follows:

CPV

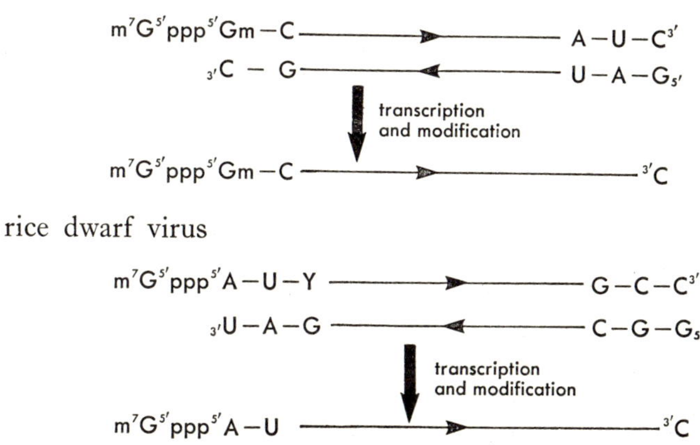

The presence of a 7-methylguanylic acid-blocking structure at the 5'-termini was also found in mRNA of other double-stranded RNA viruses (*14, 21*). The fact that it was found in CPV, in which one of the duplex carries a modified 5'-terminal structure and this strand is copied as an mRNA, was also found in our laboratory in reovirus (human type and avian type) and rice dwarf virus as follows:

reovirus

rice dwarf virus

As shown above, it was clarified that the strand containing genetic information or the positive strand has a modified structure at the 5'-terminus in the double-stranded RNA genome. This modified structure would be useful for the recognition of the starting side of mRNA synthesis by RNA polymerase, or it would work for chain separation or unwinding of double-stranded RNA.

IV. DISTRIBUTION OF CAP STRUCTURE IN RNA SPECIES

When the 7-methylguanylic acid-blocking structure was discovered by us at the 5'-terminus of CPV mRNA, Perry and Kelly (22) reported that mRNA in a mouse L cell was methylated, although it was not shown where methylation took place in an mRNA molecule. We imagined that the 5'-terminal modified structure including methylation is generally involved in mRNA. As the modified structure is a tiny part of a long mRNA molecule, it may have not been detected. Then we analyzed mRNA of vaccinia virus containing double-stranded DNA as a genome (23). This large virus contains its own RNA polymerase in its core, so that transcription can be carried out in a test tube under proper conditions as in the case of a double-stranded RNA-containing virus. *Vaccinia* mRNA synthesized *in vitro* with SAM showed two kinds of 5'-terminal modified structure, $m^7G^{5'}pppAm$ and $m^7G^{5'}pppGm$. A vaccinia virus genome contains several or more cistrons, so that some mRNAs carry $m^7G^{5'}pppGm$ while other mRNAs carry $m^7G^{5'}pppAm$ at the 5'-termini. As a representative cellular mRNA, hemoglobin mRNA was analyzed (24). Hemoglobin mRNA is synthesized preferentially in bone marrow. A primary culture of rabbit bone marrow cells was incubated overnight with [methyl ^{3}H] methionine, and the mRNA was purified and analyzed. The 5'-terminal structure was identified as $m^7G^{5'}pppA$, where A is partially methylated in the 6 position of the base and the 2'-position of ribose. Therefore, the blocking structure and methylation at the 5'-terminal part in mRNA seem to be common in eukaryotic systems. Since the blocking 7-methylguanylic acid and the first nucleotide at the 5'-terminal of RNA are facing each other in 5'-5', we called this type of structure confronting nucleotides. Since our pioneering work on the 5'-terminal modified structure of CPV mRNA was made public, many investigators have identified a similar structure at the 5'-terminus of many kinds of eukaryotic mRNA and viral RNA. This 5'-terminal modification came to be called the "Cap" of mRNA or viral RNA (25, 26). Representative contributions in the early stage other than those made by our group were on reovirus by Shatkin and Furuichi's* group (27), on vaccinia virus by Moss's group (28), on vesicular stomatitis virus by Banerjee's group (29), and on mice myeloma cell by Adams and Cory (30, 31), and so on.

Bacterial mRNA has been known to carry a different 5′-terminal structure. Imamoto, Yamamoto, and myself (32) have confirmed this by analyzing mRNAs of *E. coli* tryptophane operons. The 5′-termini of these mRNAs were pppA-, pppG-, ppA-, or ppG-. The absence of the blocking structure and methylation at the 5′-terminal part in a bacterial mRNA was also confirmed by other experiments (44).

It was surveyed whether the cap structure or the 5′-5′ confronting nucleotides structure is found in eukaryotic cellular RNAs in addition to mRNA. [^{32}P] phosphate and [^{3}H] methionine were added to a cell culture of baby hamster kidney (BHK) cells, and the total RNAs were extracted. The RNA preparations were separated by size, and the 5′-terminal part was analyzed for every RNA fraction. Besides mRNA, two kinds of nuclear small molecular weight RNA (5.7S and 5.9S) carry the cap structure. However, instead of m^{7}G in mRNA, m$_3^{2,2,7}$G was contained as a blocking nucleotide. 5.9S has m$_3^{2,2,7}$G$^{5′}$pppAm-Nm-N-, and 5.7S has m$_3^{2,2,7}$G$^{5′}$pppAm-N-N- (33). Ro-Choi *et al.* (34) reported that a low molecular weight RNA U$_2$ from the nuclei of Novikof hepatoma ascites cells has an odd terminal structure: m$_3^{2,2,7}$G$^{5′}$ppAm-Um-. Although the number of phosphates sandwiched between the nucleosides is two, this would correspond to the cap structure identified by us in 5.9S RNA.

V. CONFORMATION OF CAP STRUCTURE IN mRNA

To clarify the function of the cap structure in eukaryotic mRNA, it is necessary to know the conformation of the cap structure in eukaryotic mRNA. Since it is difficult to obtain a large amount of the cap structure from mRNA samples to perform physico-chemical experiments, the cap structure was synthesized chemically in collaboration with Tsujiaki Hata and his associates at the Tokyo Institute of Technology. The cap structure preparations, including methylated and nonmethylated materials, were characterized by ultraviolet absorption, fluorescence and circular dichroism (35–37).

---

* Dr. Y. Furuichi moved from our laboratory to Roche Institute for Molecular Biology since July 1974 after almost completion of the cap identification on CPV mRNA.

m⁷G⁵′pppA and m⁷G⁵′pppAm show a typical ultraviolet absorption spectrum, with maxima at 258 nm and 257 nm, respectively, in 0.1 M Na⁺ (pH 7) at 25°C. Molecular absorbances of these compounds are fairly lower than those calculated for each component. After the cleavage of pyrophosphate linkages by venom phosphodiesterase, absorbances of the intense peaks increased as shown in Table I. This suggests that both m⁷G⁵′pppA and m⁷G⁵′pppAm have a base-stacked conformation, and that the latter stacks more strongly than the former.

TABLE I

Ultraviolet Absorption and Hypochromicity of Cap Structure and Its Analogues (36)

Compound	λ_{max} (nm)	Molar extinction $(\varepsilon)_{max} \times 10^{-3}$	Hypochromicity (%)
GppA	256	24.0	13
GppG	250	24.6	13
m⁷GppA	260	20.6	16
GpppA	258	25.9	7
GpppG	255	24.8	8
m⁷GpppA	258	22.5	8
m⁷GpppG	255	18.7	14
m⁷GpppAm	257	20.1	20
m⁷GpppCm	263	14.9	11
m⁷GpppUm	260	17.2	10

All spectra were measured in 0.05 M phosphate buffer (pH 7.0, 0.1 M Na⁺) at 25°C. % Hypochromicity was defined as:

$$\left[1 - \frac{(\varepsilon)_{max} \text{ of compounds}}{(\varepsilon)_{max} \text{ of the phosphodiesterase hydrolysate adjusted to pH 7.0}}\right] \times 100$$

The fluorescence at 370 nm of the 7-methylguanosine residue, which was excited at 280 nm, was quenched when the base residue was involved in a stacking interaction with the adenosine residue in the cap structure m⁷G⁵′pppA. The fluorescence measurements, performed by Prof. Masamichi Tsuboi and his associates, also indicate that the base stacked structure seems to be broken at higher temperatures and at pH higher than 8.

The base-stacking state between 7-methylguanine and the confronting base in the cap structure was also supported by nuclear magnetic resonance studies (38, 39). The circular dichroism (CD) spectrum* of the cap struc-

---

* The CD work was carried out mainly by Dr. Masao Hattori with the use of a JASCO spectrometer, which can repeat spectrum scanning with a computer in Dr. T. Oshima's laboratory at the Mitsubishi Kasei Institute for Life Science.

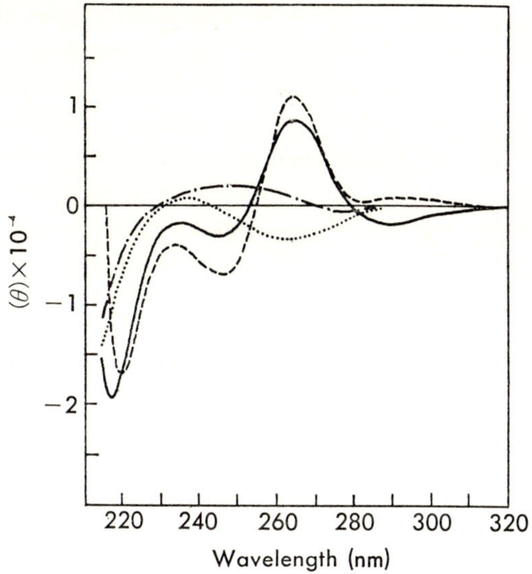

Fig. 13. Circular dichroism spectrum of the cap structure and its components (35). —— m⁷GpppA; ---- m⁷GpppAm; —·— m⁷GMP; ······ summation of AMP and m⁷GMP.

ture, m⁷G⁵′pppA or m⁷G⁵′pppAm, shows a specific curve carrying a large
positive peak at 265 nm as shown in Fig. 13. It is quite different from the
summation spectrum of each component (AMP and m⁷GMP). The peak
at 265 nm is opposite in sign and increases greatly in magnitude compared
with that of the summation spectrum. The long wavelength negative Cotton effect, which is characteristic for oligonucleotides containing a guanosine residue, is also observed. The peaks and troughs of m⁷G⁵′pppA are
highly sensitive to temperature as shown in Fig. 14. The CD spectra at
different temperatures show an isosbestic point. This suggests that the
cap structure m⁷G⁵′pppA has a base-stacked conformation at lower temperatures, and it is loosened at higher temperatures. Under the same conditions (0.1 M Na⁺, pH 7, 25°C), the CD spectrum of m⁷G⁵′pppAm is quite
similar in shape to that of m⁷G⁵′pppA except for a long wavelength band.
The magnitude of peaks at 265 nm increases appreciably compared with
that of m⁷G⁵′pppA, suggesting that methylation of a 2′-hydroxy group of
adenosine in m⁷G⁵′pppA strengthens the stacking interaction between the
N⁷-methylguanine and adenine bases. The strong positive band in the
CD spectrum of m⁷G⁵′pppA(m) increases in magnitude with increasing
concentrations of salt. As reported by Warthaw (40) single-stranded oligo-

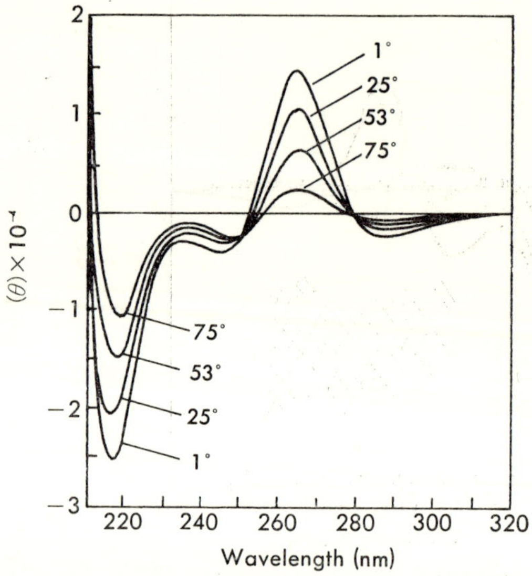

Fig. 14.   Circular dichroism spectrum of m⁷G⁵′pppA at various temperatures
(*36*). 0.11 M Na⁺; pH 6.9.

nucleotides are generally independent of the salt concentration in CD and
optical rotatory dispersion, the high dependency of CD spectra on the salt
concentration seems to be one of the characteristic properties of these cap
components having confronting bases which are linked with a polyanionic
pyrophosphate.

When changing the solution pH of m⁷G⁵′pppA, the positive band in
the CD spectrum at 265 nm is constant at pH 7, but it decreases at higher
pH (Fig. 15) (*36*). The drastic change between pH 7 and 8 is also observed
in fluorescence intensity (*37*). In pH higher than 8 the base-stacked
structure of the cap component m⁷G⁵′pppA would be lost. If the cap
structure participates in protein synthesis as described later, how does the
conformation of the cap relate to the protein-synthesizing ability of mRNA?
The rate of protein synthesis was measured by the incorporation of [³⁵S]-
methionine into polypeptides in the cell free protein-synthesizing systems
from wheat germ and reticulocyte, changing the pH of the reaction mixture. As shown in Fig. 15, the protein synthesis activity in either system
showed a narrow optimal range between pH 7 and 8, and this corresponds
to the range showing a drastic change in the conformation of the cap structure. In the pH range between 7 and 8 the m⁷G residue in the cap of

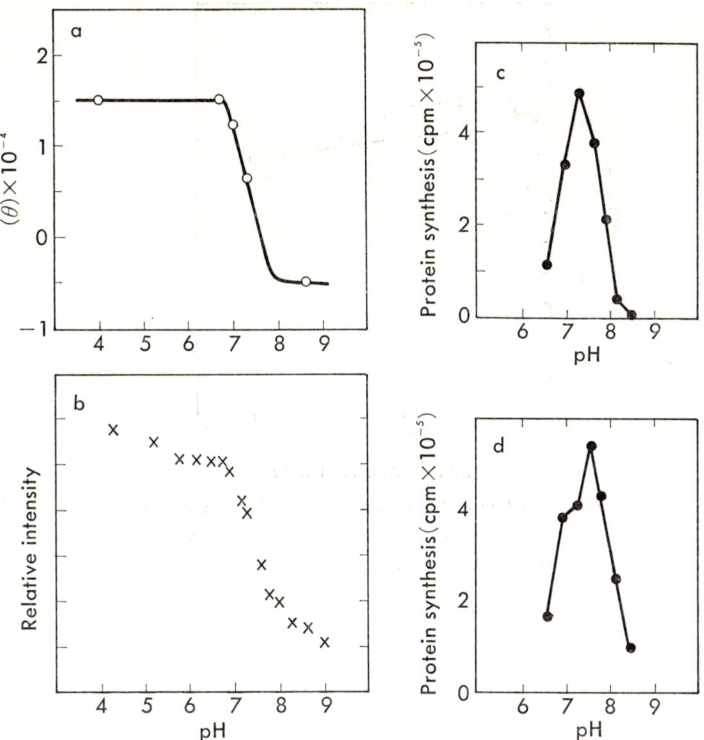

Fig. 15.   pH dependency of circular dichroism spectrum of m⁷G⁵′pppA (a), fluorescence spectrum (b) (data from (*37*)); protein-synthesizing ability of TMV RNA in wheat germ system (c); and in reticulocyte system (d). (Miura, Yamaguchi, Hattori, unpublished).

mRNA must not be in a strongly stacked with the confronting base of the first nucleotide in an mRNA, but must be in a reactive state for a factor involved in protein synthesis, especially in its initiation step.

VI.   IMPLICATIONS OF CAP STRUCTURE IN mRNA FUNCTION

1.   *Inhibition Experiments*

Since the presence of the 5′-terminal cap structure became known to be general for mRNAs in a eukaryotic system except in some cases, the cap structure is considered to function in protein synthesis. The effect of adding m⁷G⁵′pppA to the *in vitro* protein-synthesizing system was studied by incorporation of labeled amino acids into protein in the wheat germ cell-

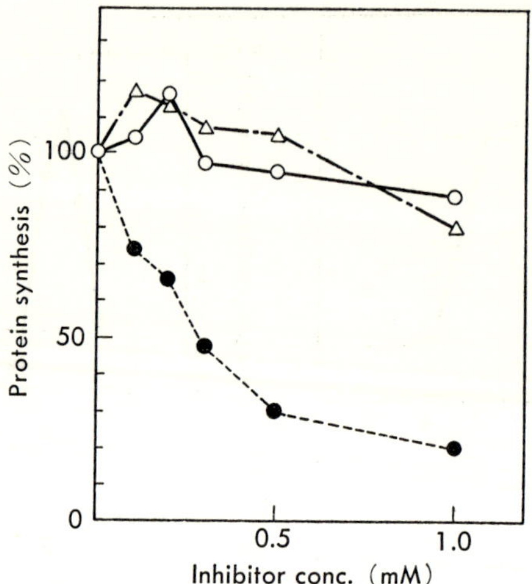

Fig. 16.   Effect of the analogs of the 5′-cap structure in an mRNA on protein synthesis in wheat germ system using TMV RNA as mRNA (*44*). ○ GpppA; ● m⁷GpppA; △ GppA.

free extract with CPV mRNA and tobacco mosaic virus RNA as the mRNA. As shown in Fig. 16, the cap structure $m^7G^{5'}pppA$ was inhibitory, whereas a nonmethylated analog $G^{5'}pppA$ was not. Addition of only pm^7G showed strong inhibition of the *in vitro* protein-synthesizing system when the mRNA carrying the cap structure was used (Fig. 17) (*41–44*). Since the methylation of guanosine in the cap of mRNA seems to be required for inhibition of protein synthesis, a variety of methylated derivatives of guanylic acid were prepared to test their inhibitory effects. We owe a part of the preparations of guanylic acid analogues to Prof. Ikehara and Dr. T. Fukui at Osaka University. Since the addition of guanylic acid (pG) to the wheat germ protein-synthesizing system showed slight inhibition, guanosine was selected as an example of lack of inhibition. This is expressed as '−' (negative) for the inhibitory effect here. Remarkable inhibition was shown with the addition of pm^7G, so it is denoted as '++'. A little inhibition, less than that of pm^7G, was shown by N^2, N^2, 7-trimethylguanylic acid ($pm_3^{2,2,7}G$). This weak inhibition is expressed as '+' (positive) for the inhibitory effect (Fig. 17). Table II summarizes the inhibition of protein synthesis by the methylated derivatives of guanylic acid. Strong inhibitors (++) carry a methyl group at the 7 position, suggesting that the 7-methyl group is especially important for inhibition of protein synthesis. However, not all of the 7-methylguanylic acid deriva-

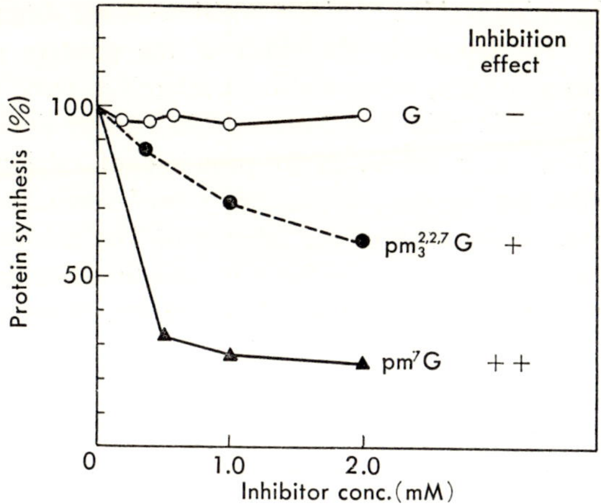

Fig. 17. Typical cases in the inhibitory effect of methylated derivatives of guanylic acid for protein synthesis in wheat germ system using TMV RNA as mRNA (*44*).

TABLE II

Inhibition Effect of the Methylated Derivatives of Guanylic Acid for Protein Synthesis in Wheat Germ Extract Using TMV RNA as mRNA (*44*)

Compound		Number of methyl group			Inhibition effect
		Position			
		2	7	8	
I	pm^{8}G	0	0	1	−
II	pm$_2^{2,8}$G	1	0	1	−
III	pm$_3^{2,2,8}$G	2	0	1	+
IV	pm^{2}G	1	0	0	+
V	pm$_2^{2,2}$G	2	0	0	−
VI	pm^{7}G	0	1	0	+ +
VII	pm$_2^{7,8}$G	0	1	1	+
VIII	pm$_3^{2,7,8}$G	1	1	1	+ +
IX	pm$_4^{2,2,7,8}$G	2	1	1	−
X	pm$_2^{2,7}$G	1	1	0	+ +
XI	pm$_3^{2,2,7}$G	2	1	0	+

tives showed strong inhibition as compounds VII, IX, and XI. In these cases, the inhibitory effect of the 7-methyl group would be cancelled by steric hindrance of configurational change by other methylation around the 7-methyl group (*44*).

The 7-methylguanine residue carries a positive charge different from other purine bases. To determine the effect of the positive charge in inhibition of protein synthesis, other similar nucleotides were prepared. Neither N^1-methyladenylic acid nor 2-methylthio-7-methylinosinic acid showed any inhibition in a wheat germ protein-synthesizing system. Thus, the methylation itself or the protonation at the 7-position in guanosine seems to be specifically required for inhibition. The nucleotide confronting pm^{7}G at the 5'-terminus of mRNA is sometimes methylated in its base moiety and the 2'-position in its ribose moiety. However, 2'-O-methylguanylic acid and 6-methyladenylic acid were not inhibitory for the wheat germ protein-synthesizing system. These results indicate that the strong inhibition of protein sythesis was caused by 7-methylguanylic acid residue in the confronting nucleotide structure (*44*).

2. Deletion of Cap from mRNA

To clarify the function of 7-methylguanylic acid in the 5'-cap of mRNA in protein synthesis, the effect of the removal of the m^{7}G residue by β-elimination from the native mRNA was tested by a few other groups (*45–49*), showing a decrease in the protein-synthesizing ability as well as in the ribosome-binding ability of mRNA. However, β-elimination removes the 3'-terminal nucleotide as well as the m^{7}G in the cap, and sometimes removal of the m^{7}G by β-elimination is incomplete. To avoid these inconveniences, an enzyme specifically attacking pyrophosphate linkages was sought. Usual pyrophosphatase splits the phosphodiester bonds in a polynucleotide chain as well as the pyrophosphate linkages. Shinshi and Miwa at the National Cancer Institute purified a pyrophosphatase from a tobacco cell culture in the process of a study of the degradation of poly (ADP-ribose), and reported that this pyrophosphatase does not split RNA and DNA at all (*50*). It was useful for the complete elimination of the capping part (m^{7}G$^{5'}$pp) from the mRNA chain without any scission in the other parts of the RNA molecule, as shown in Fig. 18 (*51–53*).

The protein-synthesizing ability of the cap-deleted mRNAs which were treated with tobacco pyrophosphatase was compared with the native mRNAs in the wheat germ extract (*53*). As shown in Fig. 19, the protein-synthesizing activity was decreased drastically by elimination of the capping structure from an mRNA. Similar results were obtained with the use of pyrophosphatase purified from potatos by Zan-Kowalczewska *et al.* (*54*), and with the use of the reovirus mRNA, synthesized without a cap (*55*). Because the removal of the modified structure was carried out al-

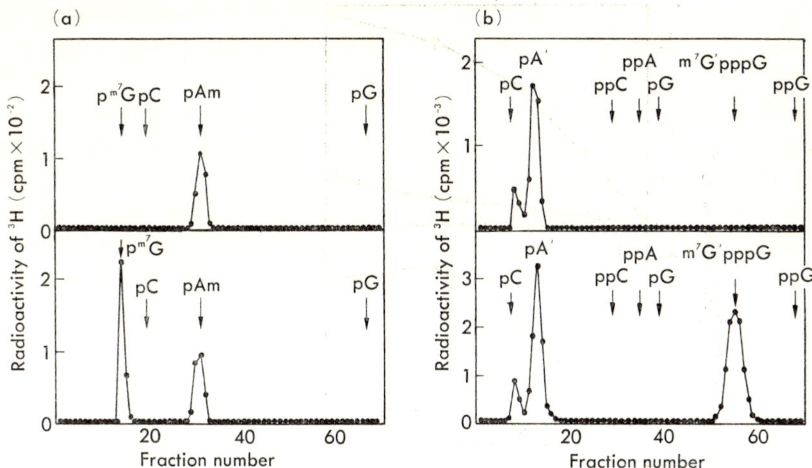

Fig. 18. Terminal nucleoside analysis of intact RNA and tobacco pyrophosphatase-treated RNA (53). Nucleotide chromatography was done by anion-
exchange resin AG 1. (a) The [methyl-³H]-labeled intact CPV mRNA (lower)
and tobacco pyrophosphatase-treated RNA (upper) were digested with nuclease P_1 and then with tobacco pyrophosphatase. (b) TMV RNA, incubated
with tobacco pyrophosphatase (upper) and not incubated with it (lower), was
labeled at the terminal nucleosides by [³H]NaBH₄ after oxidation with sodium
periodate. The RNA was digested with *Penicillium* nuclease P_1.

most completely by tobacco pyrophosphatase (51–53), it was concluded
that the confronting nucleotide structure, the cap, at the 5′-terminus of
mRNA is required for protein synthesis in all of the different types of
eukaryotic mRNA. However, the translational ability of the cap-deleted
mRNA was not abolished completely. Under optimal conditions, the tobacco pyrophosphatase-treated RNA showed 10 to 20% of the protein-
synthesizing ability of intact mRNA.

It is known that a protein molecule is synthesized on mRNA in the
direction from the 5′ end to the 3′ end, so that the 5′-terminal modified
structure and the following nucleotide sequence ahead of the initiation
codon in an mRNA may be related to the initial step of protein synthesis.
Then, the ability to form the initiation complex was tested with the cap-
deleted mRNA, which was treated with tobacco pyrophosphatase (53).
The [³H]–labeled mRNA samples, native and cap-deleted, were added to
the *in vitro* protein-synthesizing system from wheat germ, which contains sparsomycin to prevent peptide chain elongation and thus permits
detection of the initiation complex of protein synthesis. Formation of the

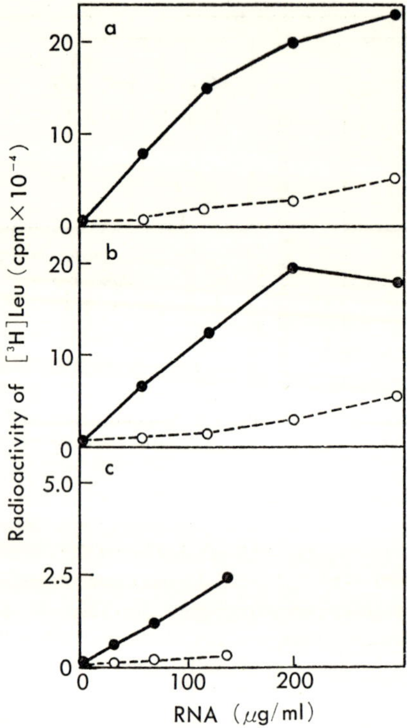

Fig. 19.   Comparison of protein-synthesizing abilities of intact mRNA (●)
and cap-deleted mRNA (○) (*53*). a: CPV mRNA. b: TMV RNA. c: globin
mRNA. Assay was carried out in a wheat germ cell-free protein synthesizing
system.

initiation complex was analyzed by sucrose density gradient centrifugation. As shown in Fig. 20, a significant proportion of the intact CPV
mRNA was recovered as the 80S initiation complex, whereas with the
cap-deleted mRNA only 15% of the amount of intact mRNA at the 80S
position was recovered. A similar loss of initiation complex formation was
obtained when 7-methylguanylic acid was added to a reaction mixture.
Thus the cap structure at the 5′-terminus of mRNA is necessary at the
step of formation of the initial complex in protein synthesis. However, the
ability of the initial complex formation was not abolished completely by
the deletion of the cap or by inhibition with the addition of pm⁷G to the
natural mRNA. The cap structure would be involved in the initiation
step of protein synthesis to form the initiation complex efficiently, but it

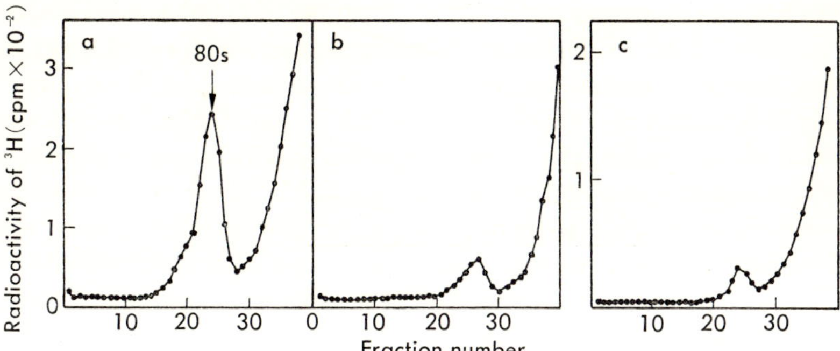

Fig. 20. Formation of the initiation complex for protein synthesis (53). Analysis of the initiation complex was done by centrifugation of a sample in a glycerol density gradient. Direction of centrifugation is from right to left. a : intact CPV mRNA. b : cap-deleted CPV mRNA. c : intact CPV mRNA was added with 7-methylguanylic acid in the protein synthesizing system.

may be dispensable. Although the cap structure is widely found in the mRNA of eukaryotic organisms and viruses, some exceptional cases are known such as in the mRNAs of satellite tobacco necrosis virus (56), encepharo myocarditis virus (57), polio virus (58, 59), and so on.

However, the strong inhibition by pm^{7}G for the initiation complex formation described in the preceding section still suggests the possibility that the cap structure in mRNA plays some direct role in the initial step of protein synthesis. Kaempfer et al. (60) showed that pm^{7}G inhibits the binding of Met-tRNA^{Met} to the initiation factor eIF-2 and mRNA binding to eIF-2 as well as the formation of the complex for the 40S ribosome and Met-tRNA^{Met}. In the experiments by Shafritz et al. (61), however, pm^{7}G inhibited the interaction of mRNA and eIF-4. Suzuki (62) showed that pm^{7}G does not inhibit the interaction between the 40S ribosome and fMet-tRNA^{Met}, but it inhibits formation of the complex of the 80S ribosome and fMet-tRNA^{Met}. He considered that a factor required for the binding of mRNA to the 40S ribosome, probably eIF-4B, would be inactivated by combining with pm^{7}G. Discrepancies in these experimental results may depend on differences in the experimental systems. However, it is likely that pm^{7}G in the cap structure interacts with some initiation factor in protein synthesis.

Owing to the progress in the nucleotide sequencing of RNA, the nucleotide sequence near the 5'-terminus has been shown on many eukaryotic mRNAs (63). The nucleotide sequences between the 5'-terminus and the first initiation codon AUG have a variety of lengths and arrangements. Recently, Shine and Dalgarno (*64*) proposed an idea for prokaryote mRNA that, because a complementary nucleotide sequence for the sequence near the 3′-terminus of 16S ribosomal RNA is commonly detected ahead of the initiation codon, the former sequence may be required for the initiation complex formation in protein synthesis. Many eukaryotic mRNAs carry the nucleotide sequence complementary to the sequence near the 3′-terminus of 18S ribosomal RNA, but some eukaryotic mRNAs do not. The analytical results obtained in our laboratory also show both cases (*65–67*). In some eukaryotic mRNAs the nucleotide sequence between the 5′-terminus and the first initiation codon can construct a folded secondary structure by forming hairpin structures with a series of base pairings, but in some other cases it is difficult to construct a stable secondary structure. According to the experiments of mRNA-ribosome binding, Kozak (*63*). considers that the secondary structure near the 5′-terminus of mRNA is required for correct binding of mRNA to the definite place in a ribosome particle after recognition for a ribosome by the cap 7-methylguanylic acid of mRNA. In eukaryotic mRNA the structure indispensable for the initiation of protein synthesis may be only the initiation codon. The cap structure, the complementary sequence to the ribosomal RNA and the specific secondary structure would enhance the effective formation of the initiation complex by cooperation with the initiation codon. These structures would be required for stable protein synthesis. A combination of these structures would determine the efficiency in the interaction of an individual mRNA with ribosome and initiation factors in the initial step of protein synthesis. Further studies are required on the molecular mechanism of mRNA function described here.

3　Stabilization of mRNA by Capped Structure

One reason for the fact that initiation complex formation is decreased by removal of the 5′-cap part in mRNA may depend on the lability of the pyrophosphatase-treated RNA. This was tested by investigating the stability of mRNA in wheat germ extract (*53*). The CPV mRNA was labeled with [methyl ^{3}H] at the cap structure and with [^{32}P]phosphate at the inner part. A portion of the labeled CPV mRNA was treated with tobacco pyrophosphatase to remove the cap m^{7}G$^{5′}$pp. Both the samples were incubated with the S-23 fraction (the supernatant fraction after 23,000 × *g* centrifugation) of the wheat germ extract under the same conditions as for the *in vitro* protein-synthesizing system. After incubation of the mixture at

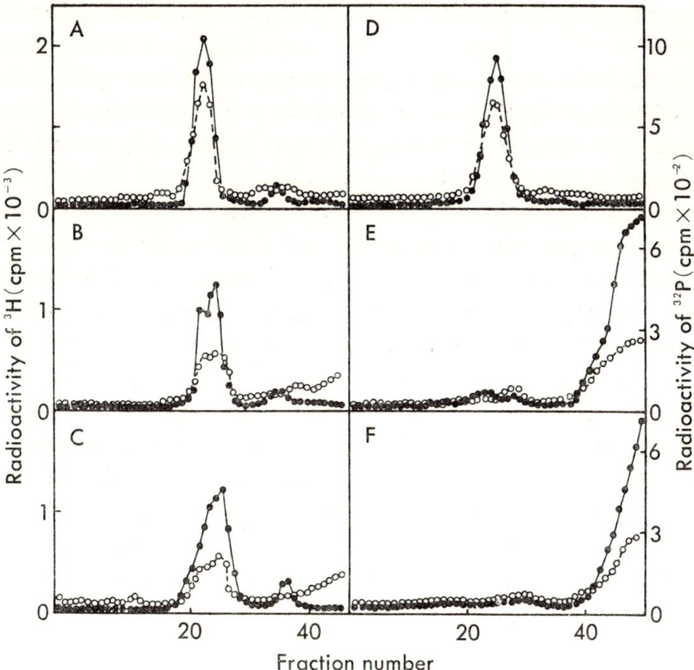

Fig. 21.   Stability of CPV mRNA during incubation with the wheat germ cell-free extract (*53*). The labeled CPV mRNA (A, B, C) and the cap-deleted RNA (D, E, F) were incubated with the S-23 fraction of wheat germ extract at 30°C for 0 min (A and D), 2 min (B and E), and 3 min (C and F). After incubation, the reaction mixture was mixed with phenol, and the extracted RNA was analyzed on a glycerol density gradient. Direction of centrifugation is from right to left. ● [methyl-³H] in the cap structure; ○ [³²P] in the inner part of RNA.

30°C for a short while, the samples were mixed with phenol to depress the enzyme activity and analyzed by glycerol density gradient centrifugation. As shown in Fig. 21, a large proportion of the intact CPV mRNA did not change in size after 2 or 3 min incubation in S-23. Although a small portion of ³²P radioactivity of CPV mRNA was transferred to the smaller RNA fraction, none of the [methyl ³H] radioactivity located at the 5′-terminus of mRNA was detected in the top fraction of the density gradient. In contrast with the intact mRNA, the cap-deleted mRNA was immediately broken down by incubation with the S-23 fraction. Both the ³H and the ³²P radioactivity moved to the top fraction of the density gradient. These results clearly indicate that the intact mRNA is resistant to nuclease attack due to the presence of the 7-methylguanylic acid blocking the 5′-terminal structure.

The analysis of the degradation product in wheat germ extract of the cap-deleted mRNA that was labeled at the 5′-terminus showed mononucleotides and nucleosides but not any oligonucleotide. This means that the first nucleotide at the 5′-terminus of the cap-deleted mRNA was released as a mononucleotide, and some of it was converted to nucleoside by phosphatase action, which may be contained in the wheat germ extract. From these results it is suggested that an exonucleolytic activity splitting RNA from the 5′- to 3′-terminus in the wheat germ extract cannot cleave the cap structure at the 5′-terminus of mRNA, and therefore cannot degrade the intact mRNA; it degrades RNA only after the cap is removed. In other words, the cap structure at the 5′-terminus of mRNA seems to protect the RNA molecule from exonucleolytic degradation; the degradation of mRNA would start by removal of the 5′-cap structure. A similar observation was made by microinjection experiments to a cell with the use of either capped or uncapped reovirus mRNA (55). The 5′-cap may be requisite for the long lifetime of mRNA and for its transfer from the nucleus to the cytoplasm in a eukaryotic cell avoiding degradation.

SUMMARY

To clarify the specific structure of the strand carrying genetic information in the duplex of genome nucleic acid, the structure of the genome and its transcript were studied on double-stranded RNA containing a virus, especially silkworm cytoplasmic polyhedrosis virus. In these virus, the genome is divided into some 10 segments. It was found that one strand of the double helix in each segment was methylated at the ribose moiety of the 5′-terminal nucleotide, showing a clear difference between the two strands. The virus carrying double-stranded RNA contains RNA ploymerase in its coat, and it can synthesize messenger RNA (mRNA) *in vitro*. Analysis of the terminal structure of a genome transcript showed that the same strand as the 5′-methylated strand in the genome duplex was copied as an mRNA. If the methyl donor S-adenosylmethionine was added to the mRNA-synthesizing system *in vitro*, the transcript mRNA was methylated at the 5′-terminus; furthermore, it was discovered that the 5′-terminal phosphates were blocked by 7-methylguanosine to become $m^7G^{5'}pppAm\text{-}G\text{-}$. One strand of the genome double helix was also blocked by 7-methylguanosine at the 5′-terminus. Therefore, in cytoplasmic polyhedrosis virus, the strand carrying protein information has a 7-methylguanylic acid blocking the structure at the 5′-terminus. Following after our discovery, a similar structure has been reported in many eukaryotic viral RNAs and eukaryotic, either cellular or viral, mRNAs, and it came to be generally called the 5′-"cap" structure of mRNA.

The 5′-cap structure was not found in bacterial mRNA. Except for mRNAs in a eukaryotic cell, the cap structure has been detected only in nuclear 5.7S RNA and 5.9S RNA, which carry the 2, 2, 7–trimethylguanosine ($m_3^{2,2,7}G$)-cap at the 5′-terminal.

The conformation of the 5′-cap region was studied by ultraviolet, circular dichroism, and fluorescent spectra with the use of a chemically-synthesized cap structure. The cap has a base-stacked structure at pH 7. Under optimal conditions for protein synthesis, between pH 7 and 8, the cap structure shows a drastic change in its conformation, when it is a reactive state.

Protein synthesis in a eukaryotic system is inhibited by the addition of a cap structure or 7-methylguanylic acid. Strong inhibition is specifically caused by 7-methylguanylic acid among the methylated analogues of guanylic acid. Initiation complex formation in protein synthesis is also inhibited by the addition of 7-methylguanylic acid. If the 7-methylguanylic acid residue in the cap of mRNA was deleted specifically by tobacco pyrophosphatase, mRNA lost its ability to synthesize protein and to form the initiation complex. However, the inhibition of protein synthesis by the addition of 7-methylguanylic acid and the loss of protein-synthesizing ability by deletion of the cap were not perfect, so that the cap structure may function cooperatively with the initiation codon, the higher order structure near the 5′-terminus of mRNA, and the sequence complementary to the ribosomal RNA to form the initiation complex efficiently. The cap structure is effective for stabilization of mRNA.

Acknowledgments

The author would like to thank the coauthors of our papers referred to in this review, especially Dr. Masahiro Sugiura, Dr. Yasuhiro Furuichi and Dr. Kunitada Shimotohno for their energetic collaboration. Our work reviewed in this article was partly supported by Grants-in-aid form the Ministry of Education, Science, and Culture of Japan and a Research Grant from the Naito Foundation.

REFERENCES

1 I. Fujii-Kawata, K. Miura, and M. Fuke, *J. Mol. Biol.*, **51**, 247 (1970).

2 K. Miura, I. Kimura, and N. Suzuki, *Virology*, **28**, 571 (1966).

3 K. Miura, I. Fujii, T. Sakaki, M. Fuke, and S. Kawase, *J. Virol.*, **2**, 1211 (1968).

4 T. Sato, Y. Kyogoku, S. Higuchi, Y. Mitsui, Y. Iitaka, M. Tsuboi, and K. Miura, *J. Mol. Biol.*, **16**, 180 (1966).

5 T. Samejima, H. Hashizume, K. Imahori, I. Fujii, and K. Miura, *J. Mol. Biol.*, **34**, 39 (1968).

6 Y. Furuichi and K. Miura, *J. Mol. Biol.*, **64**, 619 (1972).

7 Y. Furuichi and K. Miura, *Virology*, **55**, 418 (1973).

8 M. Fujimoto, A. Kuninaka, and H. Yoshino, *Agric. Biol. Chem.*, **33**, 1517 (1969).

9 K. Miura, K. Watanabe, and M. Sugiura, *J. Mol. Biol.*, **86**, 31 (1974).

10 K. Shimotohno and K. Miura, *J. Biochem.*, **74**, 117 (1973).

11 K. Shimotohno and K. Miura, *Virology*, **53**, 283 (1973).

12 K. Shimotohno and K. Miura, *J. Mol. Biol.*, **86**, 21 (1974).

13 K. Miura, Y. Furuichi, K. Shimotohno, T. Urushibara, K. Watanabe, and M. Sugiura, *Les Colloque de l'Institut National de la Santé et de la Recherche Médicale*, **47**, 153 (1975).

14 K. Miura, Y. Furuichi, K. Shimotohno, T. Urushibara, and M. Sugiura, Proc. 10th FEBS meeting, North-Holland, Amsterdam, p. 95 (1975).

15 Y. Furuichi, *Nucleic Acids Res.*, **1**, 809 (1974).

16 Y. Furuichi and K. Miura, *Nature*, **253**, 374 (1975).

17 T. Hata, I. Nakagawa, K. Shimotohno, and K. Miura, *Chem. Lett.*, 987 (1976).

18 K. Shimotohno and K. Miura, *FEBS Lett.*, **64**, 204 (1976).

19 K. Shimotohno and K. Miura, *J. Biochem.*, **81**, 371 (1977).

20 K. Yazaki and K. Miura, *Virology*, **105**, 467 (1980).

21 K. Miura, K. Watanabe, M. Sugiura, and A. Shatkin, *Proc. Natl. Acad. Sci. U.S.*, **71**, 3979 (1974).

22 R. P. Perry and D. E. Kelly, *Cell*, **1**, 37 (1974).

23 T. Urushibara, Y. Furuichi, C. Nishimura, and K. Miura, *FEBS Lett.*, **49**, 385 (1975).

24 K. Miura, Y. Furuichi, and Y. Miura, unpublished.

25 B. Griffin, *Nature*, **255**, 9 (1975).

26 A. J. Shatkin, *Cell*, **9**, 645 (1976).

27 Y. Furuichi, M. Morgan, S. Muthukrishnan, and A. J. Shatkin, *Proc. Natl. Acad. Sci. U.S.*, **72**, 362 (1975).

28 C. M. Wei and B. Moss, *Proc. Natl. Acad. Sci. U.S.*, **72**, 318 (1975).

29 G. Abraham, D. P. Rhodes, and A. K. Banerjee, *Cell*, **5**, 51 (1975).

30 J. Adams and S. Cory, *Nature*, **255**, 28 (1975).

31 S. Cory and J. M. Adams, *J. Mol. Biol.*, **99**, 519 (1975).

32 K. Miura, M. Yamamoto, and F. Imamoto, unpublished.

33 K. Shimotohno, T. Urushibara, and K. Miura, *Proc. Japan Acad.*, **52**, 563 (1976).

34 T. S. Ro-Choi, Y. C. Ro-Choi, D. Henning, J. McCloskey, and H. Busch, *J. Biol. Chem.*, **250**, 3921 (1975).

35 M. Hattori, K. Miura, K. Yamaguchi, S. Ohtani, and T. Hata, *Nucleic Acids Res.*, Special Publ. No. 5, 391 (1978).

36 M. Hattori and K. Miura, unpublished.

37 Y. Nishimura, S. Takahashi, T. Yamamoto, M. Tsuboi, M. Hattori, K. Miura, K. Yamaguchi, S. Ohtani, and T. Hata, *Nucleic Acids Res.*, **8**, 1107 (1980).

38 E. D. Hickey, L. A. Weber, C. Baglioni, C. H. Kim, and R. H. Sarma, *J. Mol. Biol.*, **109**, 173 (1977).

39 C. H. Kim and R. H. Sarma, *Nature*, **270**, 223 (1977).

40 M. M. Warshaw and I. Tinoco, Jr., *Biopolymers*, **9**, 1059 (1970).

41 E. D. Hickey, L. A. Weber, and C. Baglioni, *Proc. Natl. Acad. Sci. U.S.*, **73**, 19 (1976).

42 E. D. Hickey, L. A. Weber, and C. Baglioni, *Nature*, **261**, 71 (1976).

43 L. A. Weber, E. R. Femen, E. D. Hickey, M. C. Williams, and C. Baglioni, *J. Biol. Chem.*, **251**, 5657 (1976).

44 K. Miura, Y. Kodama, K. Shimotohno, T. Fukui, M. Ikehara, I. Nakagawa, and T. Hata, *Biochim. Biophys. Acta*, **564**, 264 (1979).

45 S. Muthukrishnan, G. W. Both, Y. Furuichi, and A. J. Shatkin, *Nature*, **255**, 33 (1975).

46 D. W. Leung, C. W. Gilbert, R. E. Smith, N. L. Sasavage, and J. M. Clark, *Biochemistry*, **15**, 4943 (1976).

47 M. Kozak and A. J. Shatkin, *J. Biol. Chem.*, **251**, 4259 (1976).

48 J. K. Rose and H. F. Lodish, *Nature*, **262**, 32 (1976).

49 R. Roman, J. D. Brooker, S. N. Seal, and A. Marcus, *Nature*, **260**, 359 (1976).

50 H. Shinshi, M. Miwa, K. Kato, M. Noguchi, T. Matsushima, and T. Sugimura, *Biochemistry*, **15**, 2185 (1976).

51 H. Shinshi, M. Miwa, T. Sugimura, K. Shimotohno, and K. Miura, *FEBS Lett.*, **65**, 254 (1976).

52 T. Ohno, Y. Okada, K. Shimotohno, K. Miura, H. Shinshi, M. Miwa, and T. Sugimura, *FEBS Lett.*, **67**, 209 (1976).

53 K. Shimotohno, Y. Kodama, J. Hashimoto, and K. Miura, *Proc. Natl. Acad. Sci. U.S.*, **74**, 2734 (1977).

54 M. Zan-Kowalczewska, M. Bretner, H. Sierakowska, E. Szczesna, W. Fillipowicz, and A. J. Shatkin, *Nucleic Acids Res.*, **4**, 3065 (1977).

55 Y. Furuichi, A. LaFiandra, and A. J. Shatkin, *Nature*, **266**, 235 (1977).

56 E. Wimmer, A. Y. Chang, J. M. Clark, and M. E. Reichmann, *J. Mol. Biol.*, **38**, 59 (1968).

57 D. Frisby, M. Eaton, and P. Fellner, *Nucleic Acids Res.*, **3**, 2771 (1976).

58 A. Nomoto, Y. F. Lee, and E. Wimmer, *Proc. Natl. Acad. Sci. U.S.*, **73**, 375 (1976).

59 M. J. Howlett, J. K. Rose, and D. Baltimore, *Proc. Natl. Acad. Sci. U.S.*, **73**, 327 (1976).

60 R. Kaempfer, H. Rosen, and R. Israeli, *Proc. Natl. Acad. Sci. U.S.*, **75**, 650 (1978).

61 D. A. Shafritz, J. A. Weinstein, B. Safer, W. C. Merrick, L. A. Weber, E. D. Hickey, and C. Baglioni, *Nature*, **261**, 291 (1976).

62 H. Suzuki, *J. Biochem.*, **84**, 1125 (1978).

63 M. Kozak, *Cell*, **15**, 1109 (1978).

64 J. Shine and L. Dalgarno, *Proc. Natl. Acad. Sci. U.S.*, **71**, 1342 (1974).

65 S. Hidaka, K. Shimotohno, K. Miura, Y. Takanami, and S. Kubo, *FEBS Lett.*, **98**, 115 (1979).
66 A. Onoue, M. Nakamura, S. Nakanishi, S. Hidaka, K. Miura, and S. Numa, *Eur. J. Biochem.*, in press.
67 S. Hidaka, K. Miura, Y. Takanami, and S. Kubo, unpublished.
68 K. Shimotohno and K. Miura, Proc. 1977 Molecular Biology Meeting of Japan, Kyoritsu Shuppan, Tokyo, p. 83 (1977).

Received for publication July 21, 1980.

Adv. Biophys., Vol. 14, pp. 239–256 (1981)

THE IONIC MECHANISM OF EXCITATION IN INTESTINAL SMOOTH MUSCLE CELLS

TAIZO SUZUKI AND HACHIRO INOMATA

Department of Applied Physiology, Tohoku University School of Medicine, Sendai, Japan

Moore and Cole stated in their review (*1*) that a new realm of physiological experimentation on the ionic mechanism of excitation in excitable cells was opened by the introduction of the ionic permeability concept (*2*) and the voltage-clamp method. The voltage clamp method, which was improved by Hodgkin and Huxley (*3–7*), permitted the quantitative study of the excitation process in membranes. The technique of the voltage-clamp method has been reviewed in detail by Moore and Cole (*1*) and others. The basic principles of this technique will be described hereafter.

When the membrane potential is displaced by V mV from the resting level (depolarized or hyperpolarized), the total current (I) from this potential displacement can be divided into a capacitative current and an ionic current (I_i)

$$I = C_m \frac{dV}{dt} + I_i,$$

where C_m is the membrane capacitance. In squid giant axons, the ion species which carry the membrane currents are Na, K, and Cl. Na current, which is a current carried by Na ions, can be expressed in terms of ionic conductance (g_{Na}).

$$I_{\mathrm{Na}} = g_{\mathrm{Na}}(E_{\mathrm{m}} - E_{\mathrm{Na}}),$$

where E_{Na} is the equilibrium potential for Na and E_{m} is the membrane potential. A similar form of equation can be set up for other ions.

When a membrane is depolarized suddenly to a predetermined level and maintained at this level, there is an immediate current surge which results from the discharge of membrane capacitance. Membrane capacitance discharges quickly and completely and the subsequent current flow is attributed to pure ionic currents through the membrane. If the membrane potential is maintained at a predetermined level during the excitation process, the change in the ionic conductance (g) of the membrane during the excitation process can be estimated by the measurement of currents through the membrane. Therefore, the critical point in membrane current experiments is how to maintain (to clamp) the membrane potential at a predetermined level. This is the reason why the voltage-clamp technique is essential for the study of the ionic mechanism of excitation.

To clamp the membrane potential at a predetermined level, two electrodes must be inserted inside the cell; the first electrode is to measure the membrane potential and the second electrode is to control the membrane potential by applying the current. The amount of current necessary to do this is equal and opposite to the ionic current generated by the cell membrane.

This technique has been applied successfully to many types of excitable cells. However, until recently the application of this technique to mammalian smooth muscle has not been successful. The ideal cell for voltage-clamp experiments would be a single cell, large enough to insert two electrodes. Most mammalian smooth muscle cells are too small in cell size to allow the insertion of two electrodes. Moreover, there are serious limitations on a voltage clamp in these tissues, because of the complexities of its morphological structure (*30–33*).

In 1969, Anderson introduced the double sucrose gap method for voltage clamping a muscle strip isolated from rat uterus (*8*). This method is particularly noteworthy because the measurement of membrane potential and the current application for clamping became feasible without using intracellular microelectrodes. The effectiveness of this method has become widely appreciated and a number of investigators have applied it to various types of smooth muscles (myometrium, *8–14, 16, 29*; taenia coli, *15–18*; *24, 27, 31*; blood vessels, *34*). Figure 1 shows how the double sucrose gap method is applied to smooth muscle preparations.

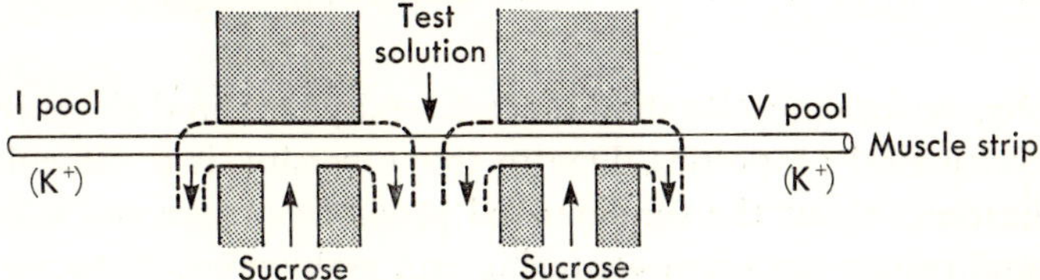

Fig. 1.   Double sucrose gap method applied to smooth muscle. Stämpfli first described the single sucrose gap method (*23*) and it was considerably developed into the double sucrose gap method by Julian *et al.* (*21, 22*). Middle test solution region (node region): region exposed to the test solution. V pool (filled with high K solution to depolarize to zero membrane potential) is used to record the potential changes with respect to the node region. Currents are injected *via* I pool (filled with high K solution).

However, the double sucrose gap method has several serious problems with voltage clamping. First, the membrane potential is not directly measured. That is, the membrane potential measured by this method is not the true membrane potential but the sum of the true membrane potentials and the voltage drop across the series resistance in the node region which is exposed to the test solution (*33*). An accurate measurement of series resistance is impossible. But, the error elicited by series resistance could be minimized by reducing the series resistance compared to the cell membrane resistance. Second, the ideal voltage-clamp is obtained only when the voltage distribution in the clamping region (node region) is kept uniform, and such a situation is impossible in the double sucrose gap method. The deviation from the ideal uniformity in voltage distribution mainly arises from the cable properties of the node region (cable loss). Of course, such a cable loss could be minimized by narrowing the width of the node region. But the narrowing of node width elicites the increase in series resistance, which in turn causes an increase in the error in the measurement of membrane potential (*24, 33*).

As mentioned above, there are some limitations of the double sucrose gap method in the accuracy of voltage control. But this method is advantageous by virtue of the small size of smooth muscle cells and the technical difficulties that occur when using intracellular microelectrodes. Much experimental data concerning the ionic mechanism underlying the generation of excitation of smooth muscle cells have been obtained by this method (*8–18, 24, 27, 31*).

I. IONIC CURRENTS IN THE EXCITATION PROCESSES

Generally, smooth muscles show great diversity both in their morphological and physiological properties. Due to this diversity, it is difficult to make generalizations about the fundamental properties of smooth muscles. Experimental results are often confusing and ambiguous. Thus, in studying smooth muscles, materials should be selected according to their purpose. The only intestinal smooth muscle that has been extensively studied with respect to excitation and contraction is the guinea pig taenia coli. The reason for using this tissue is that it is very easy to isolate at the proper diameter and length without damage. The present review will therefore be concerned only with this muscle. This tissue corresponds to the longitudinal layer of the small intestine.

1. Excitation Process in Normal Solution
1) Action potential under current clamp

As is well known, the action potential is strongly influenced by the ionic constitution of a solution. A normal action potential is obtained when the muscle strip is soaked in normal solution (physiological solution). A typical action potential of smooth muscle cells in the guinea pig taenia coli is illustrated in Fig. 2. The action potential in this figure was evoked by a long depolarizing constant current (current clamp). It is characterized by a slow depolarization, low amplitude and a rather high rate of repolarization.

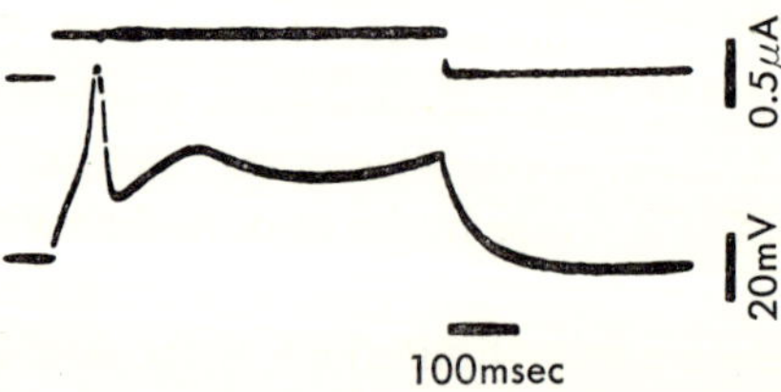

Fig. 2.   Action potential of smooth muscle cells of guinea pig taenia coli in normal Krebs solution under current clamp (recorded by double sucrose gap method). Upper trace, applied current pulse; lower trace, action potential.

2) Hyperpolarizing step potential under voltage clamp

Stepwise changes in the membrane potential under voltage-clamp conditions resulted in a family of currents. The current traces for a stepwise hyperpolarization are simple, since a hyperpolarizing step potential is not

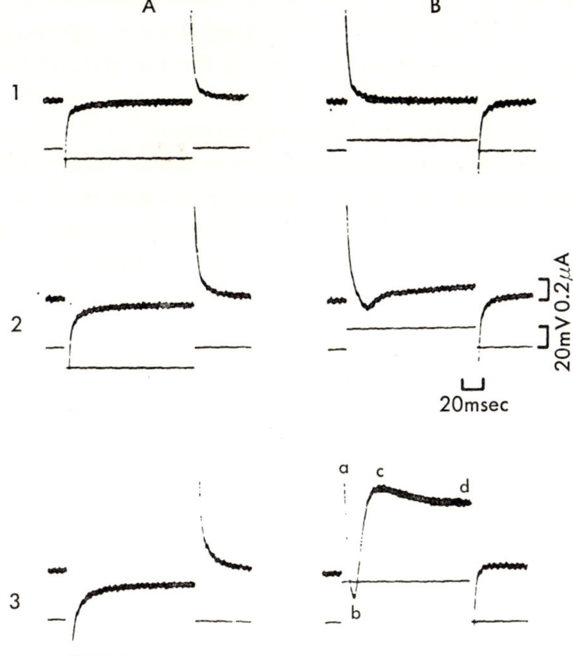

Fig. 3. Membrane currents in normal solution under voltage clamp. Column A: responses of membrane currents to hyperpolarizing steps. Column B: responses to depolarizing steps. From 1 to 3, increases in hyperpolarizing (column A) or depolarizing (column B) steps. In B-3: a, capacitative current; b, fast inward current; c, fast transient outward current; d, late outward current. These traces are before correcting for leakage and capacitative currents. Notice the small inward current in B-2 and its larger amplitude with further depolarization in B-3.

effective in evoking the action potential. As shown in Fig. 3-A, the hyperpolarizing step potential was accompanied by an initial capacitative current. This current declined exponentially with a time constant of about 5 msec. Then the inward current (shown in a downward direction) was maintained as long as the voltage steps. This inward current is termed the leakage current. The leakage current (ΔI) is nearly proportional to the hyperpolarizing step voltage (ΔV). That is, the leakage current is essentially ohmic. From $\Delta I/\Delta V$, the leakage conductance has been estimated to be about $3\,\mu$mho (Fig. 4).

The leakage current measured on the hyperpolaring side is important for making corrections of the ionic current on the depolarizing side. Because of the ohmic character of the leakage current, the correction can be made on the depolarizing side by subtracting the corresponding leakage current from the total currents which are the sum of the net current and the leakage current.

3) Depolarizing step potential under voltage clamp

Current associated with stepwise depolarization under voltage-clamp conditions is complex for the reason that the excitation process is persisted throughout this period. This consists of five components (Fig. 3-B): a) initial capacitative current; b) early inward current (shown in a downward direction); c) transient fast outward current (shown in an upward direction); d) late outward current (shown in an upward direction); and e) leakage current. Among these, the capacitative current and leakage current are excluded from the active membrane current, since these two currents are not concerned in the active process of excitation. As mentioned above, corrections for these two currents are easily performed. Therefore, the net active membrane currents consist of three components: early inward current, transient fast outward current and late outward current.

4) Early inward current

Following the capacitative current, a large inward current was observed (Fig. 3, B-3). This is the early inward current and is responsible for the spike potential. It showed a time dependence, increasing with time to reach the peak amplitude and then decreasing gradually.

It also showed a voltage dependence. With the increasing voltage of depolarization, the inward current increased its peak amplitude. By further depolarization, it reached the maximum value, then declined gradually and finally changed to outward direction. That is, the peak amplitude of the inward current is dependent on the clamped membrane potential. The peak amplitudes of inward currents as a function of clamped membrane potentials are shown in depicted in terms of the current-voltage relationship in Fig. 4.

Theoretically, the reversal potential, the voltage at which the inward current is neither inward nor outward (the voltage axis intercept, E_a in Fig. 4) is the equilibrium potential for ions that carry this current. The reversal potential is about 20 mV. Because of the overlapping of other currents (transient outward current and late outward current), this experimental reversal potential is not likely to be the true equilibrium potential. But relative equilibrium potential values under different conditions would provide some important informations when experimental conditions are not markedly different.

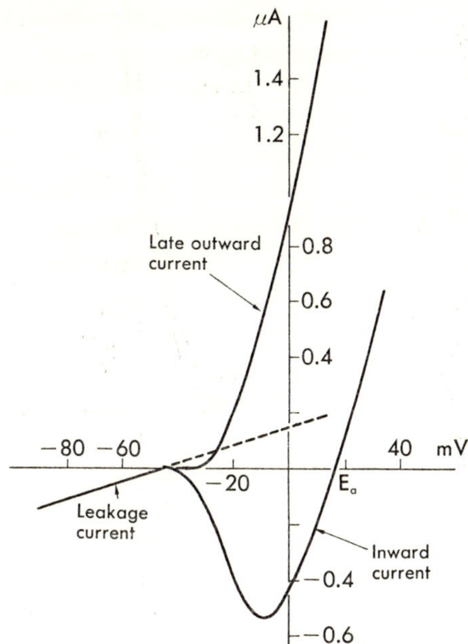

Fig. 4. Current-voltage relations in normal solution. Inward current: peak amplitude of inward current. Late outward current: steady-state value of late outward current. Leakage conductance could be estimated from $\Delta I/\Delta V$ at 10 mV hyperpolarization. E_a, reversal potential. The currents in this figure have been corrected for leakage and capacitative currents.

5) *Ionic kinetics*

Principally, the early inward current in the taenia coli is carried by Ca ions and not by Na ions (*24, 30*). This means that the spike potential of this smooth muscle is a Ca spike. The evidence in support of this idea is: 1) low $[Na]_o$ (50% reduction of $[Na]_o$) had no effect on any early inward current (Fig. 5); 2) with low $[Ca]_o$ (reduction of $[Ca]_o$ to 10% of normal), the maximum amplitude of the inward current markedly decreased and the reversal potential (E_a) shifted toward the negative; 3) Mn^{2+}, which exerts a blocking action on Ca spikes, produced a decrease in the early inward current (Fig. 6); and 4) D-600, the selective Ca blocker, lowered the inward current and shifted the reversal potential (E_a) toward the negative (Fig. 6).

6) *Transient outward current*

The early inward current was followed by a transient increase in the out-

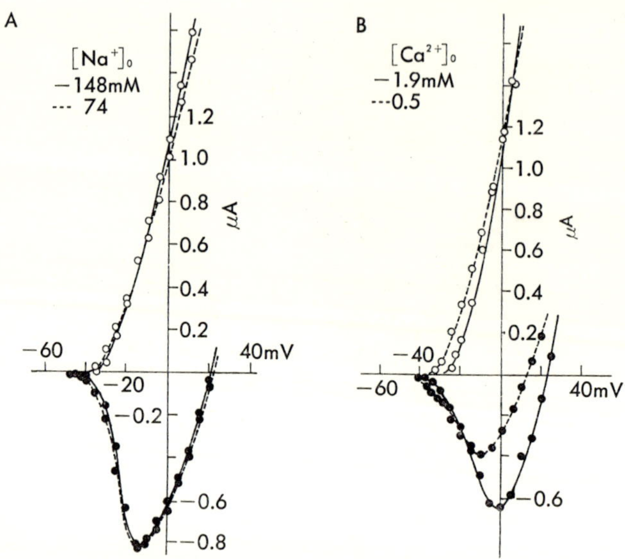

Fig. 5.   Current-voltage relations in low-Na+ (Na+: 74 mM) solution (A) and in low-Ca²+ (Ca²+: 0.5 mM) solution (B). Note that a 50% reduction of [Na]o had no effect on current-voltage relations, but low [Ca]o markedly reduced the amplitude of the inward current.

ward current. This current is the transient fast outward current. This current is specifically inhibited by tetraethylammonium (TEA) (20, 24, 27, 28). The carrier ions of this current are considered to be K ions. Similar types of transient outward currents have been described in other cells (snail neurons, 26; cray-fish X-organ cells, 25). Most of these cells have Ca spikes. From this, it seems likely that the influx of Ca ions following the inward current would induce a transient increase in K conductance, which in turn generates the transient outward current. But, at present, the question remain unsolved whether this current is the true membrane current and its nature is a subject of controversy (24).

7)   *Late outward current*

The late outward current is partly masked by the overlapping of the preceding two currents. Therefore the unmasked true late outward current can be seen when the preceding currents were suppressed by drugs. Fundamentally, the late outward current has a delayed onset, develops progressively and maintains a steady level with a clamping voltage step of about 200–400 msec. The main charge carriers of this current are K ions, but,

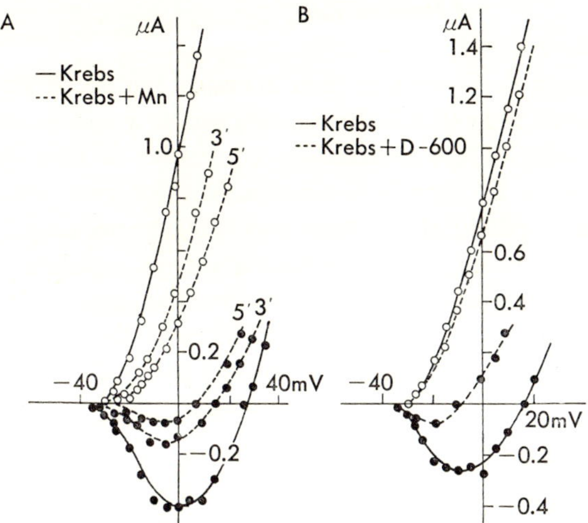

Fig. 6. Effects of Mn^{2+} and D-600 on current-voltage relations (*24*). A: Mn (2 mM) effect. —— before treatment; ------ at two different times (3 and 5 min) after treatment. B: D-600 (10^{-5} M) effect. —— before treatment; ------ after treatment. Note that D-600 suppressed the inward current selectively, while Mn^{2+} suppressed both the inward current and the late outward current.

the participation of Na ions is not negligible. In taenia coli smooth muscle cells, the resting potential is predominantly a K potential. But it is considerably less than the K equilibrium potential. This suggests that, in these cells, other ions in addition to K ions contribute to the resting potential. The most likely candidate is Na ions. At resting condition, the ratio of P_{Na} (Na permeability): P_K (K permeability) was 0.6:1. The reversal potential of the outward current (E_b) was about 15–20 mV more negative than the natural resting potential. The ratio of $P_{Na}:P_K$ for E_b was 0.05:1. It can be understood from these results that the relative increase in K permeability occurs when the late current channel is activated (*24*).

8) Voltage-current relationship of the late outward current as a function of clamping voltages

At voltages more negative than the resting potential (hyperpolarizing range), the value of $\Delta I/\Delta V$ (leakage conductance) was low. At voltages more positive than the resting potential (depolarizing range) $\Delta I/\Delta V$ was rather high compared with the leakage conductance (Fig. 4). Such a phenomenon is referred to as the rectification of membrane conductance.

2. Effects of Strontium and Barium Ions

1) Strontium ions

Strontium ions resemble Ca ions in their chemical nature. If Sr ions can carry current through the cell membrane, this action would be still present when Ca ions in solution are replaced by equimolar Sr ions. In fact, Sr ions can substitute for Ca ions in generation of the action potential (*31*). With Sr, the action potential duration was slightly prolonged (Fig. 7). This prolongation is mainly due to a slower rate of repolarization. The resting potential was slightly hypolarized by Sr. This hyperpolarization is associated with an increase in K conductance, because the leakage conductance was increased without affecting the reversal potential for the late outward current (E_b).

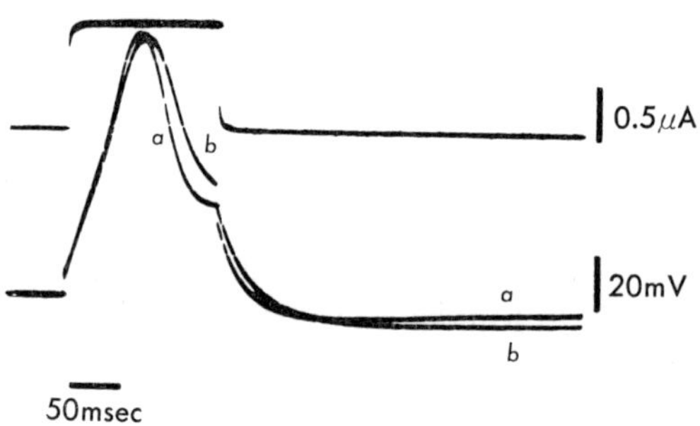

Fig. 7. Action potential in Sr solution (*31*). Trace a, in normal solution containing 2.5 mM Ca^{2+}; trace b, in solution with 2.5 mM Sr^{2+} replacing Ca^{2+}. Note the slight prolongation of spike duration with Sr^{2+}.

Under voltage-clamp conditions, following the replacement of Ca ions with Sr ions, the fast inward current due to the depolarizing step was still observed. This fact indicates that Sr ions carry the fast inward current through the Ca channel. The inward current lasted slightly longer in Sr ions than in Ca ions (*31*).

With the increase in the Sr concentration in solution, the maximum amplitude of the fast inward current was increased and the reversal potential was shifted to a more positive potential.

With Sr, the fast outward current was suppressed. The late outward current was not suppressed, but rather enhanced by Sr (Fig. 8). This enhancement is mainly due to an increase in K conductance caused by Sr (*18, 31*).

2) Barium ions

The effect of Ba ions is basically similar to that of Sr ions. By replacing

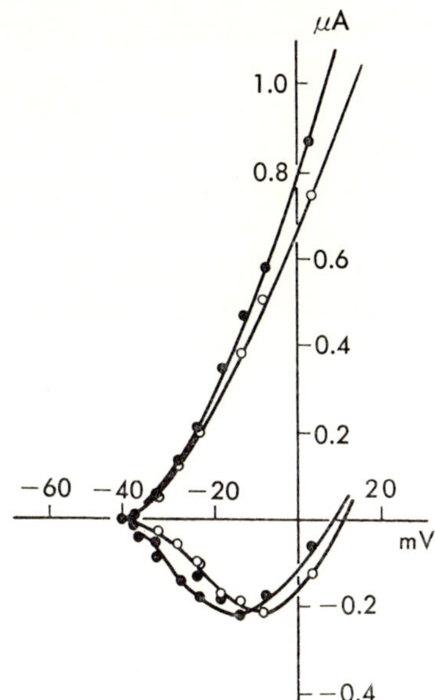

Fig. 8. Current-voltage relations in Sr solution (*31*). Open circles (◯), in normal solution containing 2.5 mM Ca^{2+}; filled circles (●), in solution with 2.5 mM Sr^{2+} replacing Ca^{2+}.

Ca with Ba, the fast inward current due to the depolarizing step was observed. This means that Ba can carry the inward current through the Ca channel. The Ba current was greater in amplitude than the Ca current. Such an augmentation in the fast inward current by Ba is due to a decrease in the overlapping outward current, since Ba ions suppressed the fast outward current. Moreover, in contrast to Sr ions, Ba ions suppressed the late outward current. This distinctive effect of Ba ions is shown in Fig. 9.

It has been shown that Ba ions produced the sustained depolarization of the action potential (Fig. 10) (*19*). Such sustained depolarization caused by Ba is mainly due to the suppression of both the fast and late outward currents.

II. INACTIVATION PROCESSES IN SMOOTH MUSCLE CELL

1. *Depolarizing Inactivation and Hyperpolarizing Inactivation*

Hodgkin and Huxley (*6*) showed that in the squid axon depolarization

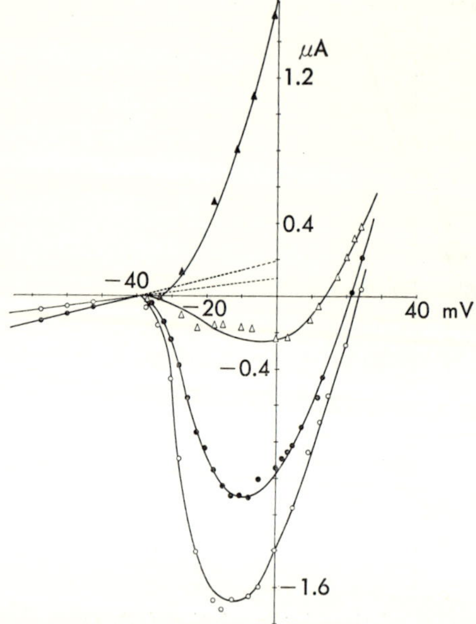

Fig. 9. Current-voltage relations in Ba solution. Filled circles (●), inward current in normal solution containing 2.5 mM Ca^{2+}; open circles (○), inward current in solution with 2.5 mM Ba^{2+} replacing Ca^{2+}; filled triangles (▲), late outward current in normal solution; open triangles (△), late outward current in Ba solution. Note the increase in the inward current and the suppression of the late outward current due to Ba.

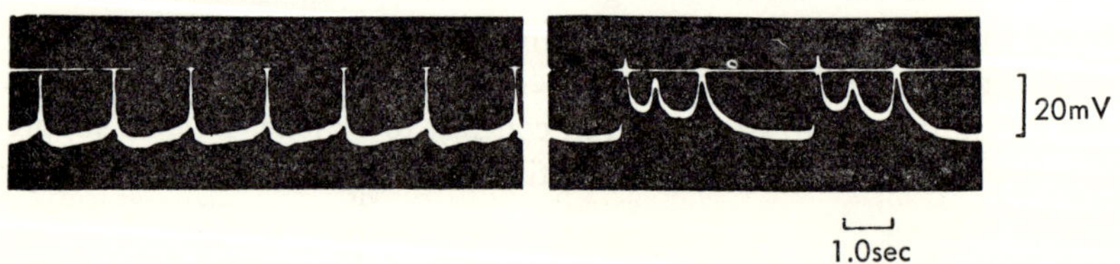

Fig. 10. Action potential in Ba solution recorded with intracellular microelectrode (*19*).

brings about two processes: 1) an activation process that rapidly increases the membrane permeability to Na ions by turning on the Na channel; and 2) an inactivation process that less rapidly inactivates the permeability changes by turning off the Na channel. Inactivation process has a direct role in determining excitability. Although smooth muscle physiologists have paid considerable attention to the analysis of the activation process of smooth muscle cells, comparatively little is known of the inactivation process of these cells.

The methods used by Hodgkin and Huxley to examine the activation process were two-step voltage-clamp techniques: the first conditioning step and the second test step. Recently, a similar two-step voltage-clamp technique has been applied successfully to intestinal smooth muscles (*18, 24, 27, 31*).

1) *Depolarizing inactivation*

There are two types of inactivation: depolarizing inactivation and hyperpolarizing inactivation. The inactivation induced by the first conditioning depolarization is referred to as depolarizing inactivation. Usually, the first conditioning depolarization, which is too small to generate any detectable activation process, is maintained for a given duration of time. The second test potential is held at a constant depolarization level which is sufficient to generate the fast inward current.

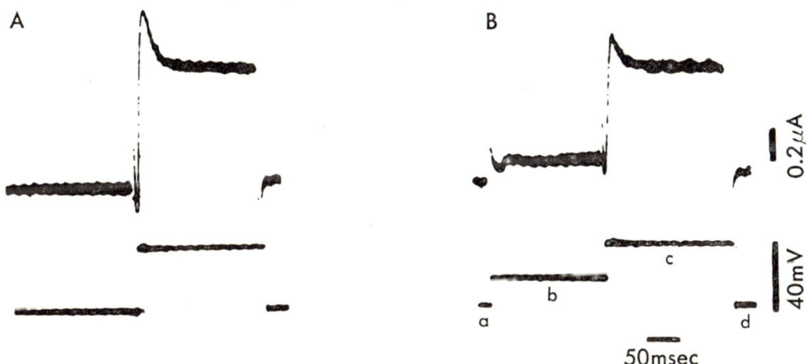

Fig. 11. Depolarizing inactivation. A: membrane currents under voltage clamp without conditioning potential. B: with conditioning potential. In B: a, preconditioning potential (holding potential); b, conditioning potential; c, test potential; d, preconditioning potential level. Preconditioning potential is usually held at the natural resting membrane potential level. Note that the peak amplitude of the inward current in B is smaller than that in A (depolarizing inactivation). Traces in this figure are before correcting for leakage and capacitative currents.

As shown in Fig. 11, the peak amplitude of the inward current in response to the test depolarization for a given conditioning depolarization (I_a') was always smaller than the amplification with no conditioning depolarization (I_a). In Fig. 12, I_a'/I_a was plotted against the duration of the

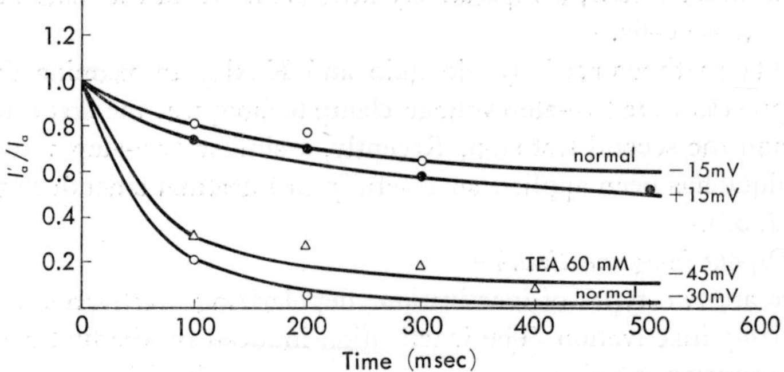

Fig. 12.   Time course of inactivation of inward current. Ordinate, I_a'/I_a (see text); abscissa, duration of conditioning potential in msec. Amplitude of conditioning potential is shown at the right in mV. Filled circles (●), in normal solution with conditioning depolarization (depolarizing inactivation); open circles (○), in normal solution with conditioning hyperpolarization (hyperpolarizing inactivation); open triangles (△), in TEA solution (TEA: 60 mM) with conditioning hyperpolarization.

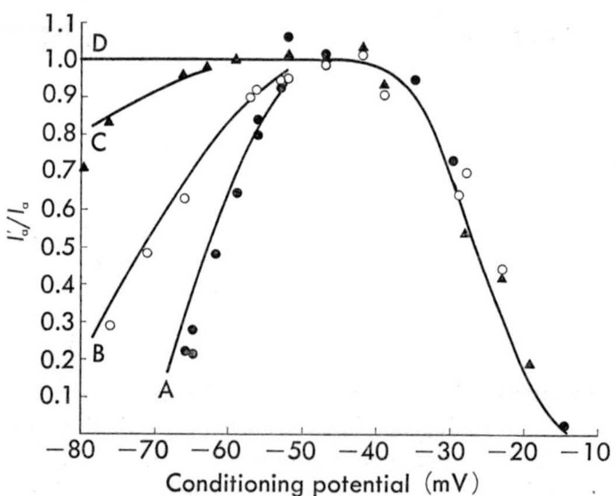

Fig. 13.   Steady-state inactivation curve of inward current and its voltage dependency. Filled circles and curve A, in normal solution; open circles and curve B, in low-TEA solution (TEA: 7 mM); filled triangles and curve C, in high-TEA solution (TEA: 60 mM). D, Hodgkin-Huxley-type response. Ordinate, I_a'/I_a (see text), abscissa, level of conditioning potential in mV. Preconditioning potential (see Fig. 11) was kept at -47 mV which is close to the natural resting membrane potential. I_a'/I_a declined both in the depolarizing range (depolarizing inactivation) and in the hyperpolarizing range (hyperpolarizing inactivation). Hyperpolarizing inactivation became less prominent with the increase in TEA concentration (A → B → C).

conditioning pulse. The conditioning depolarization tended to decrease the rate of I_a'/I_a below unity, suggesting an inactivation process (depolarizing inactivation). The depolarizing inactivation developed slowly compared with the activation process. It was dependent on the duration of conditioning depolarization (Fig. 12). In addition, it varied with the voltage of conditioning depolarization (Fig. 13). Such a depolarizing inactivation process in smooth muscles is basically similar to that in squid axons except for the slow development.

2) Hyperpolarizing inactivation

When the conditioning pulse was hyperpolarized, the intestinal smooth muscle showed a response different from that of the squid axon. The peak inward current amplitude for a given conditioning pulse (I_a') relative to the peak inward current in the absence of a conditioning pulse (I_a) was below unity (Fig. 14). This is referred to as hyperpolarizing inactivation. In squid axons (*6*), the conditioning hyperpolarization always removes the inactivation and the value of I_a'/I_a is not below unity.

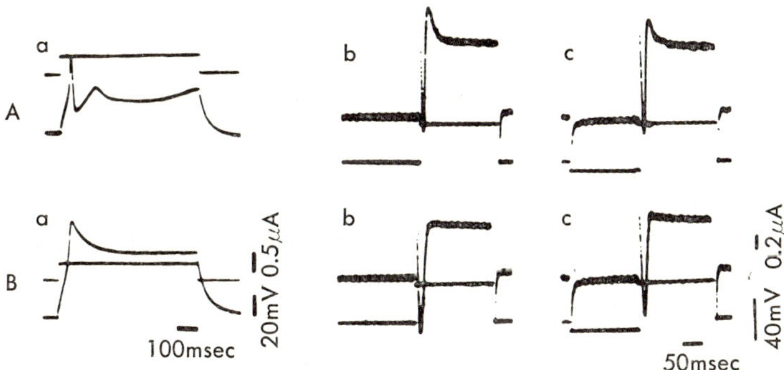

Fig. 14. Effects of TEA on the electrical activities of guinea pig taenia coli (*27*). Upper row (A): in normal solution. Lower row (B): in TEA solution. a, membrane potential under current clamp (prolonged action potential in B-a); b, membrane current under voltage clamp without conditioning potential; c, membrane current under voltage clamp with conditioning potential. Concentration of TEA: 60 mM in B-a, 7 mM in B-b and B-c. Fast transient outward current observed in normal solution (A-b and A-c) was suppressed by TEA (B-b and B-c).

Hyperpolarizing inactivation in intestinal smooth muscles is time dependent: it develops approximately exponentially over the time course and tends to a definite steady state level (see Fig. 12). The rate of decline (rate of inactivation) increased as the conditioning depolarization increased.

3) Steady-state inactivation curve

The value of I_a'/I_a at the steady state level is dependent on the amplitude of conditioning hyperpolarization, and its voltage dependency is shown in a steady state inactivation curve in Fig. 13. In intestinal smooth muscles, I_a'/I_a was maximum around the resting potential and it declined when the conditioning potential was more hyperpolarized. The steady state inactivation curve in intestinal smooth muscles deviates from that of the Hodgkin-Huxley type, particularly in the hyperpolarizing range. Similar hyperpolarizing inactivation has been described in other cells: snail neuron (*26*), Aplysia giant neurons (*35*), and crayfish X-organ (*25*). In most of these cells, the inward current was carried by Ca ions (Ca spike).

Similar hyperpolarizing inactivation was observed in Sr solution. In Sr solution, the rate of inactivation was considerably slower than that in Ca solution.

2. Ionic Kinetics of Hyperpolarizing Inactivation

The ionic mechanism of hyperpolarizing inactivation is not clear at the present time, but the results of TEA experiments will give a clue to this unsettled problem (*27*). In TEA solution, when conditioning hyperpolarization was small in amplitude (about 15 mV), I_a'/I_a slightly increased above unity, suggesting that a small conditioning hyperpolarization removes the inactivation. With a further increase in the amplitude of conditioning hyperpolarization, I_a'/I_a declined below unity. With similar conditioning hyperpolarizations, inactivation was less pronounced in TEA solution than in normal Krebs solution (Figs. 13, 14).

Moreover, in TEA solution, the steady-state inactivation curve for conditioning depolarization declined in the same way as in normal Krebs solution, but for conditioning hyperpolarization, it declined but slightly as compared with that obtained in normal Krebs solution. With an increase in TEA concentration up to 60 mM, hyperpolarizing inactivation became less prominent, but it was not abolished completely (*27*).

These results indicate that the inactivation curve of the inward current in smooth muscle cells of the guinea pig taenia coli deviates from that of the Hodgkin-Huxley type, particularly in the hyperpolarizing range, and the deviation is mainly due to the transient fast outward current. The transient fast outward current has a property of inactivation and its kinetics are similar to that of the inward current; it is dependent on the duration and voltage of conditioning hyperpolarization. Moreover, the transient fast outward current occurs nearly at the same time and overlaps with the inward current. Actually in the presence of TEA, which is effective in suppressing the transient fast outward current, the deviation of the inactivation curve in the hyperpolarizing range became less prominent as the transient fast outward current decreased. Therefore, in the guinea pig taenia coli hyperpolarizing inactivation is explained by the transient fast outward current.

SUMMARY

The ionic mechanism of excitation in intestinal smooth muscles has been reviewed. Mammalian smooth muscles are a very important group of excitable cells. There exists a considerable number of reports on the electrical and mechanical properties of these muscles. But little work has been done on the analysis of ionic mechanism of this excitable process. The recent successful application of the voltage-clamp technique using the double sucrose gap method has opened a new path for the study of the ionic mechanism in smooth muscles.

In intestinal smooth muscle cells, the fast inward current responsible for spike generation is carried principally by the influx of Ca ions and not by Na ions. The outward currents which participate in the rate of repolarization are mainly carried by efflux of K ions. Sr ions and Ba ions can carry the current through the Ca channel. That is, Sr and Ba can substitute for Ca in generation of the fast inward current.

The inactivation curve of the inward current of intestinal smooth muscle cells deviates from that of nerve cells and skeletal muscle cells in its hyperpolarizing range. Such an inactivation observed in the hyperpolarizing range is referred to hyperpolarizing inactivation. It seems likely that hyperpolarizing inactivation in smooth muscles is mainly due to the fast outward current.

The electrophysiology of smooth muscles has made a remarkable advance lately due to the introduction of the double sucrose gap method. There exist, however, some serious limitations on the voltage control in this method and more quantitative, more accurate information can be expected upon further improvements of the voltage-clamp technique.

REFERENCES

1 J.W. Moore and K. S. Cole, "Physical Techniques in Biological Research," ed. by W. Nastuk, Academic Press, New York and London, Vol. 6, p. 263 (1963).
2 K. S. Cole, *Arch. Sci. Physiol.*, **3**, 253 (1949).

3 A. L. Hodgkin, A. F. Huxley, and B. Katz, *Arch. Sci. Physiol.*, **3**, 129 (1949).

4 A. L. Hodgkin, A. F. Huxley, and B. Katz, *J. Physiol.*, **116**, 424 (1952).

5 A. L. Hodgkin and A. F. Huxley, *J. Physiol.*, **116**, 473 (1952).

6 A. L. Hodgkin and A. F. Huxley, *J. Physiol.*, **116**, 497 (1952).

7 A. L. Hodgkin and A. F. Huxley, *J. Physiol.*, **117**, 500 (1952).

8 N. C. Anderson, *J. Gen. Physiol.*, **54**, 145 (1969).

9 J. Mironneau, *J. Physiol.*, **233**, 127 (1973).

10 J. Mironneau, *Pflügers Arch.*, **352**, 197 (1974).

11 J. Mironneau, J. O. Savineau, and A. Rahéty, "Excitation-Contraction Coupling in Smooth Muscle," ed. by R. Casteels, T. Godfraind, and J. C. Rüegg, Elsevier/North-Holland, Amsterdam, p. 117 (1977).

12 G. Vassort, *J. Physiol.*, **252**, 713 (1975).

13 G. Vassort, "Physiology of Smooth Muscle," ed. by E. Bülbring and M. F. Shuba, Raven Press, New York, p. 77 (1976).

14 C. Y. Kao and J. R. McCullough, *J. Physiol.*, **246**, 1 (1975).

15 H. Inomata and C. Y. Kao, "Physiology of Smooth Muscle," ed. by E. Bülbring and M. F. Shuba, Raven Press, New York, p. 49 (1976).

16 C. Y. Kao, H. Inomata, J. R. McCullough, and J. C. Yuan, "Smooth Muscle Physiology and Pharmacology," ed. by M. Worcel and G. Vassort, Colloque Institut Nat. de la Sante et la Recherche Medicale, Paris, p. 165 (1976).

17 C. Y. Kao, "Excitation-Contraction Coupling in Smooth Muscle," ed. by R. Casteels, T. Godfraind, and J. C. Rüegg, Elsevier/North-Holland, Amsterdam, p. 91 (1977).

18 T. Suzuki and H. Inomata, *Japan J. Smooth Muscle Res.*, **14**, Suppl., 7 (1978).

19 T. Suzuki, A. Nishiyama, and K. Okamura, *Tohoku J. Exp. Med.*, **82**, 87 (1964).

20 T. Suzuki, A. Nishiyama, and H. Inomata, *Nature*, **197**, 908 (1963).

21 F. J. Julian, J. W. Moore, and D. F. Goldman, *J. Gen. Physiol.*, **45**, 1195 (1962).

22 F. J. Julian, J. W. Moore, and D. F. Goldman, *J. Gen. Physiol.*, **45**, 1217 (1962).

23 R. Stämpfli, *Experientia*, **10**, 508 (1954).

24 H. Inomata and C. Y. Kao, *J. Physiol.*, **255**, 347 (1976).

25 S. Iwasaki, Y. Satow, and T. Kuroda, *Proc. Japan Acad.*, **49**, 564 (1973).

26 N. B. Standen, *J. Physiol.*, **249**, 253 (1975).

27 H. Inomata and T. Suzuki, *Proc. Japan Acad.*, **53**, 215 (1977).

28 Y. Ito, H. Kuriyama, and Y. Sakamoto, *J. Physiol.*, **211**, 445 (1970).

29 N. C. Anderson, "Excitation-Contraction Coupling in Smooth Muscle," ed. by R. Casteels, T. Godfraind, and J. C. Rüegg, Elsevier/North-Holland, Amsterdam, p. 81 (1977).

30 D. Attwell and I. Cohen, *Prog. Biophys. Mol. Biol.*, **31**, 201 (1977).

31 H. Inomata and C. Y. Kao, *J. Physiol.*, **297**, 443 (1979).

32 T. Tomita, *Prog. Biophys. Mol. Biol.*, **30**, 185 (1975).

33 G. W. Beeler and J. A. S. McGuigan, *Prog. Biophys. Mol. Biol.*, **34**, 219 (1978).

34 C. Daemers-Lambert, "Physiology of Smooth Muscle," ed. by E. Bülbring and M. F. Shuba, Raven Press, New York, p. 83 (1976).

35 D. Geduldig and R. Gruener, *J. Physiol.*, **211**, 217 (1970).

Received for publication August 7, 1980.

CONTRIBUTOR SKETCHES

Subunit Assembly of Escherichia coli RNA Polymerase
Akira ISHIHAMA, Ph. D., has been an associate professor of biochemistry
in the Institute for Virus Research, Kyoto University since 1971. Dr.
Ishihama graduated from the Department of Biology, Nagoya University,
in 1961. He served as assistant to the Cancer Research Institute, Kana-
zawa University, from 1963 to 67, and spent two years at the Department
of Developmental Biology and Cancer, Albert Einstein College of Medicine, New York (1967–69). In 1970, Dr. Ishihama joined the Institute for
Virus Research as an instructor of genetics. Dr. Ishihama has been interested in studies on the structure, function, and molecular organization
of *E. coli* RNA polymerase and also of RNA and DNA polymerases of
animal virus.

Dynamic Analysis of Higher Order Biological Systems
Kensuke SATO, M.D., has served as Director of the Research Foundation
on Traffic Medicine, Tokyo, since 1979. Dr. Sato graduated from Niigata
University School of Medicine in 1940. He was appointed an associate
professor of physiology there, and moved to Nagasaki University School of
Medicine as professor of physiology in 1954. Dr. Sato worked in the Department of Neuroanatomy, Northwestern University, Chicago from 1957
to 58, and then studied the application of the nonlinear random theory to
electroencephalograms under Professor T. Kitagawa at the Department
of Mathematics, Kyushu University (1959–60). Dr. Sato has been
interested in the analyses of electroencephalograms and myotonograms
and has published several books on these studies.

258

Two-Dimensional NMR Spectroscopy: An Application to the Study of Flexibility of Protein Molecules
Kuniaki Nagayama, Ph. D., has been an instructor of biophysics in the Department of Physics, University of Tokyo, since 1974. Dr. Nagayama graduated from the Department of Physics, University of Tokyo, in 1968. He worked for three years in Professor K. Wöthrich's laboratory at the Swiss Federal Institute of Technology (ETH) in Zörich, from 1976 to 79. Dr. Nagayama has been engaged in physical studies of protein molecules using NMR technique.

The Cap Structure in Eukaryotic Messenger RNA as a Mark of a Strand Carrying Protein Information
Kin-ichiro Miura, D. Sc., has served as head of the Department of Molecular Genetics at the National Institute of Genetics since 1969. Dr. Miura graduated from the Faculty of Science of Gakushuin University in 1953. He studied biochemistry in the Graduate School Course of Chemistry, University of Tokyo, under the supervision of Prof. I. Watanabe. Between 1958 and 1960, he was a research associate at the Institute for Virus Research, Kyoto University, and until 1963 was a research associate at the Institute of Applied Microbiology, University of Tokyo. He worked on the heterogeneity of nucleic acids and made compositional analyses of them. From 1963 to 69, he was an associate professor at the Institute of Molecular Biology, Faculty of Science, Nagoya University, where he studied the functional structure of amino acid transfer RNA and viral nucleic acids; he especially contributed to the structural characterization of double-stranded RNA. Subsequently, he has developed his work on the mechanism of gene expression using viruses. Dr. Miura was awarded the Chunichi Culture Prize from the Central Japan Newspaper Co. for the discovery of and studies on the blocked terminal structure in a nucleic acid molecule.

The Ionic Mechanism of Excitation in Intestinal Smooth Muscle Cells
Taizo Suzuki, M.D., is a professor of physiology in the Department of Applied Physiology, Tohoku University School of Medicine. Dr. Suzuki graduated from Tohoku University School of Medicine in 1944 and was appointed as a professor of obstetrics and gynecology, Fukushima Medical College in 1956. Meanwhile he was a visiting investigator at the Rockefeller Institute for Medical Research and a visiting professor

at the State University of New York, Brooklyn. Dr. Suzuki's interest is centered on the physiology of intestinal motility.

Hachiro INOMATA, M.D., is an associate professor in the Department of Applied Physiology, Tohoku University School of Medicine. Dr. Inomata graduated from Tohoku University School of Medicine in 1961 and was an associate professor of pharmacology in the Department of Pharmacology, Faculty of Pharmacy, Tohoku University. From 1971 to 74 he was a research associate in the Department of Pharmacology, State University of New York, Brooklyn. His main interests are the ionic mechanism of excitation and inhibition in smooth muscle cells.

SUGGESTIONS TO AUTHORS

Articles for *Advances in Biophysics* should be critical and comprehensive reviews of recent research—theoretical, experimental, and applied—in the areas of biophysics. The review may be concerned in large part with work in the author's own laboratory providing that relevant contributions by other investigators are sufficiently mentioned to place the author's research in perspective.

Most of the reviews appearing in this series are written in response to invitations by the Editor, nevertheless, unsolicited manuscripts are also welcome. Prospective authors are requested to submit information such as (a) the title of the article and a short topical outline; (b) estimated length; and (c) tentative date of submission of the manuscript.

Manuscripts and editorial correspondence prior to acceptance of manuscripts should be sent to

JAPAN SCIENTIFIC SOCIETIES PRESS
6-2-10 Hongo, Bunkyo-ku, Tokyo 113, Japan.

Manuscripts should be prepared in accordance with the following general instructions:

All articles should include title, author's name, affiliation, table of contents, introduction, the body of the review, concise summary, and references. The title should be precise, informative, and suitable for accurate indexing. The introduction should be about one or two pages long and should not have a heading.

The review material should be clearly and logically arranged with main divisions and subdivisions for a general audience of biophysicists of various training, and should include a few pages at the beginning, providing background and orientation as well as introduction of specialized terms or concepts for the general reader. Some historical developments should be mentioned, but the article should not be chronological. Detailed descriptions of experiments and experimental procedure are not advisable, and should be limited to a minimum.

The article should be normally about 60 to 70 double-spaced type-written pages.

Legends for tables and figures should be typed on separate sheets, and each table and figure should have a title. Table headings should appear at the top of each table, and footnote legends should be placed directly below the table. Page numbers should be typed consecutively for text, tables, figure legends, and photograph captions.

References should be numbered in the order of appearance in the text. In the text, the number should be italicized and in parenthesis. Abbreviations such as et al., idem, ibid., should not be used in the references. Author's names should be written with initials first followed by the family name.

For examples,

E. M. Landis, *Am. J. Physiol.*, **82**, 217 (1927).

L. H. Peterson, " Pulsatile Blood Flow," ed. by E. O. Attinger, McGraw-Hill, New York, p. 263 (1964).

It is strongly suggested that recommendations by ISO, IUB, IU-PAC, and IUPAP are followed for units, nomenclature, and notations, because of the broad and numerous fields connected with biophysics.

Fifty reprints will be supplied to the author free of charge.

PUBLISHED PAPERS

VOLUME 1

Yasuji KATSUKI, Toru HASHIMOTO, and Keiji YANAGISAWA, The Lateral-
Line Organ of Shark as a Chemoreceptor

Kenji TAKAHASHI, Tsuneko UCHIDA, and Fujio EGAMI, Ribonuclease
T_1—Structure and Function

Sho ASAKURA, Polymerization of Flagellin and Polymorphism
of Flagella

Tetsutaro IIZUKA and Takashi YONETANI, Spin Changes in Hemoproteins

VOLUME 2

Yonosuke KOBATAKE, Ichiji TASAKI, and Akira WATANABE, Phase Transition in Membrane with Reference to Nerve Excitation

Isao YAMAZAKI, One-Electron and Two-Electron Transfer Mechanisms in Enzymic Oxidation-Reduction Re-
actions

Kyozo KOKETSU, The Electrogenic Sodium Pump

Tatsuo OOI and Sugie FUJIME-HIGASHI, Structure of Tropomyosin and
Its Crystal

Ei TERAMOTO, Nakako SHIGESADA, Hisao NAKAJIMA, and Koji SATO,
Stochastic Theory of Reaction Kinetics

VOLUME 3

Satoru FUJIME, Quasi-elastic Scattering of Laser Light—A New
Tool for the Dynamic Study of Biological Macromolecules

Akira TAKEUCHI and Noriko TAKEUCHI, Actions of Transmitter Sub-

Fluctuation in the Structure of a Protein Molecule

Shio MAKINO,   Interaction of Proteins with Amphiphilic Substances

VOLUME 13 : SOME NEW APPROACHES TO MUSCLE CONTRACTION

Akio INOUE, Hitoshi TAKENAKA, Toshiaki ARATA, and Yuji TONOMURA, Functional Implication of the Two-Headed Structure of Myosin

Hiroshi SHIMIZU,   Dynamic Cooperativity of Molecular Processes in Active Streaming, Muscle Contraction, and Subcellular Dynamics—The Molecular Mechanism of Self-Organization at the Subcellular Level

FORTHCOMING PAPERS

Eiichi SOEDA,   Evolutional Aspect of Papovaviruses

Masaya YAMAGUCHI,   Mathematical Model of Population and Its Development

Toru YOSHIZAWA,   Photophysiological Function of Visual Pigment

Yutaka NAITOH,   Electrophysiology of Ciliolate Protozoa

Iwao OHTSUKI and Kozo NAGANO,   Molecular Arrangement of Troponin-Tropomyosin

Haruo CHINO,   Lipid-Carrier-Protein in Insect Blood

J. T. EDSALL and H. A. McKENZIE, Water and Proteins. Part II: The Significance on the Structure of the Proteins

Tomoko OHTA,   Population Genetics of Multigene Families

Osamu GOTOH,   Prediction of Melting Profile and Local Helix Stability for Sequenced DNA

Kazuhiko KINOSITA,   Dynamic Study of Biological Membranes—A Nanosecond Fluorescence Depolarization Study

Nakaakira TSUKAHARA,   Synaptic Plasticity in the Brain

Shiro KAKIUCHI,   Calmodulin-Binding Protein

Gen MATSUMOTO and Hikoichi SAKAI, Physiological Function of Cytoskeletons in Nerve Excitation

Fumio OOSAWA,   Movement of Bacteria and of Protozoa

Shigeo YOMOSA,   Electronic Processes in Polypeptides and Proteins

Kazutomo IMAHORI,   Designs of Enzyme Molecules

Akira WATANABE,   Optical Approaches to Analysis of Neural Function

Hiroshi WATARI,   Electronic State of Non-Heme Iron Proteins

Mitsuo NISHIMURA,   Energy Transfer in Photosynthetic Membrane Systems

Toshio MITSUI,	Study of Biological Membranes by X-Ray
Shigeto MIYAJI,	Effects of Environmental Factors on the Path of Carbon in Photosynthesis
Masaki FURUYA,	Mode of Action of Membrane Bound Phyto-chrome
Yoshiki HOTTA,	Genetic Dissections of Behavior
Tomoo MASAKI,	Differentiation of Muscle Proteins
Yoshimasa KYOGOKU,	Nuclear Magnetic Resonance Studies on Bio-molecules in Solution
Yoshito KAZIRO,	Role of GTP in Protein Biosynthesis